Springer Series on

Wave Phenomena 9

Edited by L. B. Felsen

Springer Series on

Wave Phenomena

Editors: L. M. Brekhovskikh L. B. Felsen H. A. Haus
Managing Editor: H. K. V. Lotsch

Volume 1 **Mechanics of Continua and Wave Dynamics**
By L. Brekhovskikh, V. Goncharov

Volume 2 **Rayleigh-Wave Theory and Application**
Editors: E. A. Ash, E. G. S. Paige

Volume 3 **Electromagnetic Surface Excitations**
Editors: R. F. Wallis, G. I. Stegeman

Volume 4 **Asymptotic Methods in Short-Wave Diffraction Theory**
By V. M. Babič, V. S. Buldyrev

Volume 5 **Acoustics of Layered Media I** Plane and Quasi-Plane Waves
By L. M. Brekhovskikh, O. A. Godin

Volume 6 **Geometrical Optics of Inhomogeneous Media**
By Yu. A. Kravtsov, Yu. I. Orlov

Volume 7 **Recent Developments in Surface Acoustic Waves**
Editors: D. F. Parker, G. A. Maugin

Volume 8 **Fundamentals of Ocean Acoustics** 2nd Edition
By L. M. Brekhovskikh, Yu. Lysanov

Volume 9 **Nonlinear Optics in Solids**
Editor: O. Keller

Volume 10 **Acoustics of Layered Media II** Point Source and Bounded Beams
By L. M. Brekhovskikh, O. A. Godin

O. Keller (Ed.)

Nonlinear Optics in Solids

Proceedings of the International Summer School,
Aalborg, Denmark, July 31–August 4, 1989

With 139 Figures

Springer-Verlag
Berlin Heidelberg New York London
Paris Tokyo Hong Kong Barcelona

Professor Ole Keller, Ph.D.

Institute of Physics, University of Aalborg, Pontoppidanstræde 103,
DK-9220 Aalborg Øst, Denmark

Series Editors:

Professor Leonid M. Brekhovskikh, Academician

P. P. Shirsov Institute of Oceanology, Academy of Sciences of the USSR, Krasikowa Street 23,
SU-117218 Moscow, USSR

Professor Leopold B. Felsen, Ph.D.

Department of Electrical Engineering, Polytechnic University, Route 110,
Farmingdale, NY 11735, USA

Professor Hermann A. Haus

Department of Electrical Engineering & Computer Science, MIT,
Cambridge, MA 02139, USA

Managing Editor: Helmut K. V. Lotsch

Springer-Verlag, Tiergartenstrasse 17,
D-6900 Heidelberg, Fed. Rep. of Germany

ISBN-13:978-3-642-84208-5 e-ISBN-13:978-3-642-84206-1
DOI: 10.1007/978-3-642-84206-1

Library of Congress Cataloging-in-Publication Data. Nonlinear optics in solids : proceedings of the international summer school, Aalborg, Denmark, July 31–August 4, 1989 / O. Keller (ed.) p. cm.–(Springer series on wave phenomena ; 9) Includes bibliographical references and index. ISBN-13:978-3-642-84208-5
1. Solids–Optical properties–Congresses. 2. Semiconductors–Optical properties–Congresses. 3. Nonlinear optics–Congresses. I. Keller, Ole. II. Series. QC176.8.O6N66 1990 530.4'12–dc20 90-39326

Softcover reprint of the hardcover 1st edition 1990

2154/3150-543210 – Printed on acid-free paper

Preface

In recent years one has witnessed in physics a substantial increase in interest in carrying out fundamental studies in the nonlinear optics of condensed matter. At the Danish universities, this increase has been especially pronounced at the Institute of Physics at the University of Aalborg, where the main activities are centered around fundamental research within the domains of nonlinear quantum optics, nonlinear optics of metals and superconductors, and nonlinear surface optics. In recognition of this it was decided to arrange the first international summer school on nonlinear optics in Denmark at the Institute of Physics at the University of Aalborg.

This book is based on the lectures and contributed papers presented at this international summer school, which was held in the period 31 July–4 August 1989. About 60 experienced and younger scientists from 12 different countries participated. Twenty-eight lectures were given by 14 distinguished scientists from the United States, Italy, France, Germany, Scotland, England, and Denmark. In addition to the lectures given by the invited speakers, 11 contributed papers were presented. The programme of the summer school emphasized a treatment of basic physical properties of the nonlinear interaction of light and condensed matter and both theoretical and experimental aspects were covered. Furthermore, general principles as well as topics of current interest in the research literature were discussed. Among the topics included in the summer school were the following: optical bistability, principles and applications; enhanced second-harmonic generation from metallic surfaces; nonlinear effects in guiding structures, thin films and single interfaces; nonlinear optics of semiconductor quantum wells; field quantization in quantum optics and production of nonclassical light via nonlinear processes; optical phase conjugation, photorefractive effects; quantum electrodynamics in cavities, nonlinear optical effects in organic materials; light pressure on single particles, unified treatment of parametric phenomena and nonlinear ponderomotive effects in condensed matter; nonlinear optics of centrosymmetric metals and Cooper-paired superconductors; transient nonlinear optics of semiconductors; sum- and difference-frequency generation, parametric processes and harmonic generation; second-harmonic generation and nonlinear pulse generation in optical fibres; inelastic light scattering from solid surfaces, in particular Brillouin scattering; three- and four-wave mixing in disordered nonlinear media and enhanced backscattering; and critical phenomena in cooperative Raman scattering.

The invited speakers covered the following subjects: I. Abram: The nonlinear optics of semiconductor quantum wells; A.D. Boardman: Nonlinear effects in guiding structures: thin films and interfaces; J.-L. Coutaz: Surface enhanced second-harmonic generation from metals; T.J. Hall: Phase conjugation in solids; J. Hvam: Transient nonlinear optics in semiconductors; O. Keller: Nonlinear optics of centrosymmetric superconductors; P.L. Knight: Field quantization in quantum optics and production of nonclassical light; P. Meystre: Cavity quantum electrodynamics; P. Mulser: The role of radiation pressure in nonlinear optics; F. Nizzoli: Light scattering from solid surfaces; P.N. Prasad: Nonlinear optical effects in organic materials; T. Skettrup: Sum- and difference-frequency generation, harmonic generation and parametric processes; D.A. Weinberger: Second-harmonic generation and nonlinear pulse propagation in optical fibres; B.S. Wherrett: Optical bistability and digital optical computing.

In the initial phase of the planning of the summer school, the Institute of Physics cooperated with the Danish Optical Society. The school was sponsored by several organizations. The basic support came from the Danish Research Academy, which was established a few years ago in order to stimulate the education of young scientists in Denmark and to strengthen their contact to the international scientific community. Specific financial support was provided by Det Obelske Familiefond, Sparekassen Nordjylland, and the University of Aalborg.

Finally, I wish to express my thanks to the invited speakers, all the other participants, and coworkers at the Institute of Physics for their outstanding contributions in making the summer school stimulating, instructive, suggestive, and pleasant.

Aalborg, March 1990 *Ole Keller*

Contents

The Exotic World of Nonlinear Optics

O. Keller

Institute of Physics, University of Aalborg,
Pontoppidanstræde 103, DK-9220 Aalborg Øst, Denmark

It is a privilege for a physicist to work in a field which has its roots far back in the past and which has a bright future, and it is exciting if the field embraces esoteric aspects of fundamental physics as well as themes of interest to applied science. Nonlinear optics in solids dates back to the beginning of the nineteenth century when Brewster discovered the photoelastic effect in gels [1]. At the end of the century, Kerr discovered the quadratic electro-optic effect in glasses and liquids [2], and Röntgen [3] and Kundt [4] the linear electro-optic effect in quartz and tourmaline crystals. The linear electro-optic effect is often named the Pöckels effect in recognition of the important contributions made by Pöckels [5]. The above-mentioned phenomena belong to the domain of nonlinear optics since it is necessary to multiply two electric fields together to obtain the effects. The one is the high-frequency field of light, the other is the internal low (possibly dc)-frequency field associated with the deformation of a solid subjected to a mechanical stress (photoelasticity) or stemming from an externally applied electric field. In the wake of the invention of the laser thirty years ago, nonlinear optics has developed into an important and independent field of optics. By means of lasers it is possible to produce light fields of a high intensity and with a substantial degree of coherence. In turn, the use of lasers has enabled us to mix high-frequency optical fields in such manners that myriads of exotic nonlinear optical effects can be observed. Since the development of lasers possessing new outstanding properties will progress steadily in the coming years and since our understanding of the dynamics of nonlinear systems from a physical and mathematical point of view is sparse, the future of nonlinear optics will be filled with surprises and mountain tops of undreamed heights will have to be climbed. Since it is impossible in a single book to cover a substantial fraction of the subjects being studied nowadays within the domain of nonlinear optics in solids, it has been tried to bring the readers a flavour of the depth and diversity of the field by selecting a number of subjects of current interest in the research literature.

From a unified point of view, the electromagnetic interaction of light and atomic particles can be described on the basis of the wave equation for the four-vector vector potential operator $\hat{\mathcal{A}}$, i.e.,

$$\Box^2 \hat{\mathcal{A}} = -\mu_0(\hat{\mathcal{J}}^L + \hat{\mathcal{J}}^{NL}) \ , \tag{1}$$

where $\widehat{\mathcal{J}}^L$ and $\widehat{\mathcal{J}}^{NL}$ are the four-vector current density operators describing the linear (L) and nonlinear (NL) interactions, respectively. The four-vectors are given by $\mathcal{A}_4 = (\vec{A}, \mathrm{i}\Phi/c_0)$ and $\mathcal{J} = (\vec{J}, \mathrm{i}c_0\rho)$, where $\vec{A}$ and Φ are the conventional vector and scalar potentials, and $\vec{J}$ and ρ the current and charge densities. The four-dimensional nabla operator is $\square = (\vec{\nabla}, \frac{1}{\mathrm{i}c_0}\frac{\partial}{\partial t})$. Operators are denoted by a caret, $\hat{}$. It has turned out in recent years that certain phenomena in the nonlinear optics of solids can be described only if the free electromagnetic field is quantized. The space-time dependence of the quantized field can be studied via the wave equation

$$\square^2 \widehat{\mathcal{A}} = 0 \tag{2}$$

for the field operator $\widehat{\mathcal{A}}$. The field quantization leads to the introduction of the photon concept and the vacuum field state.

Among the physical phenomena requiring the field quantization to be understood is the well-known spontaneous emission resulting from the interaction between an atom and an electromagnetic field in the vacuum state. In a new subfield of physics named cavity quantum electrodynamics (cavity QED) the linear and nonlinear interactions between a single (or a few) atom(s) and a single (or a few) photon(s) in a resonator are studied. In the article by P. Meystre such novel effects as inhibited or enhanced spontaneous emission, micromaser action and quantum superposition in separated cavities are discussed. Since the spontaneous emission is triggered mainly by the zero-point fluctuations in the field which have frequencies in the vicinity of the atomic transition involved, for wavelengths below the cavity cutoff a drastic increase in the lifetime of the atom can be observed. As the lifetime is limited by the indirect coupling of the atom and the field via the wall it is of importance to consider the interaction of the cavity field with the many-body system of the wall.

In recent years the interaction between light and condensed matter surfaces has been extensively investigated, and in the present book several examples from the domain of nonlinear surface optics are presented. The approach adopted is semiclassical in the sense that the radiation field is considered to be a c-number. Much effort has been devoted to studies of second-order nonlinear effects in centrosymmetric media due to the fact that the dipole radiation is bulk forbidden. Since the centrosymmetry is broken at the outermost atomic layers the sum- and difference-frequency generation processes are surface sensitive. In the paper by myself, the nonlinear response function describing second-harmonic generation in centrosymmetric media is discussed within the framework of a nonlocal response theory. In a nonlocal formalism the light-induced nonlinear current density and the fundamental electric field, $\vec{E}$, are related via

$$\vec{J}^{NL}(\vec{r}) = \int_{-\infty}^{\infty} \overleftrightarrow{\Sigma}(\vec{r}, \vec{r}', \vec{r}'') : \vec{E}(\vec{r}'')\vec{E}(\vec{r}')d^3r''d^3r' \; , \tag{3}$$

where $\overleftrightarrow{\Sigma}(\vec{r}, \vec{r}', \vec{r}'')$ is a generalized nonlinear response function depending on the three space coordinates $\vec{r}, \vec{r}'$, and $\vec{r}''$. The nonlocal approach allows one to consider the competition between local surface effects and nonlocal bulk effects. As an example I apply the response formalism to a study of second-harmonic generation and optical rectification in a Cooper-paired superconductor. The even-order nonlinear responses of the centrosymmetric metals are so weak that one would like to invent schemes that enhance the nonlinear signal. In the article by J.L. Coutaz, a review of methods for surface enhanced second-harmonic generation from bare metallic surfaces is given. The experimental data presented are discussed on the basis of the free-electron model and the nonlinear optical surface response is treated within the framework of the two-parameter Rudnick and Stern model. In particular, enhancements obtained in attenuated total reflection configurations, via diffraction gratings and by means of surface roughness are discussed. Also the enhanced second-harmonic generation from metal islands is discussed.

Surfaces and interfaces also play a significant role for Brillouin and Raman scattering from opaque media. With main emphasis on Brillouin scattering from acoustic phonons and Raman scattering from surface phonon polaritons, F. Nizzoli reviews the field. In a section on experimental techniques for surface Brillouin scattering the single-pass and multi-pass Fabry-Perot interferometers and the tandem interferometer are discussed. The discrete and the continuous spectra of surface acoustic modes are treated, and also the vibrational spectra of unsupported and supported films are considered. Besides the elasto-optic effect, the surface rippling caused by the phonons contributes to the inelastic light scattering and sometimes even the interference between these two mechanisms. Finally, F. Nizzoli treats the theory and experiments related to Brillouin scattering from supported films. A better understanding of the nonlinear properties of optical fibre waveguides is of extreme importance for optical communication purposes. In the paper by D.A. Weinberger an interesting theme is treated, namely, that stemming from the recent observation that second-harmonic generation with a conversion efficiency approaching 5% can be obtained in fibres made of a centrosymmetric material like fused silica. Progress obtained to date in the understanding of this intriguing phenomenon is reviewed. Thus, electric quadrupole and magnetic dipole effects, the mixing model for self-organization of so-called $\chi^{(2)}$ gratings, the growth and erasure related to $\chi^{(2)}$ grating dynamics, and the evidence for defect structures are discussed.

A unified treatment of parametric phenomena and nonlinear ponderomotive effects in condensed matter is given in the interesting article by P. Mulser, who holds the view that the nonlinear optical phenomena can be considered as produced by the radiation pressure. This point of view seems most refreshing and it occurs to me that its consequences in the context of optical rectification in superconductors should be examined due to the fact that the light-induced dc-recoil of the Cooper-paired electrons leads to a frictionless dc-current. In 1861,

Maxwell argued that a light beam of flux density I, when impinging normally from vacuum upon a surface of reflectivity R, exerts a light pressure

$$P_{\mathrm{L}} = (1 + R)\frac{I}{c_0} \tag{4}$$

on the material. Taking this result as a starting point, P. Mulser presents a number of interesting historical remarks on the subject. Following the historical introduction the ponderomotive forces on single particles and in dense matter are discussed. Among the light pressure effects phenomena such as nonresonant effects, stimulated decay processes, wave pressure in moving media, and stimulated Brillouin and Stokes Raman scattering are treated. Finally, a twenty-year-old paradox is discussed.

Semiconductors play a prominent role in the nonlinear optics of solids. In the paper by I. Abram, the nonlinear optics of the new and interesting quantum wells are discussed, e.g. the observation of different types of optical nonlinearities, the physical origin of these nonlinearities and some device possibilities for optical signal processing. Among many important subjects, the optical response in the transparent regime, the resonant optical response, coherent transients, dynamical nonlinearities, and the optical Stark effect are discussed. A detailed treatment of excitonic coherent transients and of dense electron-hole populations is presented. Among possible devices the monolithic optical bistable etalon is considered. In the article by J.M. Hvam, it is described how transient optical nonlinearities are conveniently studied by time resolved excite-and-probe experiments e.g. degenerate four-wave mixing and transient laser induced grating experiments. Special emphasis is devoted to a discussion of the strong excitonic resonance enhancements of the nonlinear susceptibility observed at low temperatures in for instance CdSe. In recent years optical phase conjugation has received special attention from the scientific community. In the paper by T.J. Hall and A.K. Powell the basic principles behind phase conjugation are reviewed and the grating picture is used to demonstrate the analogy with real-time holography. The photorefractive effect is described as an example of an effect suitable for real-time holography. Interesting effects such as self-pumped phase conjugation in barium titanate are demonstrated, and illustrative examples of applications of phase conjugation are presented. Optical bistability is one of the most exciting phenomena occurring in the nonlinear optics of solids. In the article by B.S. Wherrett and D.C. Hutchings an authoritative treatment of optical bistability is given. In the wake of an interesting introductory section, the linear Fabry-Perot etalon and nonlinear refraction associated with electronic and thermal nonlinearities are discussed. The dynamics and the steady-state solutions of the nonlinear etalon and cavity optimisation are treated. Also switching power, power-time products, switching dynamics, and transphasor operation are described. For completeness, brief comments on a number of schemes for obtaining bistability, which are alternative to that of refractive bistability in Fabry-Perot cavities with reflective feedback, are given.

In recent years organic materials have emerged as an important class of nonlinear optical materials. As described in the article by P.N. Prasad, organic materials offer unique properties for both fundamental and applied research. The nonlinearities of the organic materials are related to the molecular structure of these systems and to their unique chemical binding. In the article by P.N. Prasad, the scope of the research on the organic materials is demonstrated by the variety of nonlinear effects one can obtain. The still very limited basic microscopic understanding of the optical nonlinearities of organic systems, and recent experimental studies are discussed. Following a section on measurements in soluble materials the dynamics of resonant third-order nonlinear processes and the role of free carriers and excitons are treated. Finally, related to organic systems, specific device possibilities, namely the optical waveguide and the nonlinear Fabry-Perot etalon are studied.

In the early days of the modern period of nonlinear optics, sum- and difference-frequency generation, harmonic generation, and parametric processes were extensively investigated. In the pedagogical paper by T. Skettrup, these effects are described in detail and it is demonstrated by results obtained from recent research related to so-called quasi phase matching, where the sign of the nonlinear susceptibility is reversed periodically with a period equal to the coherence length, that these subjects are still young and full of excitements.

[1] D. Brewster, Phil.Trans. A **105,** 60 (1815).
[2] J. Kerr, Phil.Mag. **50,** 337 (1875).
[3] W.C. Röntgen. Ann.Phys.Chem. **18,** 213 (1883).
[4] A. Kundt, Ann.Phys.Chem. **18,** 228 (1883).
[5] F. Pöckels, Abh. Gött. **39,** 1 (1894).

Part I

Basic Macroscopic Concepts

Second Order Nonlinear Optical Effects

T. Skettrup

Physics Laboratory III, Technical University of Denmark,
DK-2800 Lyngby, Denmark

Abstract. The nonlinear optical effects of second order are discussed. The notation of nonlinear optics is reviewed. The second order susceptibility is derived from the anharmonic oscillator model. The three coupled wave equations describing three wave mixing are presented. Second harmonic generation, phase matching, parametric amplification and oscillation are finally discussed.

1. Introduction

The purpose of the present chapter is to introduce the concepts of nonlinear optics and to review different second order effects. In the first section the notation is introduced, then the second order susceptibility is derived in a purely phenomenological way using the anharmonic oscillator model. Three wave mixing is described and applied to the cases of frequency doubling and parametric amplification and oscillation.

2. Notation

When an electromagnetic field $\bar{E}(\bar{r},t)$ is applied to a material, the response of the material is a polarization $\bar{P}(\bar{r},t)$ (i.e. the induced dipole moment per unit volume) given in general by

$$\bar{P}(\bar{r},t) = \varepsilon_o \int_{-\infty}^{\infty} \int_{-\infty}^{t} \bar{\bar{\chi}}(\bar{r},\bar{r}',t-t')\, \bar{E}(\bar{r}',t')\, d^3\bar{r}'dt' \qquad (1)$$

where ε_o is the vacuum permittivity and $\bar{\bar{\chi}}$ is the susceptibility tensor of the material. For systems exhibiting spatial invariance the dependence on $(\bar{r},\bar{r}')$ can be replaced by $(\bar{r}-\bar{r}')$. The truncation of the time integration is necessary since the response $\bar{P}(\bar{r},t)$ is only a function of the behaviour of the electric field in the past. This causality requirement leads to the well-known Kramers-Kronig relations /1/ between the real and imaginary parts of $\bar{\bar{\chi}}$ in the frequency domain. Assuming spatial invariance the Fourier transform of (1) with respect to time and space yields

$$\bar{P}(\bar{k},\omega) = \varepsilon_o\, \bar{\bar{\chi}}(\bar{k},\omega)\, \bar{E}(\bar{k},\omega) . \qquad (2)$$

Often the local response approximation is applied (i.e. the response $\bar{P}(\bar{r},t)$ is considered to depend only on the field $\bar{E}(\bar{r},t)$ at the space coordinate $\bar{r}$). In this case (2) can be written

Springer Series in Wave Phenomena, Vol. 9 **Nonlinear Optics in Solids**
Editor: O. Keller

$$\bar{P}(\omega) = \varepsilon_o \bar{\bar{\chi}}(\omega)\ \bar{E}(\omega) . \qquad (3)$$

If the material is nonlinear the susceptibility $\bar{\bar{\chi}}(\omega)$ can be expanded in powers of the electric field and (3) can be written

$$\bar{P}(\omega) = \varepsilon_o(\bar{\bar{\chi}}^{(1)}(\omega) \cdot \bar{E}(\omega) + \bar{\bar{\chi}}^{(2)}(\omega) : \bar{E}(\omega)\bar{E}(\omega) + ..). \quad (4)$$

In the literature several notations are used. Shen /2/ uses (4) directly with $\bar{E}$ and $\bar{P}$ expressed as complex quantities. This notation is simple and appealing. Yariv and Yeh /3/ use another notation with $\bar{E}$ and $\bar{P}$ expressed as real quantities. This notation is widely used and will be applied in the following. In this notation one must be careful with factors of 2 which appear, in particular, when considering harmonic generation. Following Yariv and Yeh /3/ (4) is often written in the following form:

$$P_i^r = \varepsilon_o(\chi_{ij}E_j^r + 2d_{ijk}E_j^rE_k^r + 4\chi_{ijkl}E_j^rE_k^rE_l^r + \ldots) \qquad (5)$$

where P_i^r and E_i^r are the i'th components of the field, and where summation over repeated indices is assumed. The superscript r indicates that these fields are real fields that can be expressed in terms of their complex amplitudes as follows:

$$P_i^r = \tfrac{1}{2}(P_i^\omega e^{i(\bar{k}\cdot\bar{r}-\omega t)} + c.c.) \qquad (i = x,y,z) \qquad (6)$$

where c.c. means complex conjugate.

In the following we shall only consider second order nonlinearities, i.e. the term

$$P_i^r = 2\varepsilon_o\ d_{ijk}\ E_j^r\ E_k^r \ . \qquad (7)$$

Consider two fields at frequencies, ω_1 and ω_2

$$E_j^r = \tfrac{1}{2}(E_j^{\omega_1} e^{i(\bar{k}_1\cdot\bar{r}-\omega_1 t)} + c.c.) \qquad (j = x,y,z) \qquad (8)$$

and

$$E_k^r = \tfrac{1}{2}(E_k^{\omega_2} e^{i(\bar{k}_2\cdot\bar{r}-\omega_2 t)} + c.c.) \qquad (k = x,y,z). \qquad (9)$$

Using these expressions (7) becomes

$$P_i^r = 2\varepsilon_o d_{ijk}\ (\tfrac{1}{2}E_j^{\omega_1} e^{i(\bar{k}_1\cdot\omega_1 t)} + \tfrac{1}{2}E_j^{\omega_2} e^{i(\bar{k}_2\cdot\bar{r}-\omega_2 t)} + c.c.)$$

$$x\ (\tfrac{1}{2}E_k^{\omega_1} e^{i(\bar{k}_1\cdot r-\omega_1 t)} + \tfrac{1}{2}E_k^{\omega_2} e^{i(\bar{k}_2\cdot\bar{r}-\omega_2 t)} + c.c.) . \qquad (10)$$

Due to the product in (10) terms oscillating with both $(\omega_1+\omega_2)$ and $(\omega_1-\omega_2)$ appear. Considering only the sum-frequency terms and using (6) equation (7) reduces to

$$P_i^{\omega_3=\omega_1+\omega_2} = \varepsilon_o d_{ijk}\ E_j^{\omega_1}\ E_k^{\omega_2} + \varepsilon_o d_{ijk}\ E_k^{\omega_2}\ E_j^{\omega_1} \qquad (11)$$

where it is assumed that

$$\omega_3 = \omega_1 + \omega_2 \tag{12}$$

and

$$\bar{k}_3 = \bar{k}_1 + \bar{k}_2. \tag{13}$$

To keep track of the frequencies involved a widely used notation for nonlinear susceptibilities has been introduced

$$d_{ijk} = d_{ijk}(-\omega_3, \omega_1, \omega_2) . \tag{14}$$

In this notation, for example, the difference frequency terms are expressed in terms of the $d_{ijk}(-\omega_3, \omega_1, -\omega_2)$ susceptibilities.

In (11) there can be no physical significance attached to the order in which the fields appear. Hence

$$d_{ijk}(-\omega_3, \omega_1, \omega_2) = d_{ikj}(-\omega_3, \omega_2, \omega_1) \tag{15}$$

and (11) can be written

$$P_i^{\omega_3 = \omega_1 + \omega_2} = 2\varepsilon_o \, d_{ijk}(-\omega_3, \omega_1, \omega_2) \, E_j^{\omega_1} \, E_k^{\omega_2} \tag{16}$$

with summation over repeated indices. This is the relationship between the complex amplitudes of the fields.

Equation (16) is only valid if $\omega_1 \neq \omega_2$. In case of second harmonic generation where $\omega_1 = \omega_2 = \omega$ the field components $E_j^{\omega_1}$ and $E_k^{\omega_2}$ in (10) are in fact components of the same field. Hence (10) must be written

$$P_i^r = 2\varepsilon_o \, d_{ijk}(\tfrac{1}{2}E_j^\omega \, e^{i(\bar{k}\cdot\bar{r}-\omega t)} + c.c.)(\tfrac{1}{2}E_k^\omega \, e^{i(\bar{k}\cdot\bar{r}-\omega t)} + c.c.) \tag{17}$$

resulting in

$$P_i^{\omega_3 = 2\omega} = \varepsilon_o \, d_{ijk} \, E_j^\omega \, E_k^\omega \tag{18}$$

with summation over repeated indices. (18) is valid for the complex amplitudes in case of second harmonic generation. In terms of the Kronecker delta $\delta_{\omega_1\omega_2}$ (16) and (18) can be combined

$$P_i^{\omega_3 = \omega_1 + \omega_2} = (2 - \delta_{\omega_1\omega_2}) \, \varepsilon_o \, d_{ijk} \, E_j^{\omega_1} \, E_k^{\omega_2} \, . \tag{19}$$

The 27 components of the nonlinear second order d-tensor have a number of symmetry relations among its components that restrict the number of independent elements to only a few for most crystals. If the material has inversion symmetry all the elements are zero. This can be seen from Eq. (19). Performing an inversion through a center of symmetry, the crystal remains unchanged, but both P and E change sign. When P changes sign the sign of (19) is changed, while it remains unchanged when E changes sign. Hence

$$d_{ijk} = 0 \qquad (20)$$

in centrosymmetric crystals. This is only true for bulk materials in the local response approximation. Second order effects can appear at surfaces and in case of non-local response also in centrosymmetric crystals.

Kleinman /4/ has given a useful set of symmetry conditions derived from an energy function

$$U(E) = -\frac{1}{3} d_{ijk} E_i E_j E_k \qquad (21)$$

from which the polarization can be derived as

$$\bar{P} = - \bar{\nabla}_E U. \qquad (22)$$

Since no physical significance can be attached to a rearrangement of the electric field components in (21), all the terms d_{ijk} which result from a mere rearrangement of the subscripts are equal. Hence

$$d_{ijk} = d_{kij} = d_{jki} = d_{ikj} = d_{kji} = d_{jik} \,. \qquad (23)$$

Eq. (15) is contained in (23), but is more general since (23) is only valid for the lossless case for wavelengths away from resonances in the material.

The symmetry relation (15) gives rise to the so-called contracted notation in analogy with the piezoelectric notation /5/.

$xx = 1$, $yy = 2$, $zz = 3$, $yz = zy = 4$, $xz = zx = 5$,

$xy = yx = 6$.

The resulting d_{ij} tensor forms a 3x6 matrix which operates on the electric fields according to the following rule:

$$\begin{bmatrix} P_x \\ P_y \\ P_z \end{bmatrix} = \begin{bmatrix} d_{11} & d_{12} & d_{13} & d_{14} & d_{15} & d_{16} \\ d_{21} & d_{22} & d_{23} & d_{24} & d_{25} & d_{26} \\ d_{31} & d_{32} & d_{33} & d_{34} & d_{35} & d_{36} \end{bmatrix} \begin{bmatrix} E_x^2 \\ E_y^2 \\ E_z^2 \\ 2E_zE_y \\ 2E_zE_x \\ 2E_xE_y \end{bmatrix} . \qquad (24)$$

In the usual crystals many of these terms vanish /3/.

3. Anharmonic Oscillator Model

The harmonic oscillator model is extremely useful in solid state physics for modelling the frequency response close to resonances in the material. Similarly in nonlinear optics the anharmonic oscillator model is useful.

Consider an elementary charge e with mass m bound to an oppositely charged center with a force constant $m\omega_o{}^2$ where ω_o is the angular resonance frequency. The equation of motion is then

$$mx + m\omega_o^2 x + ax^2 = eE_o \cos\omega t \tag{25}$$

where x is the displacement of the charge, and $E_o\cos\omega t$ is the applied electric field. A term ax^2 is added in order to model the second order nonlinearity.

Considering first the purely harmonic response we obtain with a = o

$$x_1(t) = \frac{e/m}{\omega_o^2 - \omega^2} E_o \cos\omega t \ . \tag{26}$$

Since the polarization P of the material is defined as the induced dipole moment per unit volume we have

$$P = Nex = \frac{Ne^2/m}{\omega_o^2 - \omega^2} E_o \cos\omega t \tag{27}$$

where N is the number of induced dipoles per unit volume. Furthermore, usual linear response theory yields (first term of Eq. (4))

$$P = \varepsilon_o \chi^{(1)} (\omega) \ E_o \cos\omega t \ . \tag{28}$$

Hence, the linear susceptibility is given by

$$\chi^{(1)}(\omega) = \frac{Ne^2}{m\varepsilon_o} \frac{1}{\omega_o^2 - \omega^2} \ . \tag{29}$$

Similarly d(i.e. $\chi^{(2)}(\omega)$)can be found from (25) in the approximation where the term ax^2 is considered small. Inserting $ax_1^2(t)$ for ax^2 in (25) where $x_1(t)$ is given by (26), the solution of (25) is given by

$$x(t) = -\frac{a}{2\omega_o^2} \left(\frac{e/m}{\omega_o^2 - \omega^2}\right)^2 E_o^2 + \frac{e/m}{\omega_o^2 - \omega^2} E_o \cos\omega t$$
$$- \frac{a}{2} \frac{1}{\omega_o^2 - 4\omega^2} \left(\frac{e/m}{\omega_o^2 - \omega^2}\right)^2 E_o^2 \cos 2\omega t \ . \tag{30}$$

It is seen that in addition to the linear term previously found both a DC-term and a second harmonic term appear. Both of these terms are small since they are proportional to the a-coefficient.

Separating x(t) in (30) into its complex amplitudes

$$x(t) = x_o + \tfrac{1}{2}(x_\omega e^{-i\omega t} + c.c.) + \tfrac{1}{2}(x_{2\omega} e^{-2i\omega t} + c.c.) \tag{31}$$

and using (27) the second harmonic polarization is given by

$$P_{2\omega} = Nex_{2\omega} = -\frac{aNe^3}{2m^2} \frac{1}{(\omega_o^2 - 4\omega^2)(\omega_o^2 - \omega^2)^2} E_o^2 \ . \tag{32}$$

Since this term generates the second harmonic signal it can be compared with (18) to yield the nonlinear susceptibility of second order

$$d = - \frac{aNe^3}{2\varepsilon_o m^2} \frac{1}{(\omega_o^2 - 4\omega^2)(\omega_o^2 - \omega^2)^2} . \quad (33)$$

This is the result of the anharmonic oscillator model. It is seen that d is, of course, proportional to the anharmonic force coefficient. It is also resonant not only at ω_o, but also at $\omega_o/2$. The corresponding absorption at this frequency is the two-photon absorption. It is also seen that d can be expressed in terms of the linear susceptibility $\chi^{(1)}$ given by (29)

$$d = - \frac{am\varepsilon_o^2}{2N^2e^3} \chi^{(1)} (2\omega) \left(\chi^{(1)}(\omega)\right)^2 . \quad (34)$$

Miller /6/ has generalised this expression in the so-called Miller's rule

$$d_{ijk}(-\omega_A, \omega_B, \omega_C) = \delta_{ijk} \chi_{ii}(\omega_A) \chi_{jj}(\omega_B) \chi_{kk}(\omega_C) . \quad (35)$$

The parameter δ_{ijk} turns out to be remarkably constant for a number of inorganic materials. Organic materials tend to have values of δ_{ijk} that are 10-100 times greater.

4. Three Wave Mixing

Only the response of the material has been considered in the previous sections. This response (the polarization) acts back via Maxwell's equations on the electric field so coupled modes are formed. In case of second order nonlinearities three waves are mixed by the coupled nonlinear wave equations. This so-called three wave mixing is responsible for the usual second order effects like sum and difference frequency generation, parametric amplification and oscillation etc.

The basic equations are Maxwell's equations for non-magnetic materials

$$\nabla \times \bar{H} = \bar{J} + \frac{\partial \bar{D}}{\partial t} \quad (36)$$

$$\nabla \times \bar{E} = - \frac{\partial}{\partial t}(\mu_o \bar{H}) \quad (37)$$

$$\nabla \cdot \bar{H} = 0 \quad (38)$$

$$\nabla \cdot \bar{D} = \rho \quad (39)$$

and the material equations

$$\bar{D} = \varepsilon_o \bar{E} + \bar{P} \quad (40)$$

$$\bar{J} = \sigma E \quad (41)$$

$$\bar{P} = \varepsilon_o \chi^{(1)} \bar{E} + \bar{P}_{ne} \quad (42)$$

where σ, the specific conductivity, is introduced in order to account for losses, and where $\bar{P}_{ne}$ is the nonlinear part of the response which for a second order effect is given by

$$P_i = 2\varepsilon_o \, d_{ijk} \, E_j \, E_k \tag{43}$$

as shown in Eq. (4). All the fields are here expressed in terms of real field amplitudes. From (36),(37),(40),(41) and (42) the usual wave equation for the electric field is obtained

$$\nabla^2 \bar{E} = \mu_o \sigma \frac{\partial \bar{E}}{\partial t} + \mu_o \varepsilon \frac{\partial^2 \bar{E}}{\partial t^2} + \mu_o \frac{\partial^2 \bar{P}_{ne}}{\partial t^2} \tag{44}$$

where

$$\varepsilon = \varepsilon_o \, (1 + \chi^{(1)}) \tag{45}$$

is the linear part of the dielectric constant of the material.

Introducing the complex amplitudes and considering for simplicity a scalar approximation the real field amplitude is given by

$$E_i^r = \tfrac{1}{2} E_i \, e^{i(k_i z - \omega t)} + \text{c.c.} \tag{46}$$

where i = 1,2,3 now indicates one of the three coupled waves, the second derivative can be written as

$$\nabla^2 E_i^r = \tfrac{1}{2} \left(\frac{d^2 E_i}{dz^2} + 2ik \frac{dE_i}{dz} - k^2 E_i \right) e^{i(k_i z - \omega t)} + \text{c.c.} \tag{47}$$

It is usually assumed that the amplitude E_i is slowly varying, i.e. on the wavelength scale of light E_i is considered nearly constant. This is the so-called SVEA-approximation (Slowly Varying Envelope Approximation)

$$|k^2 \, E_i| \gg \left| k \frac{dE_i}{dz} \right| \gg \frac{d^2 E_i}{dz^2} \, . \tag{48}$$

In this approximation the wave equation can be written

$$\frac{dE_i}{dz} e^{i(k_i z - \omega t)} + \text{c.c.} = -\tfrac{1}{2} \left(\frac{\mu_o}{\varepsilon_i} \right)^{\frac{1}{2}} \sigma_i E_i e^{i(k_i z - \omega t)} + \text{c.c.} - i \left(\frac{\mu_o}{\varepsilon_i} \right)^{\frac{1}{2}} \frac{1}{\omega_i} \frac{\partial^2 P_{ne}^r}{\partial t^2} \tag{49}$$

where

$$k_i = \omega_i (\mu_o \varepsilon_i)^{\frac{1}{2}} \, . \tag{50}$$

In Eq. (49) the P_{ne}^r-term acts as a source which generates the electric field. Obviously the three terms must oscillate synchronously, i.e. with the same frequency. Considering the case where

$$\omega_3 = \omega_1 + \omega_2 \, , \tag{51}$$

the relationship between the complex amplitudes of the second order response is given by (19)

$$P_{ne} = (2 - \delta_{\omega_1\omega_2})\varepsilon_o \, d \, E_1 \, (\omega_1) \, E_2 \, (\omega_2) \tag{52}$$

where P_{ne} is the complex amplitude defined as

$$P^r_{ne} = \tfrac{1}{2}P_{ne} \, e^{i(kz-\omega t)} + c.c. \tag{53}$$

Inserting (52) and (53) into (49) the following equations for the complex amplitudes are obtained:

$$\frac{dE_3}{dz} = - \frac{\mu_o \sigma_3 c}{2n_3} E_3 + \frac{i\alpha\omega_3\mu_o dc}{n_3} E_1E_2 \, e^{i\Delta kz} \tag{54}$$

$$\frac{dE_1}{dz} = - \frac{\mu_o \sigma_1 c}{2n_1} E_1 + \frac{i\alpha\omega_1\mu_o dc}{n_1} E_3E_2^* \, e^{-i\Delta kz} \tag{55}$$

$$\frac{dE_2}{dz} = - \frac{\mu_o \sigma_2 c}{2n_2} E_2 + \frac{i\alpha\omega_2\mu_o dc}{n_2} E_3E_1^* \, e^{-i\Delta kz} \tag{56}$$

where

$$\Delta k = k_1 + k_2 - k_3 \tag{57}$$

$$\left(\frac{\mu_o}{\varepsilon_i}\right)^{\frac{1}{2}} = \mu_o \frac{c}{n_i} \tag{58}$$

and

$$\alpha = \tfrac{1}{2}(2 - \delta_{\omega_i\omega_j}) \, . \tag{59}$$

Equations (54)-(56) form the basis of second order nonlinear optics and can be applied for all three wave mixing cases.

5. Second Harmonic Generation

In the special case of second harmonic generation we have

$$\omega_3 = 2\omega \tag{60}$$

and

$$\alpha = \tfrac{1}{2} \, . \tag{61}$$

Neglecting absorption (i.e. $\sigma_i = 0$) the coupled equations (54)-(56) reduce to

$$\frac{dE_3}{dz} = \frac{i\omega_3\mu_o dc}{2n(2\omega)} E_1^2 \, e^{i\Delta kz} \tag{62}$$

$$\frac{dE_1}{dz} = \frac{i\omega_1\mu_o dc}{2n(\omega)} E_3E_1^* \, e^{-i\Delta kz} \tag{63}$$

with

$$\Delta k = 2k_1 - k_3 = 2\frac{\omega}{c}(n(\omega) - n(2\omega)) . \qquad (64)$$

In the approximation where the fundamental wave is not depleted (i.e. E_1 is considered constant), the solution of (62) is

$$E_3(z) = \frac{i\omega\mu_o dc}{n(2\omega)} E_1^2 \, z \, e^{i\Delta kz/2} \left(\frac{\sin\frac{1}{2}\Delta kz}{\frac{1}{2}\Delta kz}\right) . \qquad (65)$$

Since the intensity of a wave is given by

$$I = \frac{n}{2\mu_o c} |E|^2 ,$$

the intensity of the second harmonic output from a crystal of thickness L is given by

$$I(2\omega) = K \, I^2(\omega) \, L^2 \left(\frac{\sin\frac{1}{2}\Delta kL}{\frac{1}{2}\Delta kL}\right)^2 \qquad (66)$$

where

$$K = \frac{2\mu_o^3 c^3 \omega^2 d^2}{n^2(\omega) n(2\omega)} .$$

It is also possible to take into account the depletion of E_1 (i.e. solve (62) and (63) exactly). The result is then

$$I(2\omega) = I(\omega)\tanh^2\left((KI(\omega))^{\frac{1}{2}} \, L \, \frac{\sin\frac{1}{2}\Delta kL}{\frac{1}{2}\Delta kL}\right) . \qquad (67)$$

When focussing the beam it is important to note that the confocal parameter of the beam must be greater than the crystal length L in order that (67) is valid. If this is not the case the efficiency of conversion is reduced as discussed by Boyd and Kleinman /7/.

6. Phase matching

In order to achieve a sufficient frequency conversion the condition $\Delta k = 0$ in (54)-(56) must be satisfied. In the case of second harmonic generation this condition becomes

$$\Delta k = k(\omega) + k(\omega) - k(2\omega) = \frac{\omega}{c}(n(\omega) + n(\omega) - 2n(2\omega)) . \qquad (68)$$

If this phase match condition is satisfied there are equal phase velocities for the fundamental and the second harmonic wave. In this case $I(2\omega) \propto L^2$ as seen from (66) or (67). If the condition is not satisfied a coherence length

$$\ell_c = \frac{\pi}{\Delta k} \qquad (69)$$

is introduced and the intensity of generated harmonic (66) can be rewritten as

$$I(2\omega) = \frac{4K}{\pi^2} I^2(\omega) \ell_c^2 \sin^2\left(\frac{\pi \ell}{2\ell_c}\right) . \qquad (70)$$

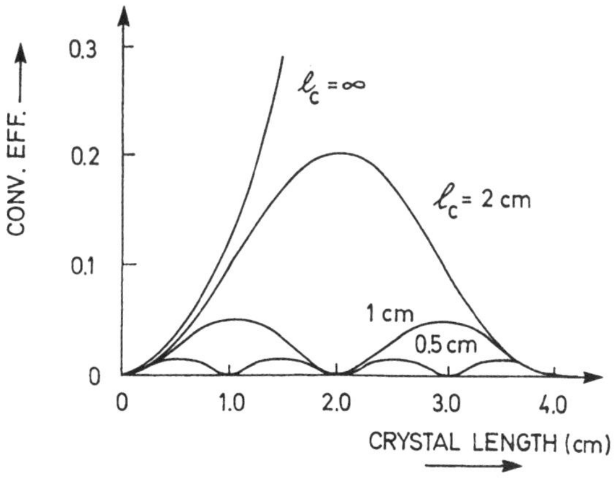

Fig. 1. Example of second harmonic conversion efficiency as a function of the length of the nonlinear crystal for various coherence lengths.

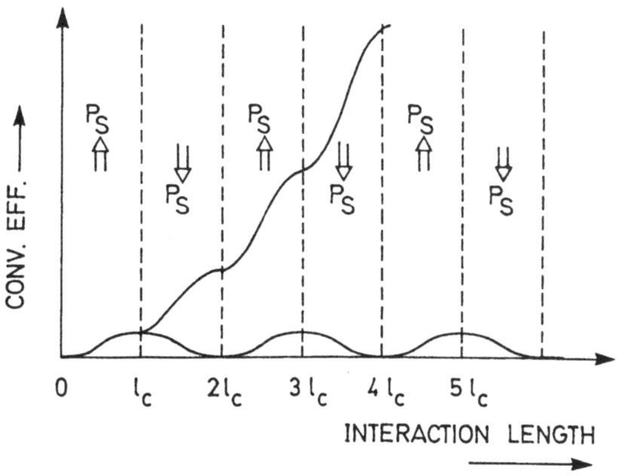

Fig. 2. Quasi phase matching. The sign of the nonlinear susceptibility is reversed periodically with the period equal to the coherence length. P_S is the spontaneous polarization of the crystal.

The conversion officiency obtained for different values of ℓ_c is illustrated in Fig. 1.

This leads to the so-called quasi phase matching scheme shown in Fig. 2, where the sign of the d-coefficient is reversed periodically by changing the direction of the spontaneous polarization in the non-linear crystal during the crystal growth /8/.

Usually the phase matching condition is satisfied using birefringent crystals. Often uniaxial crystals are applied. In this case the two refractive indices are given by

n_o ordinary index (71)

$$n_e(\theta) = \frac{n_o n_e}{(n_o^2 \sin^2\theta + n_e^2 \cos^2\theta)^{\frac{1}{2}}} \qquad \text{extraordinary index} \qquad (72)$$

where θ is the angle between the optic axis and the propagation direction and $n_e = n_e(\theta=\pi/2)$. Several possibilities now exist for satisfying (68).

Type I phase matching with positive unaxial crystals ($n_2 > n_o$):

$$n_o(2\omega) = n_e(\omega,\theta_m) \quad \text{where} \quad \sin^2\theta_m = \frac{(n_o(2\omega))^{-2} - (n_o(\omega))^{-2}}{(n_e(\omega))^{-2} - (n_o(\omega))^{-2}} \qquad (73)$$

where θ_m is the angle where phase matching occurs.

Type I phase matching with negative unaixial crystals ($n_e < n_o$):

$$n_e(2\omega,\theta_m) = n_o(\omega) \quad \text{where} \quad \sin^2\theta_m = \frac{(n_o(\omega))^{-2} - (n_o(2\omega))^{-2}}{(n_e(2\omega))^{-2} - (n_o(2\omega))^{-2}} \qquad (74)$$

Type II phase matching with positive unaxial crystal ($n_e > n_o$):

$$n_o(2\omega) = \tfrac{1}{2}(n_o(\omega) + n_e(\omega,\theta_m)) \qquad \text{where}$$

$$\sin^2\theta_m = \frac{(2n_o(2\omega) - n_o(\omega))^{-2} - (n_o(\omega))^{-2}}{(n_e(\omega))^{-2} - (n_o(\omega))^{-2}} \; . \qquad (75)$$

Type II phase matching with negative unaxial crystals ($n_e < n_o$):

$$n_e(2\omega,\theta_m) = \tfrac{1}{2}(n_o(\omega) + n_e(\omega,\theta_m)) \qquad (76)$$

where θ_m must be found by means of iteration methods.

The uniaxial crystal is cut along the phase matching direction θ_m and angle tuned. It can also be used with oblique incidence in Brewster angle and tuned by turning the crystal around its surface normal /9/. In this connection the frequency doubler deflector /10/ can be mentioned. If the fundamental beam is strongly focussed in the crystal the outgoing beam is divergent. Only a small part of this beam is phase matched since this is critically dependent on the angle. Changing the wavelength of the fundamental beam (or changing the phase match condition e.g. by electric field) the harmonic beam is deflected because the phase match angle changes.

Phase matching by angle tuning in anisotropic crystals is also called critical phase matching due to the strong angle dependence. Three problems must be considered in connection with critical phase matching.

a) Beam walk off. Since the energy of the extraordinary wave propagates in another direction than the wavefronts the crystal length must be shorter than the aperture length

$$\ell_a = \frac{a}{\rho} \tag{77}$$

where a is the beam radius and

$$\tan\rho = \tfrac{1}{2}\, n_o^2(\omega)\ \left((n_e(2\omega))^{-2} - (n_o(2\omega))^{-2}\right)\sin 2\theta_m \tag{78}$$

where ρ is the beam walk off angle.
b) Focussing. The phase matching condition (68) is only satisfied for a very narrow range of angles. The full angle $\Delta\theta$ where the second harmonic intensity has fallen to half intensity is given by /11/

$$\Delta\theta = \frac{0.44\lambda\ n_o(\omega)/\ell}{(n_o(2\omega) - n_e(2\omega))\sin 2\theta_m} . \tag{79}$$

For a crystal length $\ell \sim 1$ cm we find $\Delta\theta \sim 1$ mrad which corresponds to an unfocussed beam.
c) Spectral width. The spectral width within which more than 50% is converted is given by /11/

$$\Delta\lambda = \frac{0.44\ \lambda/\ell}{\dfrac{\partial n_o(\omega)}{\partial\lambda} - \tfrac{1}{2}\dfrac{\partial n_e(2\omega,\theta)}{\mu\lambda}} . \tag{80}$$

For a crystal length of $\ell \sim 1$ cm we find $\Delta\lambda \sim 0.1$ nm.

The strict condition (79) is somewhat relaxed for $\theta_m = 90$ degrees. This case is termed non-critical phase matching and the divergence angle corresponding to 50% drop in second harmonic intensity is now given by /11/

$$\Delta\theta = \left(\frac{0.44\ \lambda/\ell}{(n_o(2\omega) - n_e(2\omega))}\right)^{\frac{1}{2}} n_o(\omega) . \tag{81}$$

In this case $\ell \sim 1$ cm yields $\Delta\theta \sim 50$ mrad. Hence the fundamental beam can be focussed much more implying a much higher conversion efficiency.

In case of non-critical phase matching angle tuning is impossible and instead temperature tuning is applied. Since the refractive indices are temperature dependent it is often possible to find a temperature where (68) is satisfied for the anisotropic crystal.

7. Parametric effects

Parametric effects occur when one of the beams in the three wave mixing process is considerably stronger than the other two. As an example we shall here consider parametric amplification and oscillation.

The basic equations (54)-(56) of three wave mixing are rewritten as follows

$$\frac{dE_1}{dz} = -\tfrac{1}{2}\,\alpha_1\ E_1 + i\kappa_1 E_2^* E_3\ e^{-i\Delta kz} \tag{82}$$

$$\frac{dE_2}{dz} = -\tfrac{1}{2}\,\alpha_2\ E_2 + i\kappa_2 E_1^* E_3\ e^{-i\Delta kz} , \tag{83}$$

$$\frac{dE_3}{dz} = - \tfrac{1}{2}\, \alpha_3 E_3 + i\kappa_3 E_1 E_2\, e^{i\Delta kz} \tag{84}$$

where

$$\Delta k = k_1 + k_2 - k_3 \tag{85}$$

$$\alpha_i = \sigma_i \frac{\mu_o c}{n_i} \tag{86}$$

$$\kappa_i = \frac{\omega_i}{n_i} \mu_o cd \tag{87}$$

and

$$\omega_3 = \omega_1 + \omega_2 . \tag{88}$$

Only the case $\omega_1 \neq \omega_2$ is considered so $\alpha = 1$ in (59).

For simplicity the loss is neglected ($\alpha_i = 0$). We consider a nonlinear crystal where all three waves are present at the input with the fields $E_1(0)$, $E_2(0)$ and $E_3(0)$. Regarding the amplitude $E_3(0)$ of the strong field as a parameter the two coupled equations (82) and (83) can be solved

$$E_1(z)\, e^{-i\Delta kz/2} = E_1(0)\ (\cosh bz - i\, \frac{\Delta k}{2b} \sinh bz) + i\, \frac{g}{2b}\, E_2^*(0)\ \sinh bz \tag{89}$$

$$E_2^*(z)\, e^{i\Delta kz/2} = E_2^*(0)\ (\cosh bz + i\, \frac{\Delta k}{2b} \sinh bz) - i\, \frac{g}{2b}\, E_1(0) \sinh bz \tag{90}$$

where

$$g = \mu_o c \left(\frac{\omega_1}{n_1}\ \frac{\omega_2}{n_2}\right)^{\frac{1}{2}} d\ E_3(0) \tag{91}$$

and

$$b = \tfrac{1}{2}\ (g^2 - (\Delta k)^2)^{\frac{1}{2}} . \tag{92}$$

It is seen that the fields E_1 and E_2 are amplified. Consider the situation where $E_2(0) = 0$ and $\Delta k = 0$ (phase matching). Then (89) and (90) reduce to

$$E_1(z) = E_1(0)\ \cosh bz \cong \tfrac{1}{2}\, E_1(0)\ e^{gz/2} \tag{93}$$

$$E_2^*(z) = -i\ E_1(0)\ \sinh bz \cong -\tfrac{1}{2}\, E_1(0)\ e^{gz/2} . \tag{94}$$

Hence, wave one (the signal wave (ω_1) with input amplitude $E_1(0)$) is amplified with a gain g as shown in (91), but also wave two (the idler wave (ω_2) with no input intensity) is amplified.

This parametric amplification process of course takes energy from the pump wave (ω_3). A useful relation in this connection is obtained from (82)-(84) with $\alpha_i = 0$ by multiplying each of the fields of the left hand side with their complex conjugate

$$\frac{1}{\omega_1}\frac{dI_1}{dz} = \frac{1}{\omega_2}\frac{dI_2}{dz} = -\frac{1}{\omega_3}\frac{dI_3}{dz} \tag{95}$$

where

$$I_i = \tfrac{1}{2}\frac{n_i}{\mu_o c}|E_1|^2 = N_i\hbar\omega_i \tag{96}$$

is the intensity and N_i the photon flux for the i'th beam. This so-called Manley-Rowe relation /12/ states that the photon flux created in beam one is equal to that created in beam two, and that this is also equal to the photon flux annihilated from beam three.

By means of these parametric effects it is possible to construct a light amplifier. In analogy to the laser oscillator which is based on the light amplification due to stimulated emission it is possible to construct a parametric oscillator emitting coherent light based on the parametric amplifier placed in an optical resonator.

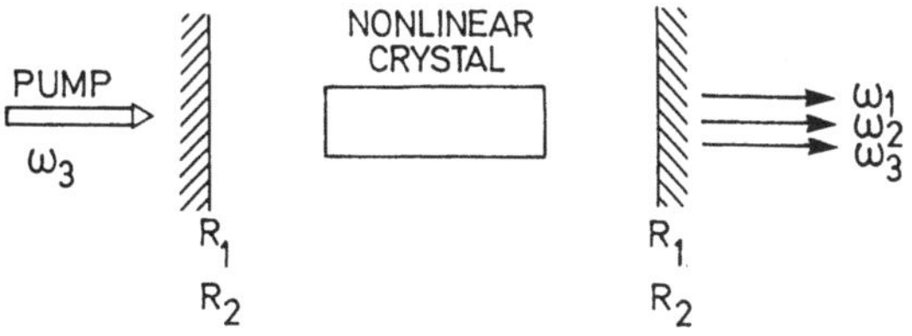

Fig. 3 Parametric oscillator. The pump wave (wave 3) pumps the nonlinear crystal so both waves 1 and 2 are amplified. If the amplification exceeds the losses, the system oscillates and emits coherent light.

The parametric oscillator is shown in Fig. 3 where the nonlinear crystal is pumped by wave three (ω_3) and is placed in an optical resonator with mirror reflectivities r_1 for wave one (ω_1) and r_2 for wave two (ω_2). Assuming that these mirror losses are the only losses in the resonator and that phase matching exists ($\Delta k = 0$) the fields for a complete round trip in the crystal of length L are given by

$$E_1(2L) = E_1(0)r_1^2\, e^{2ik_1L}\cosh gL + ir_1^2E_2^*(0)\, e^{2ik_1L}\sinh gL \tag{97}$$

$$E_2^*(2L) = -iE_1(0)r_2^{*2}\, e^{-2ik_2L}\sinh gL + E_2^*(0)r_2^{*2}e^{-2ik_2L}\cosh gL \tag{98}$$

The self-consistency requirement is

$$E_1(2L) = E_1(0)\, e^{i2\pi m} \tag{99}$$

$$E_2^*(2L) = E_2^*(0)\, e^{-i2\pi n}. \tag{100}$$

Introducing the intensity reflection coefficient R

$$r_i^2 = R_i\ e^{i\Phi_i} \qquad (101)$$

and inserting (99) and (100) into (97) and (98), non-trivial solutions only exist when the determinant is zero leading to the following oscillation conditions:

$$2k_1L = 2\pi m + \Phi_1 \qquad (102)$$

$$2k_2L = 2\pi n + \Phi_2 \qquad (103)$$

and

$$(R_1\cosh gL - 1)\ (R_2\cosh gL - 1) = R_1R_2\sinh^2 gL \qquad (104)$$

or

$$(R_1 + R_2)\cosh gL - R_1R_2 = 1\ . \qquad (105)$$

For $R_1, R_2 \cong 1$ and $\cosh gL \cong 1 + \frac{1}{2}g^2L^2$ (105) becomes

$$gL = ((1-R_1)(1-R_2))^{\frac{1}{2}}\ , \qquad (106)$$

yielding the threshold gain necessary to overcome the mirror losses.

From (91) and (96) the threshold intensity of the pump wave can be found:

$$I_{3t} = \frac{n_1n_2n_3\ (1-R_1)(1-R_2)}{2\mu_o^3c^3\omega_1\omega_2 d^2L^2}\ . \qquad (107)$$

Above threshold each pump photon is converted into a signal photon (ω_1) and an idler photon (ω_2) according to (95). Hence the output intensity is given by

$$\frac{I_1}{\omega_1} = \frac{I_2}{\omega_2} = \frac{I_{3t}}{\omega_3}\left(\frac{I_3}{I_{3t}} - 1\right)\ . \qquad (108)$$

The conditions (102) and (103) require that the oscillation frequencies must correspond to two longitudinal modes of the resonator. The frequencies ω_1 and ω_2 are selected from the conditions

$$\omega_3 = \omega_1 + \omega_2 \qquad (109)$$

and the phase matching condition

$$n(\omega_3)\omega_3 = n(\omega_1)\omega_1 + n(\omega_2)\omega_2\ . \qquad (110)$$

If as described in the preceding the mirrors have high reflectivity for both the ω_1- and ω_2-wave, both the waves are emitted from the oscillator, and it is then called a doubly resonant parametric oscillator. One serious disadvantage of this is that it has low stability /2/. Due to the conditions (102),

(103) and (109) the shift in output frequencies $\Delta\omega$ due to a small fluctuation $\Delta\ell$ in resonator length is given by

$$\Delta\omega \cong \frac{n(\omega_2)}{n(\omega_2) - n(\omega_1)} \omega_3 \frac{\Delta\ell}{\ell} . \tag{111}$$

Due to the difference in the denominator these fluctuations are several orders of magnitude greater than for an ordinary laser.

These output fluctuations can be avoided with a singly resonant parametric oscillator where the mirrors are only reflecting for the ω_1-wave. Hence $R_2 = 0$ and the threshold condition obtained from (105) becomes

$$I_{3t} = \frac{n_1 n_2 n_3 \ (1-R_1)}{\mu_{o3}^3 c^3 \omega_1 \omega_2 d^2 L^2} . \tag{112}$$

Hence it requires considerably more pump power to reach threshold for the single resonant oscillator. The ratio between the two threshold intensities is

$$\frac{(I_{3t})_s}{(I_{3t})_d} = \frac{2}{1-R} \tag{113}$$

where R is the reflection coefficient for the ω_2-wave in case of doubly resonant oscillation.

References

1 A. Yariv, "Quantum Electronics" (John Wiley & Sons, New York, 1975).
2 Y.R. Shen, "The Principes of Nonlinear Optics", (John Wiley & Sons, New York 1984).
3 A. Yariv and P. Yeh, "Optical Waves in Crystals", (John Wiley & Sons, New York 1984).
4 D.A. Kleinman, Phys. Rev. 126, 1977 (1962).
5 J.F. Nye, "Physical Properties of Crystals", (Oxford Univ. Press, London, 1960).
6 R.C. Miller, Appl. Phys. Lett. 1, 17 (1964).
7 G.D. Boyd and D.A. Kleinman, J. Appl. Phys. 39, 3597 (1968).
8 E.J. Lim, M.M. Fejer and R.L. Byer, Electron. Lett. 25, 174 (1989).
9 I. Filinski and T. Skettrup, Proc. SPIE 492, 93 (1985).
10 I. Filinski and T. Skettrup, Proc. SPIE 369, 345 (1982).
11 W. Koechner, "Solid State Laser Engineering", Ch. 10 (Springer-Verlag, Berlin 1976).
12 J.M. Manley and H.E. Rowe, Proc. IRE 47, 2115 (1959).

Part II

Cavity Phenomena

Cavity QED

P. Meystre

Optical Sciences Center, University of Arizona, Tucson, AZ85721, USA

Abstract. In cavity quantum electrodynamics (QED) [1], one alters the mode of the electromagnetic field in a resonator to obtain such novel effects as inhibited or enhanced irreversible spontaneous emission, reversible spontaneous emission, micromaser action, and "quantum collapse and revivals." This new line of investigation has numerous ramifications. In particular it impacts the generation of nonclassical fields, the study of the quantum/classical correspondence, nonlinear dynamics, and quantum measurement theory.

1. Single-Mode Spontaneous Emission

It is well known that spontaneous emission results from the interaction between an atom and an electromagnetic field in the vacuum state. In its simplest form, this interaction is described by the Hamiltonian [2]

$$\mathcal{H} = \frac{1}{2}\hbar\omega\sigma_z + \hbar\Omega a^\dagger a + \hbar(ga\sigma_+ + \mathrm{adj}) \ . \tag{1}$$

Here σ_+ and σ_- are Pauli spin-flip matrices and a, $a^\dagger$ are annihilation and creation operators of the field mode $[a,a^\dagger] = 1$. The coupling constant $g = -(\wp E_0/2\hbar)$ where $\wp$ is the atomic dipole matrix element. $\mathcal{E}_\Omega$, the "electric field per photon," is equal to $[\hbar\Omega/\epsilon_0 V]^{1/2}$, where V is the quantization volume. This Hamiltonian, which defines the Jaynes-Cummings model [3], provides a simple illustration of spontaneous emission. It also leads to dynamics that exhibit a "Cummings collapse" of the Rabi flopping induced by a coherent state, as well as "revivals" of this flopping resulting from the discrete nature of the photon field [4]. The Hamiltonian $\mathcal{H}$ can be diagonalized exactly, allowing us to determine the evolution of any observable of interest. In particular, we can compute the probability that an atom initially in the upper state $|a\rangle$, and interacting with a field initially in the Fock state $|n\rangle$, remains in that state at time t, as

$$P_a(t) = \cos^2(g\sqrt{n+1}\,t) \ . \tag{2}$$

Similarly, the probability that the atom will be in the upper state at time t if it is initially in the ground state $|b\rangle$ is

$$P_a(t) = \sin^2(g\sqrt{n}\,t) \ . \tag{3}$$

This result is sometimes referred to as quantum Rabi flopping.

One essential difference between the semiclassical and quantum Rabi problems is that in the quantum case an initially excited atom flops even in the absence of an applied field, i.e., for $n = 0$. This occurs because even though the vacuum expectation value for the field amplitude vanishes

$$\langle E\rangle = \mathcal{E}_\Omega\langle 0|a + a^\dagger|0\rangle = 0 \ , \tag{4}$$

that for the intensity does not:

$$\langle E^2\rangle = \mathcal{E}_\Omega^2\langle 0|(a + a^\dagger)^2|0\rangle = \mathcal{E}_\Omega^2 \ . \tag{5}$$

Stated another way, vacuum fluctuations exist in the electromagnetic field. These fluctuations effectively stimulate an excited atom to emit, a process called spontaneous emission. The

Springer Series in Wave Phenomena, Vol. 9 **Nonlinear Optics in Solids**
Editor: O. Keller

weak Rabi flopping that occurs for n = 0 in Eq. (2) results from spontaneous emission alone, which is neglected in the semiclassical approximation. If the atom is initially unexcited, however, no flopping occurs; spontaneous absorption does not exist.

Except for special situations, this model of spontaneous emission is unrealistic, since it involves only a single mode of the field and yields *vacuum Rabi flopping*, or *reversible spontaneous emission*, instead of the well-known exponential decay.

2. Spontaneous Emission in Free Space

In the previous section, we demonstrated that an excited atom undergoes Rabi flopping at a slow but nonzero rate because of vacuum fluctuations alone. In free space the atom interacts with a continuum of modes, which in the Born-Markov approximation leads to an exponential decay of the excited state probability. The atom-field interaction now is described by the Hamiltonian[2]

$$\mathcal{H} = \hbar\omega\sigma_z + \hbar\sum_{\mathbf{k}} \omega_k a_{\mathbf{k}}^{\dagger} a_{\mathbf{k}} + \sum_{\mathbf{k}} (g_{\mathbf{k}} a_{\mathbf{k}} \sigma_+ + \text{adj.}) , \tag{6}$$

where the sum is over all modes of the field that the atom interacts with, and $\omega_k = ck$. In this case, the Weisskopf-Wigner theory of spontaneous emission leads to the decay rate

$$\gamma = 2\pi \int d\Omega_{\mathbf{k}} \, d\omega_k \, |g_{\mathbf{k}}|^2 \, \mathcal{D}(\omega_k) \Big|_{\omega_k = \omega} , \tag{7}$$

where $\mathcal{D}(\omega_k) \, d\Omega_{\mathbf{k}} \, d\omega_{\mathbf{k}}$ is the mode density between $\omega_{\mathbf{k}}$ and $\omega_{\mathbf{k}} + d\omega_{\mathbf{k}}$ in the solid angle $d\Omega_{\mathbf{k}}$. In Eq. (7) we have replaced the sum over modes by an integral

$$\sum_{K} f(K) \to \int d\Omega \, \mathcal{D}(\Omega) \, f(\Omega) , \tag{8}$$

with

$$\mathcal{D}(\Omega) = \frac{\Omega^2 V}{\pi^2 c^3} . \tag{9}$$

Taking into account the sum over polarizations implicit in Eq. (7) leads to the well-known Weisskopf-Wigner spontaneous decay rate [5]

$$\gamma = \frac{\omega^3 |\wp|^2}{3\pi\epsilon_0 \hbar c^3} = \frac{1}{4\pi\epsilon_0} \frac{4\omega^3 |\wp|^2}{3\hbar c^3} . \tag{10}$$

The Weisskopf-Wigner theory predicts an *irreversible* exponential decay of the upper state population with no revivals, in contrast to the Jaynes-Cummings problem of the preceding section. Whereas in the Jaynes-Cummings case a quasiperiodicity results from interaction with a single mode and from the discrete nature of the possible photon numbers, in ordinary free-space spontaneous emission, the atom interacts with a continuum of modes. Although under the action of each individual mode the atom would have a finite probability to return to the upper state in a way similar to the Jaynes-Cummings revival, the probability amplitudes for such events interfere destructively when summed over the continuum of free space modes.

3. Spontaneous Emission in Cavities

In cavities with volumes comparable to the interaction wavelength, the density of states differs appreciably from that given by Eq. (10). For wavelengths below the cavity cutoff, the density of states vanishes altogether, while for wavelengths somewhat above the cutoff, the density of states may be substantially larger or smaller than the free-space value of Eq. (9). Accordingly, in such cavities, the spontaneous emission decay rate can differ substantially from the rate (10) and be either enhanced [6] or inhibited [7]. In practice, one has to consider the cavity losses, which lead to the substitution of a discrete sum over modes by an integral over a continuum. For a cavity mode density $\mathcal{D}_c(\omega)$ at the frequency of the atomic transition, and provided that the Born-Markov approximation is still appropriate in the treatment of the field modes, one finds that the atomic decay rate inside the cavity is related to the free space rate by

$$\gamma_c = \gamma \frac{\mathcal{D}_c(\omega)}{\mathcal{D}(\omega)} . \tag{11}$$

Restricting our discussion to this regime, let us consider what happens in the vicinity of a cavity resonance. For a sufficiently high cavity Q, we can approximate $\mathcal{D}_c(\omega)$ by the Lorentzian [8,9]

$$\mathcal{D}_c(\omega) = \frac{\Delta\omega_c/2}{\pi V} \frac{1}{(\Delta\omega_c/2)^2 + (\omega-\omega_c)^2} , \tag{12}$$

where $\Delta\omega_c$ is the cavity linewidth and is related to the Q factor by $Q = \omega/\Delta\omega_c$. For a cavity tuned to the atomic resonance, we find readily

$$\gamma_c \cong \gamma\, (\lambda^3/V)\, Q , \tag{13}$$

while for a cavity detuned from the atomic transition by $\omega_c - \omega = \omega$, we obtain

$$\gamma_c \cong \gamma\, (\lambda^3/V) \frac{1}{Q} . \tag{14}$$

For sufficiently high Q's, the enhancement or inhibition in spontaneous emission predicted by Eqs. (13) and (14) can be substantial.

In the limit when the cavity Q becomes very high, the Born-Markov approximation implicit in the Weisskopf-Wigner theory ceases to be valid. It is still correct to think of the modes of the electromagnetic field as a reservoir coupled to the atom, but one starts to observe effects of the colored nature of this reservoir. Zhu et al. [10] have measured the effect of the field modes acting as a colored reservoir in the response of atoms driven by a monochromatic field in a cavity.

For even higher Q's, the field modes can no longer be considered a reservoir; we reach the region of reversible spontaneous emission discussed in the previous section. In this case, we can describe to a good approximation the atom-field interaction with a Jaynes-Cummings Hamiltonian where the only field mode is that closest to resonance with the atomic transition.

Such a simplified model is often the cause of some confusion. Although the single-mode assumption is accurate and useful for a wide range of problems, questions arise that it cannot answer, such as how the atom becomes aware of the resonator boundaries. In a single-mode description, there is no propagation and no retardation, and any changes in the field at the atom are instantaneously communicated through the cavity. In a real cavity, a spontaneously decaying atom radiates a multimode field that propagates to the cavity walls and is partially absorbed and partly reflected. The reflected field acts back on the atom, carrying information about the cavity walls and about the state of the atom itself at earlier times. The picture that emerges, then, is that the atom learns of the boundary conditions with its radiated photon wave packet. Only a few cavity round trips are required for this learning process to be completed, after which the single-mode description becomes adequate [11].

The next section shows how one can take advantage of the tailored mode configuration of microwave cavities to produce masers with novel characteristics, in particular as generators on nonclassical light.

4. Micromasers

Under the best conditions, single-mode lasers and masers operating far above threshold produce light with Poisson photon statistics. The intensity fluctuations of such sources are given by $\Delta I = I^{1/2}$, where I is the intensity [12]. It has recently been realized that this property results from the fluctuations associated with the laser pump and loss mechanisms [13-15].

Micromasers are particularly simple devices on which to study how this works. They consist of exceedingly high-Q cavities in which atoms are injected at such a low rate that at most one atom at a time is present inside the resonator [16,17]. Because they operate in the regime of reversible spontaneous emission, one expects the rate of change of the mean photon number to be given by an equation of the form [18]

$$\frac{d}{dt}\langle n\rangle = \gamma N_a \sin^2 (g\sqrt{\langle n\rangle+1}\; t_{int}/2) - \gamma\langle n\rangle \ , \tag{15}$$

where N_a is the rate at which atoms are injected inside the cavity in level $|a\rangle$, t_{int} is the flying time of the atoms through the cavity, and κ the atom-field coupling constant. The first term on the right-hand side of Eq. (15) is the gain attributable to the change in atomic inversion, as deduced from the Rabi oscillations formula, the "+1" accounting for spontaneous emission into the resonator mode, while the second term describes cavity losses.

This equation is quite different in spirit from the rate equations usually encountered in laser and maser theories, which are extensions of Einstein's discussion of radiative interactions. Typically, the amplification process is described by a rate equation of the general form

$$\frac{d}{dt}\langle n\rangle = \alpha(\mathcal{N}_2 - \mathcal{N}_1)\langle n\rangle - \gamma\langle n\rangle \ , \tag{16}$$

where α is some generalized cross section proportional to the square of the dipole moment, and $(\mathcal{N}_2-\mathcal{N}_1)$ is a population inversion. More sophisticated master-equation versions of this equation exist, but they are still parameterized by the same α. This indicates that, in conventional lasers, the dipole coupling, which is proportional to both dipole moment and field strength, is replaced by a cross section that is quadratic in both. Conventional laser theory does indeed start by including quantum Rabi oscillations in the atom-field interaction, but rapidly proceeds to integrate the resulting equations over the exponential atomic level decay, leading to a result of the type given by Eq. (16). This step, which averages out the quantum Rabi phases governing the atom-field interaction, is not justified in micromasers, where the atoms do not undergo irreversible spontaneous emission. In this case, we obtain the steady-state photon statistics [13,2]

$$p_n = p_0 \prod_{k=1}^{n} \frac{\bar{n}\nu/Q + \mathcal{A}_k}{(\bar{n}+1)\nu/Q + \mathcal{B}_k} \ , \tag{17}$$

where the coefficients

$$\mathcal{A}_n = \frac{4R_a g^2}{(\omega-\Omega)^2 + 4ng^2} \sin^2[\ \sqrt{(\omega-\nu)^2 + 4ng^2}\,\tau] \ , \tag{18}$$

$$\mathcal{B}_n = \frac{4R_b g^2}{(\omega-\Omega)^2 + 4ng^2} \sin^2[\ \sqrt{(\omega-\nu)^2 + 4ng^2}\,\tau] \ , \tag{19}$$

$\bar{n}$ is the number of thermal photons in the resonator, and R_α is the rate at which atoms are injected into level α (= a or b) per cavity transit time. The normalization condition $\Sigma_n\, p_n$ = 1 determines p_0. Here, we assume that the successive atoms are injected inside the cavity without coherence between their upper and lower levels. The sinc functions in $\mathcal{A}_n$ and $\mathcal{B}_n$ are just an expression of the quantum Rabi oscillations. They cause the micromaser to exhibit a number of features that are absent in conventional lasers and masers.

For the incoherent pump mechanism considered here, the intracavity field always remains diagonal, hence the photon statistics [Eq. (17)] contain all information about the statistical properties of the steady-state field reached by the micromaser. Figure 1 shows the normalized average photon number

$$\mathcal{n} \equiv \langle n\rangle/N_a = \Sigma_n\, n p_n/N_a \tag{20}$$

as a function of the dimensionless pump parameter Θ

$$\Theta = \sqrt{N_a}\, g\tau \ . \tag{21}$$

The two curves correspond to N_a = 20 and 200 with N_b = 0, and the number of thermal photons is $\bar{n}$ = 0.1. A common feature in all cases is that $\mathcal{n}$ is nearly zero for small

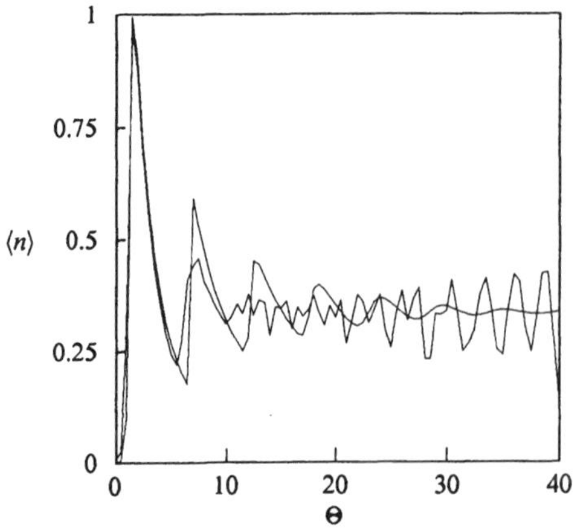

Fig. 1. Normalized average number of photons $\mathcal{n}$ from Eq. (17) vs the pump parameter Θ of Eq. (21) with $N_a = 20$ and 200, and $N_b = 0$. The number of thermal photons is $\bar{n} = 0.1$.

Θ, but a finite $\mathcal{n}$ (and $\langle n \rangle$) emerges at the threshold value $\Theta = 1$. For Θ increasing past this point, $\mathcal{n}$ first grows rapidly, but then decreases to reach a minimum at about $\Theta \cong 2\pi$, where the field abruptly jumps to a higher intensity. This general behavior recurs roughly at integer multiples of 2π, but becomes less pronounced for increasing Θ. Finally, a stationary regime with $\mathcal{n}$ nearly independent of Θ is reached. Outside the time scale of Fig. 1 there is additional structure reminiscent of the Jaynes-Cummings revivals [13, 19].

The number and, in particular, the sharpness of the features in the photon depend on N_a. At the onset of the field around $\Theta = 1$, the function $\mathcal{n}(\Theta)$ essentially does not depend on N_a if $N_a \gg 1$, but the subsequent transitions become sharper for increasing N_a. In the limit $N_a \to \infty$, this hints at an interpretation of the first transition in terms of a continuous phase transition, while the others are similar to first-order phase transitions [13, 18].

For $\Theta \ll 1$, the only solution for the field is $\langle n \rangle \cong \Theta^2 \ll 1$. The maser threshold occurs when the linearized (stimulated) gain for $\langle n \rangle \cong 0$ compensates for the cavity losses. From Eq. (15) we have

$$\gamma N_a \left. \frac{d}{d\langle n \rangle} \sin^2(g\sqrt{\langle n \rangle}\ \tau/2) \right|_{\langle n \rangle = 0} \cong \gamma N_a\ (g\tau)^2/4 = \gamma\ , \tag{22}$$

which gives precisely the threshold value $\Theta = 1$ obtained from the exact photon statistics Eq. (17). This justifies interpreting Θ as the pump parameter of the micromaser.

Figure 2 shows the normalized standard deviation

$$\sigma \equiv \frac{(\langle n^2 \rangle - \langle n \rangle^2)^{1/2}}{\langle n \rangle^{1/2}} \tag{23}$$

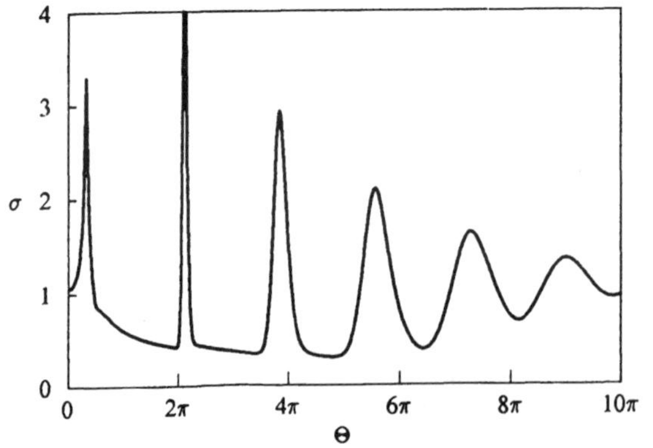

Fig. 2. Normalized deviation σ of Eq. (23) vs Θ.

of the photon distribution as a function of Θ for $N_a = 200$ and $\bar{n} = 0.1$. Above the threshold $\Theta = 1$, the photon statistics is first strongly super Poissonian, with $\sigma \cong 4$. (Poissonian photon statistics would yield $\sigma = 1$.) Further super-Poissonian peaks occur at the positions of the subsequent transitions. In the remaining intervals of Θ, σ is typically of the order of 0.5, a signature of the sub-Poissonian nature of the field.

Further nonclassial effects can be produced if the micromaser is pumped by atoms in a *coherent superposition* of their upper and lower states. Under these conditions, it is possible to generate "macroscopic" superpositions of the electromagnetic field in steady state and in the presence of dissipation (see Fig. 3). Such states are of considerable interest in the study of the quantum/classical correspondance [19,20].

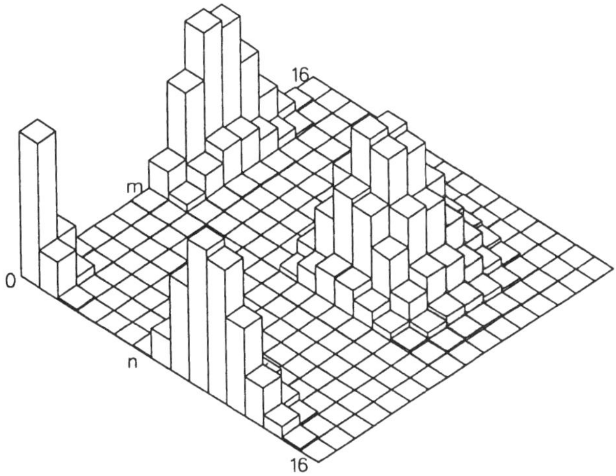

Fig. 3. Steady-state moduli of the field density matrix elements $\langle n|\rho|m\rangle$ for a micromaser pumped by a sequence of two-level atoms in state $\alpha|a\rangle + \beta|b\rangle$, with $\alpha = 0.53$ and for $N_{ex} = 10^9$. Here N_{ex} is the ratio between atomic injection rate and cavity decay rate.

The nonclassical features of the micromaser are to a large extent alien to ordinary masers and lasers: This is because micromasers possess less stochasticity and noise than macroscopic masers and lasers, where the atom-field interaction is terminated by exponential atomic decays rather than a transit time. As a result of irreversible spontaneous emission (as well as other decay mechanisms) the coherence of the quantum-mechanical light-matter interaction is averaged over and lost in conventional lasers, and the purely quantum-mechanical effects appearing in micromasers are largely lost.

5. Quantum Superpositions in Separated Cavities

The preceding section indicated how micromasers potentially can be used to generate macroscopic superpositions of the electromagnetic field. In this section, we discuss another application of cavity QED that uses a two-resonator arrangement to prepare and observe a quantum superposition between the states of two macroscopically separated systems [21]. Specifically, we consider two single-mode high-Q cavities, such as those used in micromaser experiments, separated by a macroscopic distance L on the order of millimeters.

We prepare a quantum superposition of the states of the resonator modes by sending a two-level atom initially in its excited state $|a\rangle$ through the cavities, and measuring its state as it exits the second cavity. For times such that cavity damping can be ignored the following result holds: if the atom is measured to exit the second cavity in the lower state $|b\rangle$, then the cavities are left in the coherent superposition [7]

$$|\psi\rangle = p|1,0\rangle + q|0,1\rangle \,, \tag{24}$$

where p and q depend on controllable experimental parameters such as frequency and interaction time. A simple atomic population measurement can be used to detect this coherent superposition: specifically, if a test atom is sent through the cavities initially in state (24), its state after interaction will be different from that resulting from the interaction with the same system, but prepared in the mixed state

$$\rho_{init} = |p|^2 \; |1,0\rangle\langle 1,0| \; + |q|^2 \; |0,1\rangle\langle 0,1| \quad . \tag{25}$$

Consider first the preparation of the cavities. After the atom is measured to exit the second cavity in the lower level $|b\rangle$ the field should be left in a state described by either Eq. (24) or Eq. (25), since the excitation originally in the two-level atom must wind up in either the first or the second cavity -- we neglect all sources of dissipation, including spontaneous emission during the transit time between the cavities. However, it is not obvious that it should be left in the *coherent superposition* (24). To see why, we consider the Hamiltonian of a single atom in interaction with two modes of the electromagnetic field:

$$\mathcal{H} = \hbar\omega\sigma_z + \hbar\omega_1 a_1{}^\dagger a_1 + \hbar\omega_2 a_2{}^\dagger a_2 + g_1(t)(a_1\sigma_+ + a_1{}^\dagger\sigma_-) + g_2(t)(a_2\sigma_+ + a_2{}^\dagger\sigma_-) \quad . \tag{26}$$

Here ω is the frequency of the atomic transition and ω_i the frequency of the mode of the ith cavity, with creation and annihilation operators $a_i{}^\dagger$ and a_i, and dipole coupling constant g_i. In contrast to the Jaynes-Cummings Hamiltonian Eq. (1), we now include an explicit time dependence of the coupling constants to account for the motion of the atom through the cavities. Specifically, $g_i(t) = g_i$ while the atom is in cavity i and 0 otherwise. In other words, the center of mass motion of the atom is treated classically.

We assume for simplicity that both cavities are initially in the vacuum state. The atom enters the first cavity in its excited state $|a\rangle$, so that

$$|\Psi(0)\rangle \;\; = \;\; |a,0,0\rangle \;. \tag{27}$$

At the time t_1 when the atom exits the first cavity, the state of the composite system becomes

$$|\Psi(0)\rangle \rightarrow |\Psi(t_1)\rangle = C_1|a,0,0\rangle + C_2|b,1,0\rangle \,, \tag{28}$$

and $|C_1|^2 + |C_2|^2 = 1$ and the explicit form of the C_i's are unimportant for the present discussion. Similarly, after cavity 2, we get

$$|\Psi(t_1)\rangle \rightarrow |\Psi\rangle = \mathcal{C}_1|a,0,0\rangle + \mathcal{C}_2|b,0,1\rangle + \mathcal{C}_3|b,1,0\rangle \,, \tag{29}$$

with $|\mathcal{C}_1|^2 + |\mathcal{C}_2|^2 + |\mathcal{C}_3|^2 = 1$, since no states are coupled to $|b,1,0\rangle$ in the rotating-wave approximation. The state of the atom is measured after it exits the second cavity, for example, by field ionization. If the atom is found to be in the lower level $|b\rangle$, the reduced density matrix for the field modes becomes

$$\rho_f = \mathcal{N}\; \mathrm{Tr}_{atom}\; |b\rangle\langle b|\Psi\rangle\langle\Psi| = \mathcal{N}\; [\mathcal{C}_2|0,1\rangle + \mathcal{C}_3|1,0\rangle][\mathcal{C}_2{}^*\langle 0,1| + \mathcal{C}_3{}^*\langle 1,0|] \;. \tag{30}$$

where $\mathcal{N}$ is a normalization constant such that $\mathrm{Tr}\rho_f = 1$ after the measurement. After normalization, this density matrix clearly describes the pure state

$$|\Psi_{init}\rangle = p|1,0\rangle + q|0,1\rangle \,, \tag{24}$$

$|p|^2 + |q|^2 = 1$. This demonstrates that such a preparation scheme using a *selective measurement* [22] on the atom leaves the field in a coherent superposition of the states "one photon in first cavity in first resonator and none in second" and "no photon in first resonator and one in second."

This superposition can be distinguished from a mixture of type (25) by simple population measurements. We show this by considering the situation described by the state vector

$$|\psi\rangle = pe^{i\phi}|1,0\rangle + q|0,1\rangle \quad . \tag{31}$$

This allows to obtain the mixed case (25) by integrating the pure case result over ϕ at the end of the calculation:

$$|\Psi\rangle\langle\Psi| \rightarrow \int_0^{2\pi} d\phi \; |\Psi\rangle\langle\Psi| \quad . \tag{32}$$

To analyze the quantum state of the cavities we inject a **probe** two-level atom in the state

$$|\Psi_{at}(0)\rangle = \alpha|a\rangle + \beta|b\rangle \; , \tag{33}$$

so that the combined atom-field system is initially described by

$$|\Psi(0)\rangle = [\alpha|a\rangle + \beta|b\rangle][pe^{i\phi}|1,0\rangle + q|0,1\rangle] \quad . \tag{34}$$

It is instructive to consider the states that can be reached after the probe atom exits the second cavity. We find readily

Table 1: Evolution of the atom-field initial levels.

$\|a,1,0\rangle$	→ $\|a,1,0\rangle$	→ $\|a,1,0\rangle$
		→ $\|b,1,1\rangle$
	→ $\|b,2,0\rangle$	→ $\|b,2,0\rangle$
		→ 0
$\|b,1,0\rangle$	→ $\|b,1,0\rangle$	→ $\|b,1,0\rangle$
		→ 0
	→ $\|a,0,0\rangle$	→ $\|a,0,0\rangle$
		→ $\|b,0,1\rangle$
$\|a,0,1\rangle$	→ $\|a,0,1\rangle$	→ $\|a,0,1\rangle$
		→ $\|b,0,2\rangle$
	→ $\|b,1,1\rangle$	→ $\|b,1,1\rangle$
		→ $\|a,1,0\rangle$
$\|b,0,1\rangle$	→ $\|b,0,1\rangle$	→ $\|b,0,1\rangle$
		→ $\|a,0,0\rangle$
	→ 0	

The four final states $|a,0,0\rangle$, $|a,1,0\rangle$, $|b,0,1\rangle$ and $|b,1,1\rangle$ can be reached by two channels each. For instance, for $|a,1,0\rangle$ we have $|a,1,0\rangle \rightarrow |a,1,0\rangle \rightarrow |a,1,0\rangle$ and $|a,0,1\rangle \rightarrow |b,1,1\rangle \rightarrow |a,1,0\rangle$. Each time such a situation occurs, one can expect the appearance of interference phenomena. A straightforward calculation shows that these interferences are reflected in the final atomic populations, and their measurement allows to distinguish the pure state Eq. (24) from the mixture Eq. (25). Indeed, we find that the final state of the atom-field system is

$$|\Psi_{final}\rangle = [\mathcal{A}_1 e^{i\phi} + \mathcal{A}_2]|a,0,0\rangle + [\mathcal{A}_3 e^{i\phi} + \mathcal{A}_4]|a,1,0\rangle + \mathcal{A}_5|a,0,1\rangle + [\mathcal{B}_1 e^{i\phi} + \mathcal{B}_2]|b,0,1\rangle + \mathcal{B}_3 e^{i\phi}|b,1,0\rangle + [\mathcal{B}_4 e^{i\phi} + \mathcal{B}_5]|b,1,1\rangle + \mathcal{B}_6 e^{i\phi}|b,2,0\rangle + \mathcal{B}_7|b,0,2\rangle \quad , \tag{35}$$

where the explicit dependence on the phase ϕ (see Eq. (31)) has been kept. The complex coefficients $\mathcal{A}_i$ and $\mathcal{B}_i$ are readily obtained by noting that, because $g_i = 0$ when the atom is in the (3-i)th cavity, we simply have to solve for two consecutive Jaynes-Cummings problems [8]. While the atom is in, say, the first cavity, the eigenstates of H are

$$|1,n,..\rangle = \cos\theta_n|a,n,..\rangle + \sin\theta_n|b,n+1,..\rangle \; ; \; |2,n,..\rangle = -\sin\theta_n|a,n,..\rangle + \cos\theta_n|b,n+1,..\rangle \quad , \tag{36}$$

with corresponding eigenenergies

$$E_{1n} = \hbar(\omega/2 + n\omega_1 + g_1\sqrt{n+1}\,\tan\theta_n) \; ; \; E_{2n} = \hbar(\omega/2 + (n+1)\omega_1 - g_1\sqrt{n+1}\,\tan\theta_n) \; . \tag{37}$$

Here, θ_n is given by

$$\tan 2\theta_n = \frac{25g_1\sqrt{n+1}}{\omega-\omega_1} \quad , \tag{38}$$

and the .. indicates (irrelevant) photon numbers of the second cavity. When the atom is in the second cavity, the relevant eigenstates and eigenenergies have the same form with the change of notation $E \to \mathcal{E}$, $\theta \to \Theta$, and g_1, $\omega_1 \to g_2$, ω_2.

With these results, the calculation is straightforward. Using the initial condition of Eq. (34) and the level scheme of Table 1 as a guidance, one finds, for example, that $\mathcal{A}_1$ is given by

$$\mathcal{A}_1 = \beta p \cos\theta_0\sin\theta_0[\exp(-iE_{10}\tau_1/\hbar) - \exp(-iE_{20}\tau_1/\hbar)] [\exp(-i\mathcal{E}_{10}\tau_2/\hbar)\cos^2\Theta_0 + \exp(-i\mathcal{E}_{20}\tau_2/\hbar)\sin^2\Theta_0] \exp(-i\omega_a T) \quad , \tag{39}$$

with similar forms for the other $\mathcal{A}_i$'s and $\mathcal{B}_i$'s. Here, τ_i is the transit time of the atom through the i-th cavity and T the transit time between the two cavities.

The upper state population of the atom at the exit of the second resonator is found to be

$$p_a = |\mathcal{A}_1 e^{i\phi} + \mathcal{A}_2|^2 + |\mathcal{A}_3 e^{i\phi} + \mathcal{A}_4|^2 + |\mathcal{A}_5|^2 \quad , \tag{40}$$

in the case where the field is initially in the coherent superposition (24) and, with the prescription (32),

$$p_a = |\mathcal{A}_1|^2 + |\mathcal{A}_2|^2 + |\mathcal{A}_3|^2 + |\mathcal{A}_4|^2 + |\mathcal{A}_5|^2 \quad . \tag{41}$$

in the case of the mixture (25). The interference terms attributable to the two possible channels disappear in this case, and hence a measurement of the upper state population at the exit of the resonators permits us to distinguish between a coherent and an incoherent mixture.

The two channels leading to the final state $|a,0,0\rangle$ originate from states where the atom is in the lower state $|b\rangle$ and the two channels leading to $|a,1,0\rangle$ originate from states where the atom is initially in the upper state. This indicates that it is not necessary for the atom to be initially in a coherent superposition, Eq. (33). For example, it is sufficient to inject atoms in the ground state $|b\rangle$ to distinguish between the coherent superposition (24) and the mixture (25). Also, it is straightforward to show that spontaneous emission of the probe atom between the two cavities does not impede the capability of this scheme to distinguish between a coherent superposition and a mixture.

6. Outlook

The preceding section showed an example of a situation where selective measurements are used to prepare a specific state of a system. This is just one of many examples that have been worked out recently to illustrate the impact of cavity QED on a better understanding of quantum measurement theory.

Until recently, most atomic experiments turned out ensemble averages, as if the same measurement were repeated on a large number of identically prepared atoms. The outcome of such experiments is given by the quantum-mechanical density matrix. With the advent of cavity QED, one is now in a position to perform experiments on a single quantum system, as opposed to an ensemble. In electron or ion traps, the single system is an isolated electron or ion, while in micromasers, it is a single (or few) mode(s) of the electromagnetic field. To properly describe the dynamics of such single quantum systems, it is essential to explicitly describe the measurements performed to monitor them [23-25]. This requires coupling the system under observation to a meter system. As we have seen in the preceding section, the measurement process typically produces a back action on the system and influences its subsequent dynamics. Thus, the observed dynamics are both measurements induced and measurements dependent [24]. This is quite different from the classical situation, where any observation of the system would involve some intervention from the outside, but the structure of the theory is such that the effects of measurements can easily be isolated and/or ignored.

We have illustrated how measurements on a single quantum system allow the preparation and study of unusual states, for example, macroscopic superpositions. They are also of considerable interest in the study of the quantum/classical correspondence. In particular, in the case of dissipative systems, one is confronted with the apparent paradox that the quantum system typically evolves toward a unique steady state, while its classical counterpart need not, but may instead exhibit instabilities and chaos. As already mentioned, conventional quantum mechanics yields predictions about ensemble averages, whereas classically the dynamics is typically interpreted in terms of single realizations. The study of repeated quantum measurements on single quantum systems, as has become possible in cavity QED, suggests a way out of this dilemma.

Acknowledgments

This work is supported by ONR Contract N00014-88-K-0294, NSF Grant PHY-8902548, and the Joint Services Optics Program.

References

1. For a pedagogical review, see S. Haroche and D. Kleppner, Physics Today **42**, 24 (January 1989).
2. See e.g., P. Meystre and M. Sargent, III, *Elements of Quantum Optics*, Springer Verlag, Heidelberg (1990), in press.
3. For a review of the Jaynes-Cummings model, see e.g., S. M. Barnett, P. Filipowicz, J. Javanainen, P. L. Knight, and P. Meystre, in *Frontiers in Quantum Optics*, E. R. Pike and S. Sarkar, Eds. (Adam Hilger, Bristol 1986), p. 485.
4. J. H. Eberly, N. B. Narozhny, and J. J. Sanchez-Mondragon, Phys. Rev. Lett. **44**, 1323 (1980).
5. V. Weisskopf and E. Wigner, Z. Phys. **63**, 54 (1930).
6. E. M. Purcell, Phys. Rev. **69**, 681 (1946).
7. D. Kleppner, Phys. Rev. Lett. **47**, 233 (1981).
8. S. Haroche and J. M. Raimond, in *Advances in Atomic and Molecular Physics*, Vol. 20, D. Bates and B. Bederson, eds. (Academic Press, 1985).
9. D. P. O'Brien, P. Meystre, and H. Walther, in *Advances in Atomic and Molecular Physics*, Vol. 21, D. Bates and B. Bederson, eds. (Academic Press, 1985).
10. Y. Zhu, A. Lezama, T. W. Mossberg, and M. Lewenstein, Phys. Rev. Lett. **61**, 1946 (1988).
11. J. Parker and C. R. Stroud, Jr. Phys. Rev. **A35**, 4226 (1987).
12. M. Sargent III, M. O. Scully, and W. E. Lamb, Jr. *Laser Physics* (Addison-Wesley, Reading 1974).

13. P. Filipowicz, J. Javanainen, and P. Meystre, Phys. Rev. **A34**, 3077 (1986).
14. S. Machida, Y. Yamamoto, and Y. Itaya, Phys. Rev. Lett. **58**, 1000 (1987); S. Machida and Y. Yamamoto, Phys. Rev. Lett. **60**, 792 (1988).
15. M. Marte and D. F. Walls, Phys. Rev. **A37**, 1235 (19888); M. A. M. Marte, PhD Thesis, University of Waikato (1988), unpublished; F. Haake, S. M. Tan, and D. F. Walls, preprint (1989).
16. D. Meschede, H. Walther, and G. M{ller, Phys. Rev. Lett. **54**, 551 (1985).
17. M. Brune, J. M. Raimond, P. Goy, L. Davidovich, and S. Haroche, Phys. Rev. Lett. **59**, 1899 (1987); L. Davidovich, J. M. Raimond, M. Brune, and S. Haroche, Phys. Rev. **A36**, 3771 (1987).
18. See A. Guzman, P. Meystre, and E. M. Wright, Phys. Rev. **A**, in press, for a justification of this equation.
19. E. M. Wright and P. Meystre, Optics Lett. **14**, 177 (1989).
20. J. J. Slosser, P. Meystre, and S. L. Braunstein, Phys. Rev. Lett. **63**, 934 (1989); J. J. Slosser, P. Meystre, and E. M. Wright, submitted to Optics Lett.
21. S. Haroche and P. Meystre, unpublished.
22. E. B. Davies, *Quantum Theory of Open Systems* (Academic Press, London 1976); K. Krauss, *States, Effects, and Operations: Fundamental Notions of Quantum Theory* (Springer Verlag, Berlin 1983).
23. W. E. Lamb, Jr., in *Chaotic Behavior in Quantum Systems, Theory, and Applications*, Ed. by G. Casati (Plenum, N.y. 1985); W. E. Lamb, Jr., in *New Techniques and Ideas in Quantum Measurement Theory*, (The New York Academy of Sciences, New York 1986).
24. P. Meystre, Opt. Lett. **12**, 669 (1987); P. Meystre and E. M. Wright, Phys. Rev. **A37**, 2524 (1988); J. Krause, M. O. Scully, and H. Walther, Phys. Rev. **A36**, 4547 (1987).
25. For the related problem of quantum jumps in ion traps, see H. Demhelt, Bull. Am. Phys. Soc. **20**, 60 (1975); R. J. Cook and J. F. Kimble, Phys. Rev. Lett. **54**, 1023 (1985); W. Nagourey, J. Sandberg, and H. Demhelt, Phys. Rev. Lett. **56**, 2727 (1986); J. Javanainen, Phys. Rev. **A33**, 2121 (1986); A. Schenzle, R. G. deVoe, and R. G. Brewer, Phys. Rev. **A33**, 2127 (1986); C. Cohen-Tannoudji and J. Dalibard, Europhys. Lett. **1**, 441 (1986) and references therein.

Superradiance in the Dicke Model

E.P. Kadantseva, W. Chmielowski, and A.S. Shumovsky

Laboratory of Theoretical Physics for Nuclear Research,
P.O. Box 79, SU-141980 Dubna, USSR

Abstract. We derive an exact differential operator equation for the operator level populations. Particular cases of one- and two-photon processes are considered for which the equations are integrated in quadratures.

The theory of superradiance is based on the Dicke model whose Hamiltonian for the simplest point-like one-mode system in a perfect resonator is of the form [1-2]:

$$H = \omega\, a^{+} a + \omega_{0} R_{3} + g\, \{ R^{+} a^{m} + (a^{+})^{m} R^{-} \}\,, \qquad (1)$$

where R_3 is the operator of level populations of a system of two-level atoms with the transition frequency ω_0; the operators R^+ and R^- describe a transition to an upper (lower) level with absorption (emission) of m photons; a^+ and a - are the operators of creation and annihilation of photons with frequency ω ($\hbar = 1$); g - is the atom-field coupling constant. Using the commutation relations for bosons and the Heisenberg equations we may be verify that the following constants of motion do exist

$$\hat{n} + m\, R_{3} = \text{const} \;\; (\equiv \hat{M}_{1}), \qquad (2)$$

$$R^{+}R^{-} + R_{3}^{2} - R_{3} = \text{const} \;\; (\equiv \hat{M}_{2}). \qquad (3)$$

Taking account of (2) and (3) we find from (1)

$$\frac{d^{2}}{dt^{2}} R_{3} = \Delta\omega\, (\hat{H} - \omega \hat{M}_{1}) - (\Delta\omega)^{2} R_{3} + 2g^{2} \Big\{ (\hat{M}_{2} - R_{3} - R_{3}^{2})* \prod_{k=1}^{m} (\hat{M}_{1} - mR_{3} + k) - (\hat{M}_{2} + R_{3} - R_{3}^{2}) \prod_{k=0}^{m-1} (\hat{M}_{1} - mR_{3} - m - k) \Big\}$$

$$= P_{m+1}(R_{3})\,. \qquad (4)$$

Springer Series in Wave Phenomena, Vol. 9 **Nonlinear Optics in Solids**
Editor: O. Keller

P_{m+1} is a polynomial of degree m+1 in the variable R_3. Thus, the dynamics in the Dicke model can be described by the exact constant of motion (2) without any assumptions on the initial state of the system and the way of switching-on the interaction between atoms and the field. Consider the commutator

$$[R_3, i\frac{d}{dt} R_3] = g(R^+a^m + (a^+) R^m) = H - \hat{M}_1(\equiv \hat{M}_3), \quad (5)$$

where $\hat{M}_3$ is a time-independent operator proportional to the atom-field coupling constant g. This parameter is small and if the right-hand side of (5) is neglected ($\hat{M}_3 \simeq 0$), then multiplying the rhs and lhs of (5) by ${}^{d}/_{dt}R_3$ we may pass to an equation of the form

$$\frac{d}{dt} R_3 = 2 \sqrt{\int P_{m+1}(R_3)\, dR_3 + C}\ , \quad (6)$$

where C is a new time-independent operator. Assuming g to be small, solutions to the exact equation (5) may be found by the perturbation theory in the atom-field coupling constant, the solution to eq. (6) being a zeroth approximation. A qualitative character of the dynamic behaviour of the system can be determined if in (4) we perform averaging over the initial state of the system $\rho_0 = |0;\alpha\rangle\langle 0;\alpha|$. We take advantage of a mean-field approximation when $\langle R_3^2\rangle_t \simeq \langle R_3\rangle_t^2$. Then, consider particular cases:

1. One-photon process (m=1).

Equation (5) for m=1 assumes the form

$$\ddot{x} = g^2\{6x^2 - 2(2M_1 + \kappa + 1)x - M_2 + 2\kappa(x_0 + \lambda)\}\ , \quad (7)$$

where $M_i \equiv \langle\hat{M}_i\rangle$, $\kappa = 1/2(\Delta\omega/g)^2$, x_0 is a number of the level populations at t=0, λ is a coefficient for the interaction of a field with an atomic system. The first-order equation is

$$(\dot{x})^2 = g^2[4x^3 - 2(2M_1 + \kappa + 1)x^2 - 2(M_2 - 2\kappa(x+\lambda))x + C]\ . \quad (8)$$

Equation (8) is integrated in quadrature . Let initial conditions be such that for t=0 the function of level

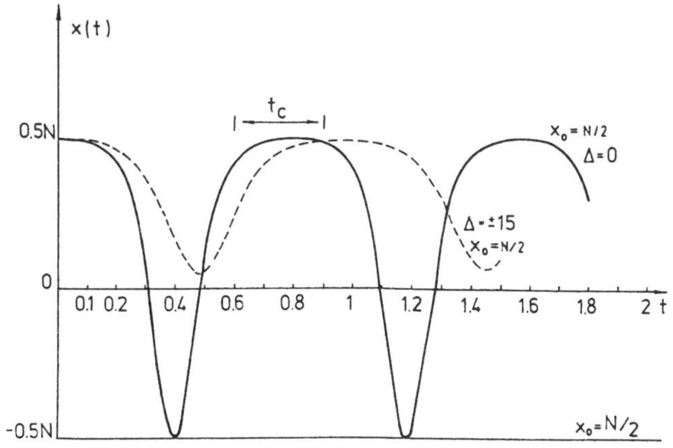

Fig. 1. Level population as a function of time for m=1.

populations assumes one of the values x_1, x_2, x_3 that are zeros of equation (8) and $x_1>x_2>x_3$. The solution to eq. (8) takes on the form [3]

$$x_i(t) = x_i + \frac{(x_i - x_j)(x_i - x_k)\,\mathrm{sn}^2(gt\sqrt{x_1-x_3}, k')}{(x_1 - x_3)-(x_i - x_3)\,\mathrm{sn}^2(gt\sqrt{x_1-x_3}, k')}, \qquad (9)$$

where $k'^2=(x_2-x_3)/(x_1-x_3)$. That solution coincides with the results of other authors [4]. It describes a double-periodic process of the energy transfer between photon and atom subsystems in a perfect resonator (see Fig. 1).

2) Two-photon process (m = 2)

For m = 2 equation (4) assumes the form

$$\ddot{x} = -2g^2\{16x^3-6(2M_1+1)x^2+2(M_1^2+M_1-2M_2+1+\tfrac{\delta}{2})x + (2M_1+1)M_2-\delta(x_0+\kappa) . \qquad (10)$$

The corresponding first-order equation is

$$\dot{x} = -4g^2\{4x^4-2(2M_1+1)x^3+(M_1^2+M_1-2M_2+1+\tfrac{\delta}{2})x^2 + [(2M_1+1)M_2-\delta(x_0+\kappa)]x+C . \qquad (11)$$

This solution is much more complicated than the solution to eq. (8), and therefore we restrict ourselves to a numerical solution (Fig. 2). It is then seen that the

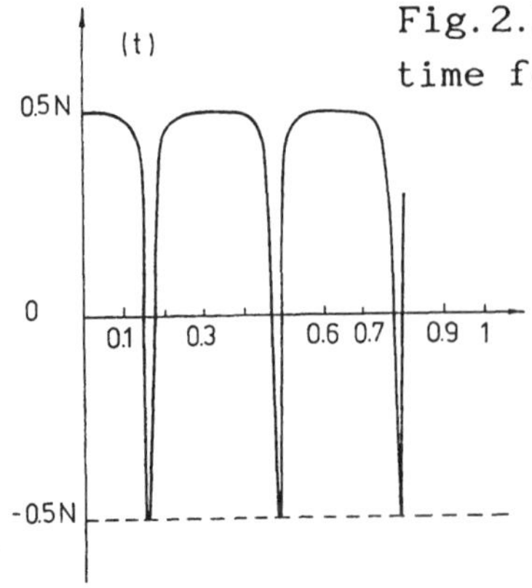

Fig. 2. Level population as a function of time for m=2.

higher the magnitude of photon multiplicity, the longer the time of stay of an atom in an excited state and the lower the time of delay t_D.

Thus, the presented approach allows the description of periodic processes of emission of pulses of the electromagnetic field in a perfect resonator alternating with periods of absorption of the field for an arbitrary photon multiplicity of the transition. Superradiance pulses possess a characteristic asymmetry with respect to the delay time (position of the intensity maximum). We emphasize that the experimentally observed asymmetry of pulses is usually attributed different additional physical assumptions [3]. As can seen from the time dependence of level populations, a system of two-level atoms is in an upper state for a longer time than in a lower state, because the process of de-excitation starts as a rather slow spontaneous decay of individual excited states and only gradually, through self-induction of correlations, the process becomes cooperative. So, the time of stay in the upper state, t_c, can be identified with the time of collectivization [3]. The frequency tuning out from a resonance leads , on the one hand, to an increase in the time of collectivization and , on the other hand, to a decrease in the intensity of radiation caused only by a partial reverser of the initial inverted state in the process of generation (Fig. 1).

R e f e r e n c e s

1. L. Allen, J.H. Eberly - Optical Resonance and Two-level Atoms (Wiley - Interscience, New York, 1975).
2. N.N. Bogolubov(Jr.), A.S. Shumovsky - Superradiance, JINR, Dubna, 1987.
3. E. P. Kadantseva, W. Chmielowski, A. S. Shumovsky - Dynamics in Dicke model (in press).
4. Sunil Kumar, C.L. Mehta - Phys. Rev., A21 (1980) 1573.

Part III

Metals, Superconductors, and Ponderomotive Effects

Surface Enhanced Second Harmonic Generation from Metals

J.-L. Coutaz

Laboratory of Electromagnetism, Microwaves and Optoelectronics (LEMO), ENSERG, BP 257, F-38016 Grenoble Cedex, France

1. Introduction

Second Harmonic Generation (SHG) at a metallic surface was first reported by F. Brown *et al.* [1] in 1965. At the same time, Jha [2] established that this phenomenon is a surface effect. Indeed, in bulk centrosymmetric media, even order nonlinear (NL) effects, in particular SHG, are forbidden. However, there is, by geometry, a break of the symmetry at the surface. The second harmonic (SH) signal is generated because of the discontinuities occurring at the surface. A second source term is due to the decay of the pump field in the medium. These initial ideas are well described in a famous paper by Bloembergen *et al.* [3] published in 1968.

As SHG from a metal is a surface effect, it is very sensitive to any change occurring at the surface, in particular to an increase of the pump field amplitude because of surface effects, or to a modification of the surface configuration, for example by contamination by chemical adsorbates. In 1974, Simon *et al.* [4] reported, for the first time, on enhancement of the second harmonic (SH) signal by excitation of a surface plasmon (SP) in a Kretschmann prism coupler configuration [5]. At the same time, Chen *et al.* [6] demonstrated the effect on the SH signal of a monolayer of adsorbates deposited onto the metal surface. It is notable that these pioneering works attracted attention only several years later.

A radical evolution in the NL optical studies of surfaces arose from the discovery of the giant Raman effect by Fleischmann *et al.* [7] in 1974: molecules adsorbed at a rough metallic surface scatter a Raman signal about 10^6 larger than the same molecules in solution. Even if all the mechanisms responsible for this spectacular effect were not fully understood at this time, it was certain that the increase of the pump field in the vicinity of the surface was a cause of the enhancement [8,9]. The surface amplification of the pump field is due to the excitation of electromagnetic (EM) resonances, localized at submicronic rough patterns, or propagating along the surface. In the following, we will call the latter resonance SP, and the former "localized" SP. Therefore, it is possible to connect the origins of the enhancement in Simon's experiment [4] with Fleischmann's observation [7] because of the similitude between prism devices and roughness: both act as couplers between an incident beam and the EM resonance. A third kind of coupler can also be used to excite the SP, the diffraction grating.

Of course, this EM origin of the enhancement could be used to amplify any NL surface effects. Among them, SHG is of great interest, because the enhancement of this effect is only of EM origin. In the 80's, considerable work has been devoted to the study of enhanced SHG at a metal surface. Three kinds of couplers have been used to enhance SHG at a metal surface: as written above, the prism coupler by

Springer Series in Wave Phenomena, Vol. 9 **Nonlinear Optics in Solids**
Editor: O. Keller

Simon *et al.* [4], then in 1981, Chen *et al.* [10] reported SHG at a rough surface, and more recently, Coutaz *et al.* [11] demonstrated enhancement of SHG by excitation of SP at a metallic diffraction grating. SHG from a flat metal (without enhancement effect) has also attracted great attention, because of its sensitivity to the surface state. This opens a wide domain of applications towards the optical analysis of surfaces and the study of chemical surface effects [12].
The purpose of this paper is to give **a review of surface enhanced SHG from bare metallic surfaces.** However, before dealing with enhancement, one has to solve the major problem of SHG from a flat metal surface. This is a lively debate, and there exist still now different theories [13,14,15,16]. We will give in Sect. 2 the approach proposed by our group [14,15], using the free electron gas model to represent the optical response of the metal. In Sect. 3, we will show how to increase the pump field amplitude at the metal surface. Enhancement of SHG will be presented in Sect. 4 for prism configurations, and in Sect. 5 for diffraction grating devices. Section 6 will be devoted to the effect of roughness on SHG.

2. Second Harmonic Generation at a Metallic Surface [15]

2a NL polarization inside metals [14]

We will use here the free electron gas model to describe the NL optical response of the metal. It means that we neglect the contribution of intraband transitions and the effect of bounding potential structure, which is effective in the immediate vicinity of the surface. Even if this model is simple and very crude, it can be successfully employed to give a fairly good description of the concerned phenomena.
In the following, the time dependence of the waves at frequency ω will be $e^{-j\omega t}$, the subscripts 1 and 2 represent quantities taken at frequencies ω and 2ω respectively. The geometry of the problem is depicted in fig. 1, where there is a symmetry of translation in the z direction ($\partial/\partial z=0$). The upper medium is the vacuum, its dielectric constant is 1 at all frequencies.
The expression for the NL polarization inside the metal was derived by Bloembergen and his co-workers [3] twenty years ago:

$$\mathbf{P}^{NL} = \gamma\ \nabla(\mathbf{E}_1.\mathbf{E}_1) + \beta\,\mathbf{E}_1\,(\nabla.\mathbf{E}_1) \tag{1}$$

where $\beta=\varepsilon_0/2m\omega^2$ and $\gamma=\beta\,(1-\varepsilon_1)/4$. (2)

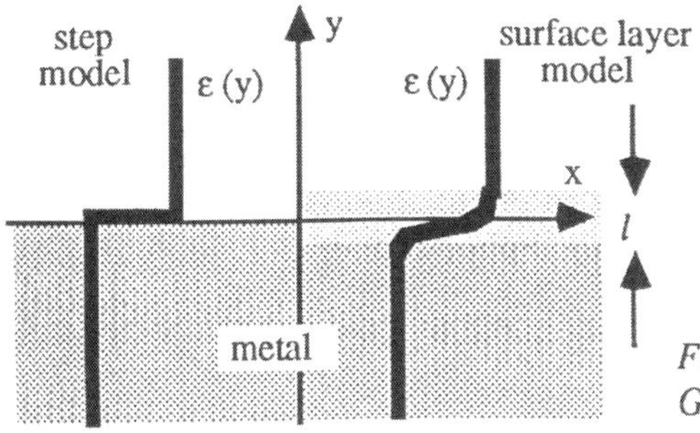

Fig. 1:
Geometry of the metal surface

Within this model, the dielectric susceptibility ε_1 of the metal is given by the Drude relation:

$\varepsilon_1 = 1- \omega_p^2/\omega_1^2$ (ω_p is the plasma frequency). (3)

Consider first that the surface corresponds to an abrupt variation of the medium: we know from the usual boundary conditions, derived from Maxwell equations, that the normal component of the electric field at ω is discontinuous through the surface. This component, as well as ε_1, can then be written as a step function. These considerations yield a mathematical ambiguity. Indeed, the term $\gamma \nabla(\mathbf{E}_1.\mathbf{E}_1)$ along the y direction is equal to $\gamma\, \partial/\partial y(\mathbf{E}_1.\mathbf{E}_1)$. Because of the discontinuity of $\mathbf{E}_1$ through the surface, $\partial/\partial y(\mathbf{E}_1.\mathbf{E}_1)$ is proportional to a Dirac function. The product of a Dirac function and a step function (here γ) is not mathematically defined. An identical problem holds for $\beta\, \mathbf{E}_1(\nabla.\mathbf{E}_1)$.
However, an expression for the NL polarization, valid even at the surface, can be derived [14] from the bulk expression (1). For the sake of simplicity, we will consider here only the flat surface case, but this treatment can be extended to any shape of surface [14]. To understand it, consider a transition layer of thickness l at the metal surface (fig. 1), where the parameters of the medium vary continuously from the bulk to the vacuum values. It should be noted [16,17] that, inside this layer, ε_1 varies from a negative (bulk metal) to a positive value (vacuum). So there exists a location where ε_1=0. The conservation of $D_{1y}=\varepsilon_1 E_{1y}$ leads to an unphysical infinite value for E_{1y}. This difficulty is mathematically removed with the formalism of principal values of Cauchy for $1/\varepsilon_1$, or the finite part of Hadamard for $1/\varepsilon_1^2$ [14]. An alternative is to introduce the optical losses of the metal through a complex dielectric constant. Inside this layer, the components of the NL polarization are (from (1))

$$P_u^{NL} = \beta\,[E_{1u}(\frac{\partial E_{1x}}{\partial x} + \frac{\partial E_{1y}}{\partial y}) + (\frac{1-\varepsilon_1(y)}{4})\frac{\partial}{\partial u}(E_{1x}^2+E_{1y}^2+E_{1z}^2)], \qquad u = x \text{ or } y,$$

$$P_z^{NL} = \beta\,[E_{1z}(\frac{\partial E_{1x}}{\partial x} + \frac{\partial E_{1y}}{\partial y})].$$

Our method is here to find the limit of $\mathbf{P}^{NL}$ in the transition layer when l tends towards 0. First, E_{1x} and E_{1z} are continuous through the layer, so

$$\lim_{l\to 0} \frac{\partial E_{1x}}{\partial y} = \lim_{l\to 0} \frac{\partial E_{1z}}{\partial y} = 0.$$

Secondly, because of the geometrical symmetry of the problem, the $\partial/\partial x$ operator, applied to $\mathbf{E}_1$, leads to a finite quantity, whose limit is equal to 0 when l tends to 0. Therefore:

$$\lim_{l\to 0} P_u^{NL} = \lim_{l\to 0} \beta E_{1u} \frac{\partial E_{1y}}{\partial y}, \qquad u = x \text{ or } z,$$

$$\lim_{l\to 0} P_y^{NL} = \lim_{l\to 0} \beta \frac{3-\varepsilon_1(y)}{4}\frac{\partial E_{1y}^2}{\partial y}.$$

Let us determine: $\lim_{l\to 0} \dfrac{\partial E_{1y}}{\partial y} = \dfrac{\partial}{\partial y}(\lim_{l\to 0} E_{1y}) = J\,[E_{1y}]\,\delta_S$

where J[..] is for jump of. Identically, we find

$$\lim_{l \to 0} \frac{\partial E_{1y}^2}{\partial y} = \frac{\partial}{\partial y} (\lim_{l \to 0} E_{1y}^2) = J\,[E_{1y}^2]\,\delta_S.$$

We have also to calculate $\lim_{l \to 0} \frac{\varepsilon_1 \partial E_{1y}^2}{\partial y}$:

$$\varepsilon_1 \frac{\partial E_{1y}^2}{\partial y} = 2\,\varepsilon_1 E_{1y} \frac{\partial E_{1y}}{\partial y} = 2\,D_{1y} \frac{\partial E_{1y}}{\partial y}.$$

D_{1y} is continuous through the surface layer, so

$$\lim_{l \to 0} \left(\frac{\varepsilon_1 \partial E_{1y}^2}{\partial y}\right) = 2\,D_{1y} \lim_{l \to 0} \frac{\partial E_{1y}}{\partial y} = 2\,D_{1y}\,J\,[E_{1y}]\;\delta_S.$$

Thus we have demonstrated that there exists at the metal surface a NL surface polarization $\mathbf{P}^{NL}$ given by

$$\mathbf{P}_S^{NL}\,\delta_S = \beta\,(\varepsilon_1 - 1)\,E_{1my}(0)\,\left[\mathbf{E}_{1m}(0) + (\varepsilon_1 - 1)\,\mathbf{E}_{1my}(0)/4\right]\delta_S \qquad (4)$$

where $\mathbf{E}_{1m}(0)$ corresponds to the electric field inside the metal at y=0. This NL polarization is associated with a surface current density $\mathbf{I}_{2s}$ by the following relation:

$$\mathbf{I}_{2S}\,\delta_S = \frac{\partial\,\mathbf{P}^{NL}\,\delta_S}{\partial t}.$$

Inside the bulk metal, the origin of the NL polarization comes only from the evanescent decay of the pump field, since $\nabla.\mathbf{E}_1$=0. This bulk NL polarization is

$$\mathbf{P}_B^{NL} = \beta\,(1 - \varepsilon_1)\,\nabla(\mathbf{E}_1 . \mathbf{E}_1)/4. \qquad (5)$$

We are now able to write the total NL polarization of the metal:

$$\mathbf{P}_B^{NL} = \mathbf{P}_B^{NL} + \mathbf{P}_S^{NL}\,\delta_S. \qquad (6)$$

We then know the source terms for the SH field.

2b Boundary conditions at 2ω frequency

The above treatment is valid for the free electron gas model. It is obvious that a more accurate model is needed for a complete understanding of SHG at metallic surfaces. Examples of sophisticated models are presented by O. Keller in this book [13]. The general feature of all kinds of theories for NL polarization inside metals is that this NL polarization is described by equation (6). It means that the NL polarization is characterized by a bulk and a surface term. We will now derive the boundary conditions for the SH field in view of such a NL polarization. Let us write the Maxwell equations for the SH field:

$$\nabla \times \mathbf{H}_2 = -2j\omega\, \varepsilon_0 \varepsilon_2 \mathbf{E}_2 - 2j\omega\, \mathbf{P}^{NL} \tag{7}$$

$$\nabla \times \mathbf{E}_2 = 2j\omega\, \mu_0 \mathbf{H}_2. \tag{8}$$

As $\mathbf{P}^{NL}$ includes a distributive term (the surface term), equation (7) has to be written in the sense of distributions (see appendix A):

$$\{\nabla \times \mathbf{H}_2\} + \mathbf{n}\times J[\mathbf{H}_2]\,\delta_S = -2j\omega\, \varepsilon_0 \varepsilon_2 \mathbf{E}_2 - 2j\omega\, \mathbf{P}_B^{NL} - 2j\omega\, \mathbf{P}_{S\parallel}^{NL}\delta_S - 2j\omega\, \mathbf{P}_{Sy}^{NL}\delta_S \tag{9}$$

where {..} means that the function has to be taken as a usual function. **n** is the unit vector normal to the surface, and the subscript ‖ means parallel to the surface.
This relation (9) is true if functional and distributive terms of each side of the expression are respectively equal. In the left side of (9), there is no component normal to the surface, which compensates $-2j\omega \mathbf{P}^{NL}{}_{sy}\,\delta_S$ in the right part of (9). It means that we have to introduce new purely distributive terms into this equation through the EM fields. It is easy to show that making $\mathbf{H}_2$ a distribution does not lead to a mathematical solution. The only way is then to write $\mathbf{E}_2$ as a distribution:

$$\mathbf{E}_2 = \mathbf{E}_{2B} + \mathbf{E}_{2S}\,\delta_S. \tag{10}$$

So equation (9) can now be written

$$\{\nabla \times \mathbf{H}_2\} = -2j\omega\, \varepsilon_0 \varepsilon_2 \mathbf{E}_{2B} - 2j\omega\, \mathbf{P}_B^{NL} \tag{11}$$

$$\mathbf{n} \times J[\mathbf{H}_2]\,\delta_S = -2j\omega\, \varepsilon_0 \varepsilon_2\, \mathbf{E}_{2S}\,\delta_S - 2j\omega\, \mathbf{P}_S^{NL}\,\delta_S. \tag{12}$$

We perform a similar substitution in (8):

$$\{\nabla \times \mathbf{E}_{2B}\} = -2j\omega\, \mu_0 \mathbf{H}_2 \tag{13}$$

$$\nabla \times (\mathbf{E}_{2S}\,\delta_S) = -\mathbf{n} \times J[\mathbf{E}_{2B}]\,\delta_S. \tag{14}$$

It can be shown [14,15] that (14) is equivalent to

$$\mathbf{E}_{2S}\,\delta_S = E_{2S}\,\mathbf{n}\,\delta_S \tag{15}$$

$$J[E_{2Bx}] = [\nabla . E_{2S}]_x. \tag{16}$$

We substitute $\mathbf{E}_{2s}\delta_s$ by (15) in equation (12) which gives

$$E_{2S}\,\delta_S = -P_{Sy}^{NL}\,\delta_S / \varepsilon_0 \varepsilon_2 \tag{17}$$

$$-\mathbf{n} \times J[\mathbf{H}_{2\parallel}]\,\delta_S = 2j\omega\, \mathbf{P}_{S\parallel}^{NL}\,\delta_S \tag{18}$$

and projecting (18) on the x and z axes:

$$J[H_{2x}] = 2j\omega\, P_{Sz}^{NL} \tag{18a}$$

$$J[H_{2z}] = -2j\omega\, P_{Sx}^{NL}. \tag{18b}$$

To summarize, we obtain the following relations:
- (11) and (13) are the usual Maxwell equations for the SH fields, without distributive terms.

- The surface NL polarization generates (17) a surface SH field $\mathbf{E}_{2s}$ normal to the surface.
- Contrary to the linear case, the component of the SH magnetic field $\mathbf{H}_2$ parallel to the surface is discontinuous (18): this discontinuity comes from the surface NL polarization.

To go on with the determination of the SH field, we have now to introduce the expression for the NL polarization in the relations (11), (13), (17) and (18). Let us first focus on relation (17), which describes the contribution of P^{NL}_{sy}. As we have shown for similar relations in §2a, the right hand side of this relation (17) is not mathematically defined, because of the product of a Dirac function ($\mathbf{P}^{NL}_s\delta_s$) and a step function $1/\varepsilon_2$. We will use the same formalism as previously (surface layer) to remove this ambiguity. Using the definition of P^{NL}_y, we derive the limit of P^{NL}_y/ε_2 when l tends to 0:

$$\lim_{l\to 0} P_y^{NL}/\varepsilon_2 = \lim_{l\to 0} \beta \frac{3-\varepsilon_1(y)}{4\,\varepsilon_2(y)} \frac{\partial E_{1y}^2}{\partial y}.$$

According to the Drude model, $\varepsilon_2 = (3+\varepsilon_1)/4$. After a tedious, but straightforward calculation, it is shown that

$$E_{2S}\,\delta_S = \lim_{l\to 0} P_{Sy}^{NL}/\varepsilon_0\varepsilon_2 = -\beta U \varepsilon_1^2 E_{1my}^2(0)\,\delta_S/\varepsilon_0$$

$$\text{with}\quad U = \frac{4}{9}\ln\left(\frac{\varepsilon_1}{\varepsilon_2}\right) + \frac{1}{3\varepsilon_1^2}\left(\frac{\varepsilon_1-1}{3-\varepsilon_1}\right).$$

Substituting E_{2s} in (16), we find

$$J\left[E_{2Bx}\right] = \left[\nabla.\mathbf{E}_{2S}\right]_x = \frac{\partial E_{2S}}{\partial x} = -\frac{\beta\varepsilon_1^2 U}{\varepsilon_0}\frac{\partial E_{1my}^2(0)}{\partial x}.$$

Because of the invariance of translation along z, $\partial/\partial z=0$, then $J[E_{2Bz}] = 0$. Let us summarize again the boundary conditions at 2ω:

$$J\left[H_{2x}\right] = 2j\omega\, P_{Sz}^{NL} \qquad (18a)$$

$$J\left[H_{2z}\right] = -2j\omega\, P_{Sx}^{NL} \qquad (18b)$$

$$J\left[E_{2Bx}\right] = -\beta\frac{\varepsilon_1^2}{\varepsilon_0}U\frac{\partial E_{1my}^2(0)}{\partial x} \qquad (19)$$

$$J\left[E_{2Bz}\right] = 0 \qquad (20)$$

We are now able to determine the EM field at SH frequency, both in the vacuum and in the metal. However, in the case of reflexion on a metallic surface, the reflected SH field is the only measurable quantity, which is why we will now focus on the calculation of its characteristics.

2c Reflected SH field

We will calculate only the reflected SH field of TM polarization: the resolution for the TE case is similar. The propagation equation for H_{2z} is (everywhere, except at the surface)

$$\{\Delta H_{2z}\} = 2j\omega\, \varepsilon_0 \varepsilon_2\, \{\nabla \times \mathbf{E}_{2B}\} + 2j\omega\, \{\nabla \times \mathbf{P}_B^{NL}\}.$$

The last term of this equation is null in the metal, because $\mathbf{P}^{NL}{}_B$ is proportional to a gradient (see equation 5). Substituting $\{\nabla \mathbf{x}\, \mathbf{E}_{2B}\}$ by its definition (13) we get

$$\{\Delta H_{2z}\} + (2\omega/c)^2 \varepsilon_2 H_{2z} = 0. \qquad (21)$$

This relation is a Helmholtz equation: even inside the metal, there is no forced magnetic field. In the geometry we are concerned with, (21) can be rewritten as

$$\frac{\partial^2 H_{2z}}{\partial x^2} + \frac{\partial^2 H_{2z}}{\partial y^2} + \left(\frac{2\omega}{c}\right)^2 \varepsilon_2 H_{2z} = 0. \qquad (22)$$

The solutions of (22) are

$H_{2vz} = H_{2v} \exp(j(k_{2x}\, x + k_{2vy}\, y))$ in the vacuum,
$H_{2mz} = H_{2m} \exp(j(k_{2x}\, x - k_{2my}\, y))$ in the metal.

These two relations, together with (11), enable us to derive the expressions for E_{2x} and E_{2y}. The SH field amplitudes are then determined with the help of the boundary conditions equations:

$J[H_{2z}] = H_{2mz} - H_{2vz}$
$J[E_{2Bx}] = E_{2mx} - E_{2vx}$.

This leads to the expression of the SH field propagating in the vacuum:

$$H_{2vz} = \frac{2\omega\, \varepsilon_0 \varepsilon_2\, J\,[E_{2Bx}] - k_{2my} J\,[H_{2z}] + 2\omega\, P_{Bx}^{NL}(y=0)}{k_{2vy}\, \varepsilon_2 + k_{2my}}. \qquad (23)$$

A similar demonstration could be done for the TE SH field case. We would find

$$E_{2vz} = -2\omega\mu_0\, J[H_{2x}] / (k_{2vy} + k_{2my}). \qquad (24)$$

(23) and (24) are the general equations defining the SH field generated by reflexion on the flat surface of a centrosymmetric medium, provided the SH polarization can be written as the sum of a bulk and a surface term. Let us notice that the TE SH field E_{2vz} is generated only by a surface term, but the TM field is radiated also by the bulk metal through $P^{NL}{}_{Bx}$.
To calculate the SH field as a function of the pump field, we have first to substitute in (23) and (24) the values of the jumps of the SH field (i.e. (18 a), (18 b) and (19)) and then to introduce the pump field, defined by

$$\mathbf{E}_{1i}(\mathbf{r}) = \mathbf{E}_{1i} \exp(j(k_{1x}\, x - k_{1vy}\, y)) \qquad (25)$$

with $E_{1ix} = E_{1i} \cos\phi \cos\theta$, $E_{1iy} = E_{1i} \cos\phi \sin\theta$, $E_{1iz} = E_{1i} \sin\phi$, where ϕ is the angle between $\mathbf{E}_{1i}$ and the plane of incidence.

Depending on the state of polarization of the pump beam, the characteristics of the SH beam are quite different. For example, when the pump beam is TE polarized, only the bulk $\mathbf{P}^{NL}$ contributes to SHG. On the other hand, with a TM pump beam, both bulk and surface $\mathbf{P}^{NL}$ generate a SH field. In addition the TE SH field comes only from the surface NL polarization (see equation 24).

Recently, Guyot-Sionnest and Shen [18] have demonstrated an additional term $\mathbf{P}^{NL}_{add}$ in (1), originating in higher order in the multipole expansion of (1):

$$P^{NL}_{add,i} = -\Sigma_{j,k,l} \left(\frac{\partial \chi^{Q}_{jikl}}{\partial j}\right) E_{1l} E_{1k}.$$

Obviously, this expression exhibits a surface term through $\partial/\partial y$, but, because of the product $E_{1l}E_{1k}$, it is now impossible to separate bulk and surface contributions. Fortunately, this additional contribution is very weak for a clean metal surface [18] and can be neglected here.

In the more general case of centrosymmetric media, or if the metal surface is not clean, it is admitted [19] that one can describe the surface contribution by $\mathbf{P}^{NL}_s = \chi_s \mathbf{E}_1 . \mathbf{E}_1 \delta_s$. Due to the tensorial form of χ_s [17,20], $\mathbf{P}^{NL}_s$ exhibits components over all the x,y,z axes, which are generated by the terms of the product $\mathbf{E}_1.\mathbf{E}_1$, involving all the possible states of polarization for $\mathbf{E}_1$. In this case, the above considerations do not hold. For example, one can no longer separate surface and bulk contributions [21,18], even if one is able to select the polarization of both pump and SH beams.

2d Comparison with experiments: the need for phenomenological parameters

Experimentally, SHG from a metallic surface has been known since the pioneering work of F. Brown *et al.* [1]. Many measurements of SHG efficiency have been reported, using silver mirrors [22,3,23] or ATR configurations [24,25]. If one uses the expressions derived in the preceding section, or similar theories [3], to describe the experimental results, one will find a good qualitative agreement, but the calculated values of the SH intensity are about 20 times larger than the measured ones [23]. This leads us to modify [23] the expression (1) of the NL polarization by the adjunction of phenomenological parameters A and B, in such a way that

$$\mathbf{P}^{NL} = A\gamma\, \nabla(\mathbf{E}_1 . \mathbf{E}_1) + B\beta\, \mathbf{E}_1(\nabla . \mathbf{E}_1). \qquad (26)$$

This modification is in fact equivalent to weighting the relative contributions of the bulk and surface terms, but keeping the mathematical expression close to the Bloembergen relation, since this last one allows us to get a mathematically well-defined solution, as demonstrated previously. The coefficient A modifies only the weight of the "surface" contribution, since $B\beta\, \mathbf{E}_1 (\nabla.\mathbf{E}_1)$ is null inside the metal. On the other hand, B weights the "bulk" contribution with regard to the change of $\mathbf{P}^{NL}$ due to A. It means in fact that in the vicinity of the surface, the response of the metal can no longer be described by a free electron gas. Let us notice that these coefficients have to be determined phenomenologically from measurements. A calculation similar to the previous one can be done again, using expression (26) instead of (1). One finds

$$H_{2vz} = \frac{2\omega\ \varepsilon_0\ \varepsilon_2\ J^*[E_{2Bx}] - k_{2my}\ B\ J[H_{2z}] + 2\omega B\ P^{NL}_{Bx}(y=0)}{k_{2vy}\ \varepsilon_2 + k_{2my}} \tag{27}$$

$$E_{2vz} = \frac{-2j\omega\ \mu_0 B\ J\ [H_{2x}]}{k_{2vy}\ \varepsilon_2 + k_{2my}}, \tag{28}$$

where $J^*[E_{2Bx}]$ is equal to $J[E_{2Bx}]$ (see (19)), but with U_{AB} instead of U:

$$U_{AB} = \frac{1}{3}\,[(2A + B)\,U + (B - A)(\varepsilon_1 - 1)]. \tag{29}$$

A similar modification has been introduced by Rudnick and Stern [26], and used afterwards by many other authors [27,16,24]. As we showed in §2a, the surface NL polarization could be associated with a surface current density. Rudnick and Stern modified the value of this surface current density, the normal component by "a" and the tangential component by "b". This model has been extended by Sipe *et al.* [27], who determined the NL reflectivity of a metal surface in different configurations (mirror, prism): they found that b=-1 and $|a|\approx 1$, using Green function formalism (which is an alternative to the distribution method we developed here) and introducing an effective plasma frequency in the selvedge to arbitrary remove the difficulties which appear in the determination of $P^{NL}{}_{sy}$ (see §2b). The details of the calculations of the boundary conditions are not explicitly given in [27]. However, Sipe's theory and the approach given here are based on similar concepts, and they give close results as seen in fig. 2: experimental SHG efficiency of a silver mirror [23] is plotted together with theoretical fits from both theories. The best fits are for A=2.2, B=0.577 and a=-1.04, b=-1 (the difference between (A,B) and (a,b) comes in fact from their respective definitions).
The coefficient "a" has been carefully measured in many different configurations by Quail and Simon [24], who found a=0.9 at λ=1.06 µm for silver, this value being independent of the index of refraction of the above medium. The value of this phenomenological parameter "a" depends strongly on the conditions of fabrication of the metal surface and of its cleanness. This explains the different values reported here (a=0.9 and a=-1.04), which are however both in good agreement with the predictions [27] of the hydrodynamic model ($|a|\approx 1$).

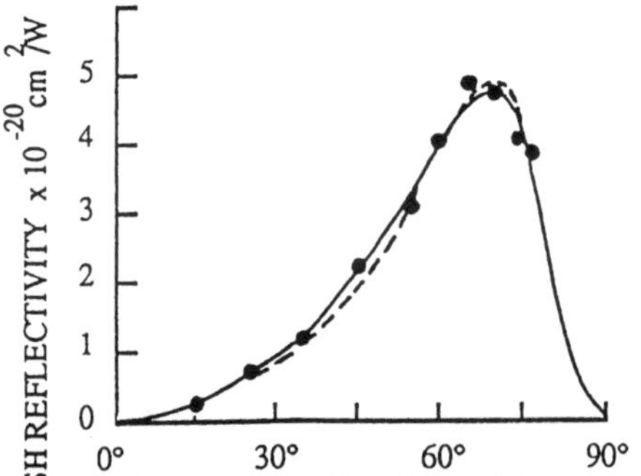

Fig. 2:
NL reflectivity of a silver mirror at 1.06 µm pump wavelength. The curves correspond to theoretical fits: Sipe's theory [27] (dashed), theory developed here [23] (continuous).

Many theoretical attempts have been made to determine the value of "a" [28,29,30,31]. It seems that these calculations, using density functional [29,30] or Thomas-Fermi theories [31], lead to very large values of "a", apparently in contradiction with the measurements reported here. Some authors claim that this discrepency is due to the bad cleanness of the real surface, and that a layer of adsorbates on the metal can greatly reduce the measured value of "a". Finally one should note that recently large experimental values of "a" have been published (a=-4 [25], a=-10 [32]).
It is certain that more theoretical work is required for the determination of the phenomenological parameters. But, even if they are not sophisticated enough to derive the coefficients, the hydrodynamic theories allow one to fit experimental results of SHG from flat metal surfaces. We will show later on that they lead also to a good prediction of some experiments of surface enhanced SHG.

3. Enhancement of the Pump Field

The nonlinear polarization at the metal surface is proportional to the square of the pump field amplitude in the metal, whatever the theory used to determine $\mathbf{P}^{NL}$. For a given metal surface, the only way to enhance the SH signal is then to increase the pump field amplitude at the surface. In this section, we will show how to reach this goal, by the excitation of SP. Another way to enhance the SH signal is to modify the metal surface, in order to change the electron dynamics in the vicinity of the surface [33,34]. According to the previous section, this leads to modification of the value of $\mathbf{P}^{NL}{}_s$ through χ_s. This effect has been reported many times, especially when another metal is deposited at the surface [6,35,36]. The reader will find supplementary information in the previous references [33,34,35,36]. We will concentrate here on the description of the effect of propagative SP excitation at the metal surface. Information on local (nonpropagative) SP will be given in §6.

3a Surface plasmon waves [37,38,39]

From the EM point of view, a SP corresponds to an EM wave propagating at the interface between 2 media, with an exponential decay of the wave amplitude in both media. Solving the Maxwell equations in the interface geometry leads to the conditions of SP existence:
- The wave is TM polarized.
- The product of the dielectric constants of both the media should be negative: let us consider vacuum as the first medium. Since ε(vacuum)=1, the real part of ε(material) should be negative. It means that this material could be: a metal, if the light frequency is below the plasma frequency of the metal; a dielectric or a semiconductor, in a spectral range of absorption. We will continue here to study the case of a metal-vacuum interface: ε is the dielectric constant of the metal.

- The SP dispersion relation is given by $$k_{SP}(\omega) = \frac{\omega}{c}\sqrt{\frac{\varepsilon}{1+\varepsilon}} \qquad (30)$$

where k_{SP} is the longitudinal wavevector amplitude of the SP.

The real part of ε is negative, so $k_{SP}(\varepsilon)>\omega/c$. The SP wavevector amplitude is then larger than the one of an incident plane wave in the vacuum. As for every guided wave, it means that it is impossible to directly excite a SP with a parallel light beam.

3b Optical excitation of SP

The use of an optical coupler allows one to match the longitudinal wavevectors of the incident wave and of the SP. Two kinds of couplers are used in practice: prism couplers (also often called attenuated total reflexion (ATR) devices) and diffraction gratings. It is also possible to excite SP through surface roughness. Indeed, the roughness can be modelled by the random superposition of many gratings [8].
The prism is used in the total reflexion configuration. The longitudinal wavevector of the beam, inside the prism, is large because of the high index of refraction of the prism. The wave, under the prism base, is evanescent and a metallic surface, brought into the vicinity of the prism base, is illuminated by the evanescent wave. Its longitudinal wavevector, because of momentum conservation through the interface, could be larger than those of the SP at the metal surface. Phase matching is obtained by tuning the angle of incidence of the beam in the prism. If the propagation of the SP is not too much perturbated by the prism, the resonant angle is deduced from

$$n_p \sin(\theta) = \sqrt{\frac{\varepsilon}{\varepsilon+1}} \quad (31)$$

This original configuration was proposed by Otto [40] (fig. 3a). A variation of this device is the so-called Kretschmann geometry [5], where a metallic layer is deposited on the prism (fig. 3b). As the optical field is evanescent in the metal, SP can be excited at the lower face of the metal. This configuration is often used in practice since it is very easy to control with great precision the metal thickness and its parallelism during the fabrication process. A hybrid device (fig. 3c) has been proposed by Sarid [41], with which it is possible to excite a SP at both borders of the metal layer embedded in two media. The corresponding EM field in the metal layer can be described by two modes of propagation, one of symmetrical distribution (relative to the y direction), the other one antisymmetrical. The latter mode has only a small fraction of its field in the metal, and then it suffers low attenuation: it is called a *long range* SP. The other mode is more attenuated, it is the *short range* plasmon. For all kinds of prism couplers, the EM field map can easily be calculated with usual formalisms, as multilayer theories [42].

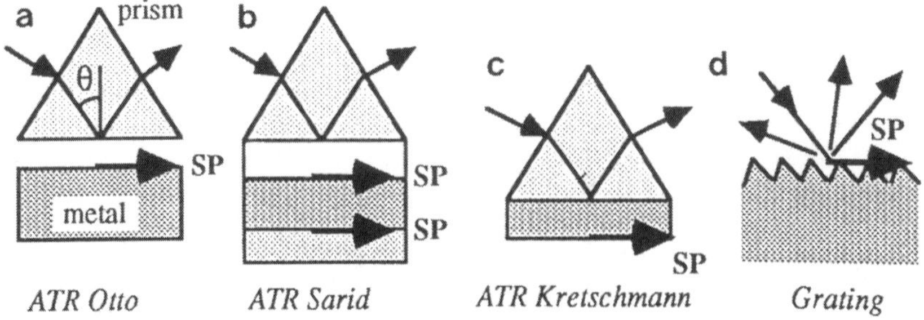

Fig. 3 : SP couplers

The grating coupler (fig. 3d) is based on the fact that a diffraction grating, due to its space periodicity, acts as a longitudinal wavevector generator [43]. The diffracted fields can then have longitudinal wavevectors larger than the incident one. Some of these fields are evanescent and then can correspond to SP at the grating surface. Here again, the wavevector matching is achieved by tuning the angle of incidence. As long as the SP propagation is not too much perturbated by the wavy shape of the surface, the excitation condition is described by

$$\sin(\theta) + \frac{m\lambda}{d} = \sqrt{\frac{\varepsilon}{\varepsilon+1}}, \tag{32}$$

where d is the periodicity of the grating and m is an integer corresponding to the order of diffraction. Unlike the prism coupler, the calculation of the field map is now more complicated, especially in the grating grooves, where an expansion of the field in plane waves is not rigorously allowed. However, there exist several exact methods to calculate the diffracted field [44,45]. Among them, the integral method is well adapted to solving both linear and nonlinear (SHG) problems [46]: this method is briefly described in appendix B.

The main difference between prism and grating couplers is that the latter can couple light to the SP through different orders of diffraction (see equation 32). It means that if we are looking at the specularly reflected beam versus the angle of incidence, we will observe only one minimum with the prism, but several, if allowed, with the grating. These minima correspond to a kind of Wood anomaly in the diffraction efficiency of gratings [44,43].

3c Enhancement of the field at the surface

As for the excitation of any resonant system, the efficiency of the optical coupling is maximum when the radiative losses are equal to the absorption losses. The absorption is determined by the optical losses of the metal, the radiation losses arise from the decoupling of the SP energy by leakage through the coupler. If the optical absorption is set for a metal, the radiative losses can be varied by changing the coupling constant of the device, i.e. the metal thickness or the groove depth. It means that there exists an optimum value of this constant, for which the coupling is maximum: this has been checked for both structures [47,48]. For the resonant excitation of SP, the lack of light in reflexion corresponds to energy coupled to the SP: the field amplitude E_{SP} at the metallic surface is thus enhanced as compared to the field E_R at a metal mirror in a usual reflexion experiment. The magnitude of the enhancement depends on the optical losses of the metal: silver, which possesses the lowest absorption coefficient in the visible range [49], is the most often used metal. For example, the enhancement, defined as the ratio of E_{SP}/E_R, reaches 10 in a Kretschmann configuration with a 600 Å thick silver layer for λ=5000 Å.

4. SHG in ATR Configuration

The first experimental reports of surface enhanced SHG in an ATR (Kretschmann) configuration were published by H. J. Simon *et al.* in the 70's [4]. Since this pioneering work, the same group has reported many results for different ATR configurations [24,50,51], involving SHG at metal-crystal interfaces, where both

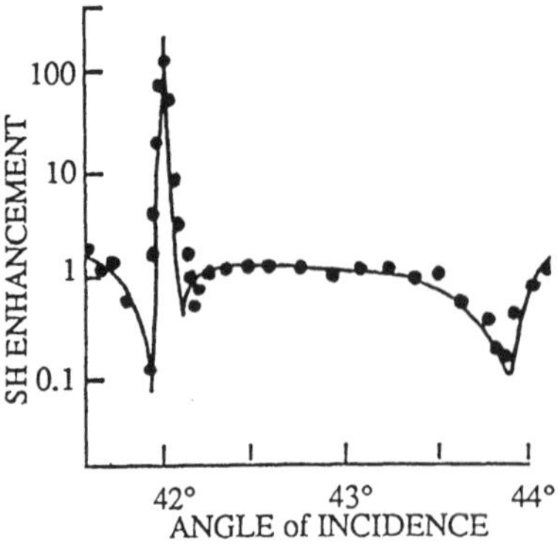

Fig. 4:
Enhancement of the SH signal in an ATR Kretschmann configuration with 500 Å of silver and a BK7 prism, from [24].

metal and crystal participate in the SHG. Similar experiments have been published more recently by Chen *et al.* [52]. The SP enhanced SHG at a metal-crystal interface has also been studied by De Martini *et al.* [53], who claimed they detected enhancement due to the NL excitation of a SP at the harmonic frequency. Simon's group has in addition studied the enhancement due to the excitation of short and long range SP [54,55,56].

Typical results obtained by Simon's group [24] are shown in fig. 4 for a 500 Å thick layer of silver evaporated on a BK7 prism. The main peak, close to 42°, corresponds to the excitation of a SP at ω that leads to an enhancement of $5 \cdot 10^2$ as compared to the flat mirror case. In the wings of the resonance, one observes dips due to interferences between SH sources at both surfaces of the metal layer. A large dip is also seen at the excitation angle of SP at the harmonic frequency. In this case, the SH field is generated only at the prism-metal interface, because the pump field is damped in the layer (no resonance at pump field). However, the SH light is coupled to the SP at the harmonic frequency, causing a leakage of energy towards the metal-air interface, and then a decrease is observed in the reflected SH intensity. Sipe's theory [27,24] describes very well the experimental results, *without any unknown parameters*, since the a and b parameters have been measured elsewhere, in an experiment of SHG in reflexion [24].

When a crystal is put against the metal layer, SHG comes from both crystal and metal [50,51,53]. Simon's group has demonstrated enhancement of SHG due to SP excitation at the pump frequency, but no detectable effect of the nonlinearly excited SP at the harmonic frequency. Using the birefringence of the crystals, for example quartz, it is possible to detect transmitted SHG in an ATR configuration, provided that the critical angle at the SH frequency is larger that the SP angle at the pump frequency [50,51].

Theoretically, Sipe *et al.* [27,16] have devoted numerous publications to describing the SHG enhancement in the Kretschmann device, using the hydrodynamic theory of SHG. As written previously, this theory gives a good description of the experimental results [24]. More recently, Weber [57] computed the influence of the metal layer thickness, using the same model to describe the NL response of the metal. As expected, he demonstrated the existence of an optimum metal thickness, for which he determined an enhancement of $6 \cdot 10^3$ for silver. As shown in §3, this optimum value corresponds to the offset to the optical losses in the metal by the radiative losses through the prism. The situation is however more complicated in the NL case, because two resonant phenomena are involved: the coupling of the pump field to the SP and the decoupling of the SH field. It explains why the optimum metal thickness in SHG is different from the optimum value in linear optics (respectively ≈ 500 Å and ≈ 600Å of silver thickness for λ=1.064 μm).

Excitation of the long range SP leads to high enhancement: this was theoretically predicted for different device configurations, including the excitation of counterpropagating long range SP [58,59,60]. To produce the long range SP, Quail *et al.* [54,55,56] used the ATR configuration: the gap between the prism and the silver layer is filled up with an index matching liquid. On the other side of the metal layer, there is a quartz crystal on which the metal was evaporated. The enhancement is over two orders of magnitude larger than the enhancement due to the excitation of a single-surface plasmon. In addition, it is possible to match the velocities of the short range SP at ω and long range at 2ω by tuning the thickness of the metal layer. Quail and Simon [56] observed the SH signal when the silver thickness (220Å) does not allow the phase-matching of both kinds of SP and measured an enhancement due to the excitation of the short range SP at ω. Increasing the silver thickness to the optimum value (285Å) enables the simultaneous excitation of short range and long range SP at ω and 2ω respectively and the previous enhancement is increased by two orders of magnitude.

Although , to our knowledge, no experimental report on SHG in Otto's ATR configuration has been published, this device has been extensively studied theoretically by Chen and Burstein [61], who calculated an enhancement of $5 \cdot 10^3$ for silver. They also suggested the possibility of generating a SH beam by NL mixing of counterpropagating SP. The theory of this phenomenon has been studied by Fukui *et al.* [62], in the framework of the Rudnick and Stern model for SHG [26]. Chen *et al.* [63] performed the experiment, using a Kretschmann ATR configuration. The metal is silver, and quartz is the medium below.The two counterpropagating SP are excited by two symmetrical input beams. The SH signal, generated by both metal and quartz, propagates in a direction perpendicular to the metal surface.

A different SP coupling process has been used by Sasaki *et al.* [64] to enhance SHG. Indeed, when a metal layer covers an optical dielectric waveguide, the guided wave can be coupled to a SP at the upper surface of the metal layer. The remarkable feature of the mode of propagation of the entire structure is the possibility of matching the phase velocites at ω and 2ω. The SH signal is generated by the metal layer. When phase matching is realized, a high SH signal is detected. Unfortunately, no SH conversion factor is given in [64].

5. SHG at a Metallic Diffraction Grating

First experimental results on surface enhanced SHG using silver gratings have only recently been published [65,11]. Coutaz *et al.* [11,66] reported on enhancement of SHG due to the excitation of SP in the first order of diffraction of the grating. They studied the angular dependence of the phenomenon, as well as the role played by the grating groove depth h, for which they demonstrated the existence of an optimum value h_{opt} leading to the greatest enhancement. They showed also that the enhancement depends on the grating period d. Other experimental results come only from Simon's group, who studied mainly the influence of the shape of the grating

groove [67]. In another publication, they described the SHG enhancement in a quartz-silver grating device, where SH light is generated by the quartz and the silver metal [68]. They also used the grating to produce long range and short range SP enhanced SHG [69,70]. A chinese group [71,72,73] reported on SP enhanced SHG with a GaAs-Al grating at the CO_2 laser wavelength. Here, SH is only generated by the GaAs crystal, and the contribution of the metal is neglected.
A grating-diffraction effect produced by a regular array of silver particles has been observed by Wokaun *et al.* [74] in a SHG experiment. The enhancement is due to electrostatic resonance in each particle. As this effect is concerned with local SP excitation, we will discuss it in the next section.

5a Theory of SHG at a metallic grating

If experimental results on SHG at a metal grating are mainly due to two groups, the theoretical study of this phenomenon has been the subject of numerous publications. A perturbation theory was first described by Agarwal and Jha [75,76]. The SH diffracted intensity is found to vary as the square of the groove depth, which is in contradiction with what we deduced in §3b, i.e. the existence of an optimum groove depth for which the SH signal is maximum. More sophisticated methods have since been used. Reinisch and Nevière calculated the SH diffracted field with the help of the differential theory [77,78,79]. They first gave the expression of the direction for the diffracted SH orders, and predicted the existence of the optimum groove depth h_{opt}. Farias and Maradudin [80] used the method of reduced Rayleigh equations to obtain the amplitudes of the diffracted SH fields. More recently, D. Maystre *et al.* [46,15] presented the resolution of the SH diffraction problem with the help of the integral theory. The outline of this theory is given in appendix B.
The rigorous theories [79,46,15] predict that the SH signal can be enhanced as compared to the mirror case by the excitation of a SP at the pump frequency. The second prediction is the existence of an optimum groove depth for which the enhancement is maximum. Due to the wavelength dispersion of the grating, the directions of propagation of the diffracted orders at ω and 2ω are different. It is easy to show [74,77] that the dispersion relation for the SH field is given by

$$n(2\omega)\sin(\theta_m) = n(\omega)\sin(\theta_i) + m\frac{\lambda}{2d} \qquad (33)$$

where n(ω) is the index of refraction of the above medium at frequency ω. When the above medium is not dispersive (for example vacuum), the SH NL diffraction pattern is the same as the one obtained, in linear optics, if the grating is illuminated by an incident beam at frequency 2ω: the 2ω diffracted beams propagate in the same directions as the diffracted pump beam for even orders m, but in different directions for odd orders. In practice, the detection of the odd-order SH beams is easier, since one does not have to separate harmonic and pump beams.

5b Comparison with experiment

Experiments were performed by Coutaz *et al.* [11] in order to check the predictions of the theory. The grating is made by holography, i.e. exposing a photoresist layer in an interference fringes pattern. After developing, the exposed resist is removed, and

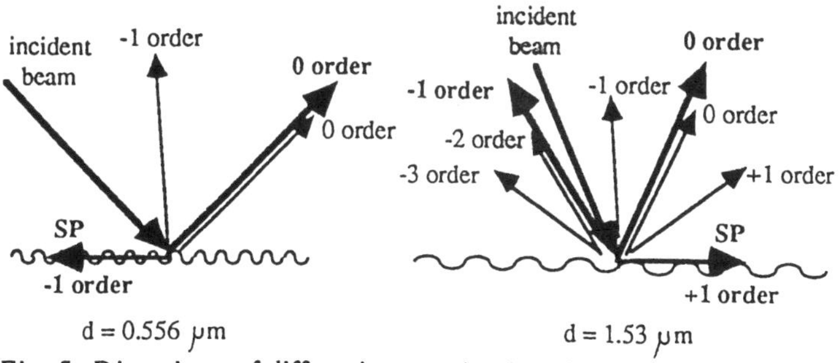

Fig. 5: Directions of diffraction at ω(—) and 2 ω(—).

the sinusoidal wavy surface is covered by an evaporated thick silver film. The groove depth was determined by comparing the measured and calculated linear efficiencies of diffraction. The value was then checked by observation with a scanning electron microscope. The grating period is d=0.556 μm, the pump beam is TM polarized and its wavelength is 1.06 μm. Under these experimental conditions, the SP at ω corresponds to the -1 order of diffraction (see fig. 5).

The characteristics of the 0 and -1 orders of diffraction at 2ω were measured as a function of the angle of incidence, exhibiting enhancement due to the excitation of the SP at ω. Fig. 6a shows the maximum of the SH signal achieved for each grating, as a function of the groove depth h of the grating, for the 0 and +1 orders of diffraction. The maximum value 36 of the enhancement is clearly seen, corresponding to an optimum groove depth $h_{opt} \approx 300$ Å. It should be emphasized that it corresponds to a very shallow modulation of the grating, since $h_{opt}/d \approx 5\%$. The same experiment has been performed again [66] for another grating of large period, d=1.53 μm. With this kind of grating, the SP at pump frequency corresponds to the +1 order of diffraction, and the SH signal enhancement was measured for the 0 and -1 orders at 2ω (fig. 6b).The measured maximum enhancement is 2500, showing the role played by the grating period in the phenomenon. The optimum modulation is still very weak (6.5%).

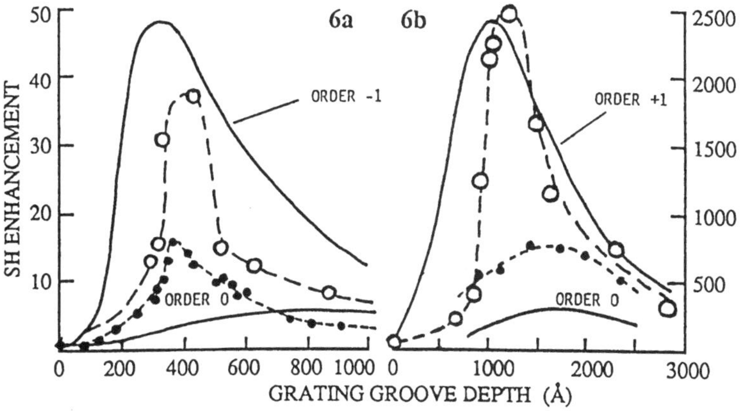

Fig. 6: Enhancement of the SHG by excitation of SP at grating surfaces versus the groove depth. a: period d=0.556 μm; b: period d=1.53 μm (from [11,15]).

The increase of the enhancement together with the increase of the grating period could be explained by the following: the diffracted pump beam in the metal is written as [67]

$$E_1(x,y) = \Sigma_p E_{1p}(y) \exp(j k_{px} x) \qquad \text{where } k_{px} = k_{ix} + p\frac{2\pi}{d}.$$

Then, the NL polarization is

$$P^{NL}(x,y) \approx |E_1(x,y)|^2 \approx \Sigma_p \exp(j k_{px} x) \, \Sigma_q E_{1q}(y) E_{1p-q}(y).$$

We calculate the field at 2ω:

$$E_2(x,y) \approx P^{NL}(x,y) \approx \Sigma_p E_{2p}(y) \exp(j k_{px} x)$$
$$\text{where } E_{2p}(y) = \Sigma_p E_{1q}(y) E_{1p-q}(y).$$

Let us suppose the SP is excited in the k order of diffraction at ω, then the field component E_{1k} is mainly enhanced, but also, by EM coupling, all the neighbouring orders, i.e. E_{1k-1} and E_{1k+1}, etc. The most enhanced orders at 2ω are $E_{2,2k} \approx E_{1k}E_{1k}$, then $E_{2,2k-1} \approx E_{1k}E_{1k-1}$ and $E_{2,2k+1} \approx E_{1k}E_{1k+1}$. Because E_{1k} is evanescent (SP), $E_{2,2k}$ is also evanescent and then the most enhanced propagating orders at 2ω are $E_{2,2k+1}$ or $E_{2,2k-1}$. When the grating period is reduced, the coupling between diffracted orders is increased, and then the amplitude of the E_{1k-1} and E_{1k+1} orders. But, simultaneously, the number of propagating diffracted orders increases, and so the optical losses by radiation. The competition between the two opposite effects leads to a value of the grating period for which the SH signal is maximum.

Continuous lines in fig. 6 are the theoretical enhancements calculated with the integral theory [15]. In the calculation, no unknown parameter is used, since the A and B parameters were deduced from a SHG experiment with a silver mirror [23], and the dielectric constants of silver at ω and 2ω were measured by ellipsometry. The experimental data, for the nonspecular orders (m≠0), are well represented by the theory: indeed, for d=1.53 μm, the theory predicts an enhancement of 2300 for h_{opt}=900 Å, and they have been measured respectively as 2500 and 1000 Å. However, for the specular order (m=0), there is a discrepancy between experiment and theory, which appears in the same manner for both families of gratings (d=0.556 and d=1.53 μm): the calculated enhancement is lower than the measured one, by a factor of ≈3.

A similar observation was made by Quail and Simon [67] with a grating of d=833 μm and h=220 Å. They calculated the SH diffracted signal with the help of the formulism developed by Farias and Maradudin [80], using Rayleigh expansions to describe the grating diffraction and Sipe *et al.'s* [27] theory to calculate the NL polarization in the metal. Because of the weak grating modulations considered here, the Rayleigh expansion is used in its range of validity. The agreement with the observed signal is very good for the -1 and 0 orders at 2ω, but there is a large discrepancy for the +1 order, since the calculated enhancement is about 10^{-3} smaller than the measured one. Quail and Simon [67] suggested that this problem could arise from the shape of the grating grooves. They computed the SH diffracted signal,

assuming a grating profile function that is 97% sinusoidal with a 3% admixture of the third spatial harmonic. The -1 and 0 order signals are not very sensitive to this change, but the calculated +1 order is enhanced, leading to a good agreement with the measured value. By the same procedure, Simon et al. [69,70] were also able to describe experimental results they obtained with a thin layer of silver deposited on a grating grooved in quartz crystal: in this way, they excited long range surface plasmons at the silver surface, and they measured enhanced SHG from both silver and quartz. They reported an enhancement of 4 orders of magnitude. Thus, it is demonstrated that a very weak perturbation of the sinusoidal groove profile leads to a redistribution of the energy in the different orders of diffraction at 2ω. This effect is well known in linear optics (blaze effect), but it seems more effective in NL optics. It explains also why Coutaz *et al.* [11,66] measured enhancements larger than the calculated ones for the specular order: the gratings they used have a quite truncated sinusoidal waveform [11].
Since the enhancement depends greatly on the groove profile, this profile can be optimized to lead to larger enhancements. This idea has been exploited by Nevière *et al.* [79], who computed the enhancements for gratings of the same period (1.53 μm) but of different groove profiles: for a sinusoidal profile, the enhancement is 2370, and it becomes 10^5 for a trapezoidal profile!

6. Enhancement by excitation of local SP

Another way to enhance the EM field at the metal surface is to excite the local SP. Indeed, it has been known for a very long time that sharp geometrical metallic structures (e.g. bumps, pits, tips, particles...) act to confine the motion of electrons associated with them, leading to resonances of the collective oscillations of these electrons. Such a resonance corresponds to what we called local field resonance, or local SP [81]. The dimensions of the sharp structures are usually very small compared to the beam wavelength, thus the propagation effect can be neglected and the calculation of the field in the vicinity of the structure can be performed within the electrostatic formalism. The simplest way to excite local SP is to illuminate a rough surface, which is in fact a collection of small sharp patterns. Rough surfaces are easily prepared by electrolytic oxidation and reduction cyclings. Some authors have also fabricated microscopic particles of metal deposited on a surface.
Local SP have been widely used to enhance NL surface effects, mainly Raman scattering of adsorbates [82], but also Raman scattering in semiconductors [83], two-photon luminescence [84], stimulated Raman scattering [85], etc. We will focus here on SHG from bare rough surfaces, where the enhancement is only of EM origin.

6a Theories of SHG with local SP

Two methods enable us to calculate the enhancement due to roughness. The first one consists of representing the rough surface by a flat metal covered by submicronic particles, of spheroidal or ellipsoidal shapes. This is in fact a good description of the surface of metallic electrodes, which exhibits microscopic bumps, after oxidation-reduction cyclings. Assuming the particle size is small compared to the wavelength, the field E_{in} inside the particle is calculated within the electrostatic approximation.

For an insulated ellipsoid in vacuum, E_{in} is given by

$$E_{in} = \frac{E_0}{1 + (\varepsilon - 1) A} \tag{34}$$

where E_0 is the incident field, and A is a shape factor depending on the principal axes f_a and f_b of the ellipsoid [83]. A=0.33 for a sphere and it decreases when the sphere becomes elongated.

E_{in} reaches a maximum when $\varepsilon = 1-1/A$, but E_{in} does not become infinite because the resonance is damped by the optical losses of the metal. This resonance is usually called localized SP. Because of the dispersion of ε, there exists an optimum value of the frequency for which E_{in} is maximum. For a silver ellipsoid with $f_a/f_b=2$, the resonance is obtained for $\lambda \approx 4100$ Å. The field E_{out} at the surface of the particle is deduced from equation (34) with the aid of usual boundary conditions. At the tip of the ellipsoid, E_{out} increases rapidly as the ratio f_a/f_b increases [81]. This phenomenon is referred to as the lightning rod effect [87]. For SHG, the enhancement can be approximated by $L^4(\omega)L^2(2\omega)$, where $L(\omega)=E_{in}/E_0$ is the local field factor [10]. Using a more complicated model, involving image field effect in the conducting flat surface [88], Boyd *et al.* [89] derived the enhancement for various metals, relative to silver. However, we have to note that the actual rough surface cannot be readily compared with an isolated ellipsoid. Many attempts have been made to approximate the real situation by two or several interacting spheroids, or more complicated models. In this case, the prediction of the magnitude of the enhancement is subject to many adjustable parameters, even in the electrostatic approximation [90].

The second way [91,92] to determine the enhancement is to consider the rough surface as a whole. The roughness is described by a surface profile function which is a random variable. The determination of the scattered field is achieved following the Green's function techniques described by Maradudin and Mills [93]. Using Sipe's theory [16] for the metal NL polarization, Arya [91] found an enhancement of 10^4 for a rough silver surface, characterized by an average roughness amplitude of 200Å and a transverse correlation length of 800Å. On the other hand, Deck and Grygier [92] determined an enhancement of 2 orders of magnitude and claimed that the SH scattered beam is diffracted from the metal in a narrowly defined direction.

6b Enhanced SHG at rough metallic surface

The first experimental report on enhanced SHG at a rough metal surface is due to Chen *et al.* [10]. These authors used a metal surface roughened by chemical etching in a solution of KCl in water. For silver, the surface roughness consists of 500 Å particles. The SH light is scattered nearly isotropically in angular distribution. In addition to the SH frequency, a broad spectrum is detected, which may be of luminescence origin. An enhancement of 10^4 is reported for a rough silver-air interface (here the enhancement corresponds to the ratio of the SH signal intensity, integrated over the 2π solid angle, to the collimated SH signal generated by a flat surface). Using a similar technique, Murphy *et al.* [94] reported an enhancement of 700 for silver etched by a solution of NH_4OH. Boyd *et al.* [89] measured the enhancement for a wide set of materials, including metals, semi-metals and semiconductors. To compare all these materials with each other, they used a rough glass substrate, on which they evaporated the material. After the measurement, they removed the metal layer, and used the same substrate again in a similar way to test another material. Thus they claimed that the roughness is identical for each tested material. They reported enhancements of $2 \cdot 10^3$ for silver. For the other materials, except gallium and aluminium, the enhancements are weaker, all the results being well described by the local field effect model.

Formation of pores at a silver film surface may give rise also to localized resonances [95,96]. These pores appear between grains during the growth of "cold metal films", which are metal layers evaporated on a substrate at low temperature (≈100 K). For a silver cold film, Aktsipetrov *et al.* [97] reported enhancements of 2-3 orders of magnitude. When the cold film is heated to room temperature, the effect of enhancement disappears irreversibly, because the pores fill up by self-diffusion of silver [98].

Chen *et al.* [11] demonstrated also that the enhancement of the SH signal, for an electrolyte-silver interface, depends on the electrolytic cycling. This is of great interest for probing electrochemical processes at surfaces, and this powerful tool has since been used many times [99,100,101]. In this kind of experiment, it appears that SHG is mainly due to the contribution of chemical adsorbates deposited on the silver during the electrolytic cycling. The roughness of the surface enhances the efficiency of SH conversion and makes it easy to detect. But, as the scope of this paper is restricted to SHG from bare metals, we let the interested reader find further information on SHG from adsorbates on rough surfaces in some review articles [9,102,90,12].

6c Enhanced SHG from metal islands

Remarkable experiments on surface NL optics have been reported by a group at Bell Laboratories, involving the Raman effect [103] and SHG [74]. A regular array of metal islands is deposited on a sapphire substrate by a microlithographic technique. The SHG efficiency is measured [74] as a function of the mass thickness d of the islands, defined as the mass deposited by unit area over the bulk metal density (fig. 7). For low mass thickness, the dimensions of the metal particles are small (d<100Å) and the particles are well separated. Local SP are then excited by the laser beam in each particle. A local SP resonance is observed for d=36 Å, corresponding to a maximum of enhancement of 10^3. This observed maximum confirms the predictions

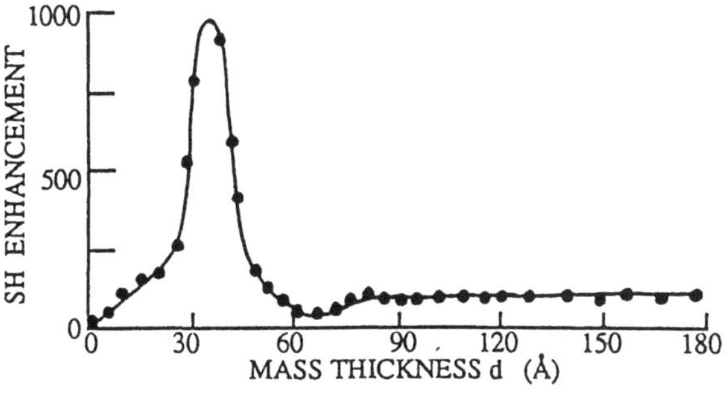

Fig. 7: SHG from Ag island films as function of mass thickness (from [74])

of the local field effect theory. For larger mass thicknesses (d>100 Å), the metal islands coalesce into a continuous film, and the enhancement reaches a plateau of 10^2. This plateau is attributed to the excitation of propagating SP. When the array of particles is perfectly regular, the NL grating effect (§5) is observed.

Russian physicists [104] have studied the influence of the shape of the metal islands, showing that the lightning rod effect may lead to large enhancements. They fabricated conical metal islands, with dimensions (base of the cone ≈0.5 µm, height 2-3 µm) large enough to avoid local SP resonance inside the island. The lightning rod effect leads to an enhancement of 10^5.

Golubtsov *et al.* [105] have deposited a very thin metal film on the base of a prism. The pump beam impinges through the prism at angles exceeding the total reflexion angle. The metal islands are then illuminated by an evanescent field, which leads to an enhancement of $8.5 \cdot 10^3$.

7. Concluding Remarks

We have tried to show that enhancement of SHG may be a very good tool for scientists studying surfaces:

- It is easy to make use of it, whatever the chosen coupling device. In the ATR configuration, one has just to control the thickness of the metal layer during the evaporation process. With a diffraction grating, the major parameter is the groove depth, which is determined by the exposure time in the holographic fabrication. In the case of a rough surface, the amplitude of the roughness depends on the electrolytical cycling, for rough electrodes, or of the mass of evaporated metal for ultrathin rough metal layers. In any case, the main parameter of the pump field enhancement is easily controlled during fabrication.
- It makes the experiment easier to perform, especially the detection of the signal, because the magnitude of enhancement can be very large, in any case larger than 2 orders of magnitude. Such a large increase has been cleverly used by Boyd *et al.* [106] to detect a monolayer of adsorbates on a rough surface with a CW laser diode of low power (20 mW). This spectacular experiment demonstrates the interest in enhancement. However, one should not forget that, even if the enhancement is large, the measured SH intensity remains very small, because of the low efficiency of the phenomenon. For a flat silver surface, the efficiency of SHG is about 10^{-20} cm^2/W at 1.06 µm pump wavelength [94,24,23]. It corresponds to only 1000 reflected SH photons for an incident beam of 10^{17} photons (i.e. a YAG-laser beam of 2 MW/cm^2, 1 cm beam diameter and 10 ns pulse duration).

The large enhancements reviewed in this article have been actually achieved with very low perturbations of the surface flatness. Indeed, the optimum grating modulation is about 5% and typical roughnesses correspond to surface protusions of 100s of angstroms in size. So weak a modulation of the surface leads to large enhancement of the SH signal.

It is notable that there exist theories which allow a fairly good description of the enhanced SHG phenomena. It means that the NL EM diffraction problem has been solved. But the weak point of the theory is the description of the SHG phenomenon itself. Many physicists are now working on this problem, from a solid state point of view [13]. One has however not to forget that the real metal surfaces fabricated by the experimentalists are never perfectly clean and flat. This fact is of prime importance since the SHG is extremely sensitive to the surface state.

Appendix A: Some useful vectorial relations in the theory of distributions

We would like to give here some basic expressions of the distribution theory, to enable the reader to follow the calculations of Sect. 2. Let $\mathbb{R}$, $\mathbb{C}$ and $\mathbb{F}$ be respectively the field of real numbers, the field of complex numbers and a vector space of functions. We call each application of $\mathbb{F}$ on $\mathbb{C}$ which relates a function to a complex number the functional T :

$$\forall\, \phi(x) \in \mathbb{F} \qquad < T, \phi > = c \qquad c \in \mathbb{C}.$$

Let $\mathbb{D}$ be the space of functions which are differentiable an infinite number of times and whose support is bounded. The functional T, defined on $\mathbb{D}$, is a distribution if T is linear and continuous. We define a regular distribution T_f associated with the function f by

$$\forall\, \phi(x) \in \mathbb{D} \qquad < T_f , \phi > = \int f(x)\, \phi(x)\, dx \; = < f(x), \phi(x) >$$

and the singular distribution δ by

$$\forall\, \phi(x) \in \mathbb{D} \qquad < \delta , \phi > = \phi\ (0).$$

In three dimensions, we write

$$\forall\, \phi(\mathbf{r}) \in \mathbb{D} \qquad < \delta(\mathbf{r}) , \phi(\mathbf{r}) > = \phi\ (0).$$

Let S be a surface and a(r) a function which is defined on S. The distribution $a\delta_s$ is

$$\forall\, \phi(\mathbf{r}) \in \mathbb{D} \qquad < a\ \delta_s , \phi > = \int_S a\ \phi\ dS.$$

The derivative of a distribution is

$$\forall\, \phi(x) \in \mathbb{D} \qquad < \frac{dT}{dx}, \phi > = - < T, \frac{d\phi}{dx} >.$$

Let U be a regular function everywhere, except on S. One shows that

$$\frac{\partial U}{\partial x} = \{ \frac{\partial U}{\partial x} \} + n_x\ J\ [\ U\ ]\ \delta_s$$

where {...} means that the derivation is made in the usual sense of the theory of functions and n_x is the x component of the normal **n** to the surface. From this relation, we can deduce the expressions of some vectorial operators valid in the sense of distributions:

$$\nabla\ U = \{\ \nabla\ U\ \} + \mathbf{n}\, J\ [\ U\ ]\ \delta_s$$

$$\nabla\ \mathbf{U} = \{\ \nabla\ \mathbf{U}\ \} + \mathbf{n}\ .\ J\ [\ \mathbf{U}\ ]\ \delta_s$$

$$\nabla \times \mathbf{U} = \{ \nabla \times \mathbf{U} \} + \mathbf{n} \times J [\mathbf{U}] \delta_s$$

To learn more about the powerful theory of distributions, the reader will find a pedagogical introduction by Petit in appendix A of [44].

Appendix B: The integral method [45,44]

To determine the amplitude of the field diffracted by a grating, one has to solve the Maxwell equations, and find a solution that obeys the boundary conditions at the grating surface. Above and below the groove region, the media are homogeneous, so the field can be described by a plane wave expansion, the Rayleigh expansion. But inside the modulated region, the medium is inhomogeneous by geometry. The exact field is no longer represented by a plane wave expansion. If the grating modulation (i.e. the groove depth over the periodicity) is small, one can approximate the exact solution by a plane wave expansion (Rayleigh hypothesis).
However, a rigorous way to determine the solution, among others [45], is the integral method, which is well adapted to the case of metallic gratings, and which, in addition, allows one to solve the NL problem with the same formalism. We will give here only the scheme of this method. The interested reader will find complete information in [45,44].

B1 The integral method in linear optics

We denote by R^+ (R^-) the region above (below) the grating surface. The grating surface is determined by y=f(x). Let F be the diffracted field ($F=E_Z$ for TE polarization, $F=H_Z$ for TM) and F_i the incident field. F obeys the following relations:

$$\Delta F + k^2 \varepsilon F = 0 \qquad \text{where } \varepsilon = 1 \text{ in } R^+,\ \varepsilon = \varepsilon(\text{metal}) \text{ in } R^-, \tag{B1}$$

and, at the grating surface

$$F^+ + F_i\,(x,f(x)) = F^- \tag{B2}$$

$$\partial F^+/\partial n + \partial F_i(x,f(x))/\partial n = Cp\, \partial F^-/\partial n \qquad \text{with } C_p=1 \text{ (TE)},\ C_p=1/\varepsilon \text{ (TM)} \tag{B3}$$

where n is the coordinate normal to the grating surface. In addition, F satisfies a radiation condition for $y \rightarrow \pm\infty$. In order to find F, it is useful to define a fictitious field U(x,y) which is equal to F(x,y) in R^+, which is a solution of the propagation equation in R^- (R^- being filled with vacuum), and which is continuous everywhere. The subtlety of the method introduced by D. Maystre [107] is that the field F can be described everywhere as a function of only one variable, the jump, through the grating surface P, of the derivative of U normal to the surface. Let us define the function $\phi(x)$ related to this jump by

$$\phi\,(x) = [\frac{\partial U^+}{\partial n}(x) - \frac{\partial U^-}{\partial n}(x)]\,h\,(x) \qquad \text{with } h\,(x) = (1+(\frac{\partial f}{\partial x}(x))^2)^{1/2}. \tag{B4}$$

Note that a fictitious surface current density $I_S(x)$ could be associated to $\phi(x)$ by $I_S(x)= j\phi(x)/h(x)\omega\mu_0$. The field U(x,y) in R^+ is calculated with the Helmholtz-

Kirchhoff theorem:

$$U(x,y) = \int_0^d A(x,y,x')\,\phi(x')\,dx'. \tag{B5}$$

Let us notice that, due to the periodicity of the grating, it has been possible to find a Green function A(x,y,x') in such a way that the integration is performed over only one period d of the grating. From (B5), we can derive the limit U^+ of U(x,y) just above the grating:

$$U^+ = U(x,f(x)) = \int_0^d A(x, f(x), x')\,\phi(x')\,dx' \equiv G\{\phi\}. \tag{B6}$$

In (B6), we have introduced the operator G which is applied to ϕ. In the following, we will use the same notation: for example, Z{X} means that an operator Z is applied to a function X through a relation similar to (B6). Identically to the derivation of (B6), one finds

$$\frac{\partial U^+}{\partial n} = \left[\frac{\phi}{2} + N\{\phi\}\right] / h(x). \tag{B7}$$

Since in R^+, F(x,y)=U(x,y), we then know the total field in R^+ by adding the incident field $F_i(x,y)$. Let us define the limit value of F_i just above the grating:

$$\psi_0 = F_i(x, f(x)) \tag{B8}$$

$$\phi_0 = h(x)\,\frac{\partial F_i}{\partial n}(x, f(x)). \tag{B9}$$

From the boundary conditions (B2) and (B3), we derive the value of the field F just below the grating:

$$F^- = F^+ + \psi_0 = U^+ + \psi_0 = G\{\phi\} + \psi_0 \tag{B10}$$

$$h(x)\,C_p\,\frac{\partial F^-}{\partial n} = \frac{\partial F^+}{\partial n} + \phi_0 = \frac{\partial U^+}{\partial n} + \phi_0 = \frac{\phi}{2} + N\{\phi\} + \phi_0. \tag{B11}$$

To determine the field in R^-, we define a field V which is equal to F in R^- and to 0 in R^+. The Helmholtz-Kirchhoff theorem leads to the knowledge of V(x,y) from F^- and $\partial F^-/\partial n$:

$$V(x,y) = -\int_0^d A_\varepsilon(x,y,x')\,\frac{\partial F^-}{\partial n}(x')\,h(x')\,dx' - \int_0^d B_\varepsilon(x,y,x')\,F^-(x')\,dx'. \tag{B12}$$

As compared to (B5), there is an additional second term in (B12), which comes from the discontinuity of V(x,y) through the surface. From (B12), we determine the limit of F just below the grating:

$$\frac{F^-}{2} = V(x,f(x)) = -\int_0^d A_\varepsilon(x,f(x),x')\,\frac{\partial F^-}{\partial n}(x')\,h(x')\,dx' - \int_0^d B_\varepsilon(x,f(x),x')\,F^-(x')\,dx'$$

$$\equiv -\left(\frac{1}{C_p}G_\varepsilon\{h\frac{\partial F^-}{\partial n}\} + N^*_\varepsilon\{F^-\}\right) = -\left(\frac{1}{C_p}G_\varepsilon\{\phi_0 + \frac{\phi}{2} + N\{\phi\}\} + N^*_\varepsilon\{G\{\phi\} + \psi_0\}\right). \tag{B13}$$

The field F^- is linked by (B10) to the field just above the grating. Substituting its value in (B13) leads to the integral equation

$$G\{\phi\} + \frac{1}{C_p}G_\varepsilon\{\frac{\phi}{2} + N\{\phi\}\} + N^*_\varepsilon\{G\{\phi\}\} = -\psi_0 - G\{\phi_0\} - N^*_\varepsilon\{\psi_0\}. \tag{B14}$$

Let us notice that the only unknown function in (B14) is ϕ. Indeed, the right-hand term of (B14) corresponds to the contribution of the incident field. The operators G, G_ε, N, N_ε^* can be written as Fourier expansions, because of the periodicity of the grating. Solving (B14) leads to ϕ, and through (B5) and (B12) to the knowledge of F above and below the grating.

B2 Application to the SH diffraction problem

To solve the NL problem, we use an analogy which exists between the linear and NL equations, in the case of SHG from a metallic grating. Let us consider only the TM SH polarization, therefore $\mathbf{H}_2 = \mathbf{H}_{2z}$. We will use similar notations as those defined in §B1.

As we demonstrated in §2c, the propagation equation for H_{2z} in R^+ and R^- is (21):

$$\{\Delta H_{2z}\} + k_2^2 H_{2z} = 0 \tag{B15}$$

where: $k_2 = \frac{2\omega}{c}\sqrt{\varepsilon_2}$, $\varepsilon_2 = 1$ in R^+ and $\varepsilon_2 = \varepsilon_2(\text{metal})$ in R^-.

We know from (18) the value $J[H_{2z}]$ of the jump of H_{2z} through P, thus

$$H^-_{2z} = H^+_{2z} + J[H_{2z}]. \tag{B16}$$

The H_{2z} field is linked to the NL polarization by the following Maxwell equation:

$$\nabla \times \mathbf{H}_{2z} = -\frac{\partial}{\partial t}(\mathbf{D}_2 + \mathbf{P}^{NL}).$$

For simplicity, let us here consider only a flat surface, with its normal along y. The projection over x of the above relation gives

$$\frac{\partial H_{2z}}{\partial y} = -2j\omega\,[\varepsilon_2 E_{2x} + P_x^{NL}].$$

As this relation is true in both R^+ and R^-, we obtain by subtraction

$$\frac{\partial H^+_{2z}}{\partial y} - \frac{1}{\varepsilon_2}\frac{\partial H^-_{2z}}{\partial y} = -2j\omega\left(J[E_{2x}] - \frac{P_x^{NL}}{\varepsilon_2}\right).$$

It can be shown [46], that using a curvilinear coordinates (X,Y,z) system, this last relation can be generalized to an arbitrary shape surface by

$$\frac{\partial H^{+}_{2z}}{\partial n} + 2j\omega \left(J\,[E_{2X}] - \frac{P^{NL}_{X}}{\varepsilon_2} \right) = \frac{1}{\varepsilon_2} \frac{\partial H^{-}_{2z}}{\partial n} . \qquad \text{(B17)}$$

Thus, the SH diffracted field obeys relations (B15,B16,B17) similar to those which define the linear diffracted field (B1,B2,B3). The NL source terms replace the contribution of the incident field, which was the source for the linear diffracted field. The resolution of the system (B15,B16,B17) is similar to that of (B1,B2,B3). The full computation of the SH diffracted field involves the following steps:
- determination of the pump field in the metal by the integral method,
- calculation of the NL source terms $J[E_2]$, $J[H_2]$, and P^{NL}_{Bx} from the values of the pump field in the metal,
- determination of the SH diffracted field with the integral method, the SH density curent being the source terms.

References

[1] F. Brown, R.E. Parks, A.M. Sleeper, Phys. Rev. **14**, 1029 (1965)
[2] S.S. Jha, Phys. Rev. Lett. **15**, 412 (1965)
[3] N. Bloembergen, R. Chang, S. Jha, C. Lee, Phys. Rev. **174**, 813 (1968)
[4] H.J. Simon, D.E. Mitchell, J.G. Watson, Phys. Rev. Lett. **33**, 1531 (1974)
[5] E. Kretschmann, H. Raether, Z. Naturforsch. **23a**, 2135 (1968)
[6] J.M. Chen, J.R. Bower, C. Wang, C.H. Lee, Opt. Commun. **9**, 132 (1973)
[7] M. Fleischmann, P.J. Hendra, A.J. McMillan, Chem. Phys. Lett. **26**, 163 (1974)
[8] M. Moskovitz, Rev. Mod. Phys. **53**, 783 (1985)
[9] D. Ricard, Ann. Phys. Fr. **8**, 273 (1983)
[10] C.K. Chen, A.R.B. de Castro,Y.R. Shen, Phys. Rev. Lett. **46**, 145 (1981)
[11] J.L. Coutaz, M. Nevière, E. Pic, R. Reinisch, Phys. Rev. **B32**, 2227 (1985)
[12] Y.R. Shen, Nature **337**, 519 (1989)
[13] O. Keller, in *Nonlinear Optics in Solids*, edited by O. Keller, Springer-Verlag, Berlin (1989)
[14] D. Maystre, M. Nevière, R. Reinisch, Appl. Phys. **A39**, 115 (1986)
[15] R. Reinisch, M. Nevière, A. Akhouayri, J.L. Coutaz, D. Maystre, E. Pic, Opt. Engineering **27**, 961 (1988)
[16] J.E. Sipe, G.I. Stegeman, in *Surface Polaritons*, edited by V.M. Agranovich and A.A Maradudin, North Holland, Amsterdam (1982)
[17] P. Guyot-Sionnest, W. Chen, Y.R. Shen, Phys. Rev. **B33**, 8254 (1986)
[18] P. Guyot-Sionnest, Y.R. Shen, Phys. Rev. **B38**, 7985 (1988)
[19] Y.R. Shen, *The Principles of Nonlinear Optics*, Wiley Interscience, New-York (1984)
[20] J.E. Sipe, D.T. Moss, H. M. van Driel, Phys. Rev. **B35**, 1129 (1987)
[21] J.E. Sipe, V. Mizrahi, G.I. Stegeman, Phys. Rev. **B35**, 9091 (1987)
[22] H. Sonnenberg, H. Heffner, J. Opt. Soc. Am. **58**, 209 (1968)
[23] J.L. Coutaz, D. Maystre, M. Nevière, R. Reinisch, J. Appl. Phys. **62**, 1529 (1987)

[24] J.C. Quail, H.J. Simon, Phys. Rev. **B31**, 4900 (1985)
[25] W. Zheng, L. Li, S. Dong, Z. Zhang, paper ThA4, IQEC, Tokyo (1988)
[26] J. Rudnick, E.A. Stern, Phys. Rev. **B4**, 4274 (1971)
[27] J.E. Sipe, V. So, M. Fukui, G.I. Stegeman, Phys. Rev. **B21**, 4389 (1980)
[28] M. Corvi, W.L. Schaich, Phys. Rev. **B33**, 3688 (1986)
[29] M.G. Weber, A. Liebsch, Phys. Rev. **B35**, 7411 (1987)
[30] A. Liebsch, Phys. Rev. Lett. **61**, 1233 (1988)
[31] A. Chizmeshya, E. Zaremba, Phys. Rev. **B37**, 2805 (1988)
[32] K.J. Song, E.W. Plummer (unpublished) cited in [34]
[33] M.G. Weber, A. Liebsch, Phys. Rev. **B36**, 6411 (1987)
[34] B.N.J. Persson, L.H. Dubois, Phys. Rev. **B39**, 8220 (1989)
[35] H.W.K. Tom, C.M. Mate, X.D. Zhu, J.E. Crowell, Y.R. Shen, G.A. Somorjai, Surf. Sci. **172**, 466 (1986)
[36] K.J. Song, D. Heskett, H.L. Dai, A. Liebsch, E.W. Plummer, Phys. Rev. Lett. **61**, 1380 (1988)
[37] H. Raether, *Excitation of Plasmons and Interband Transitions by Electrons*, Springer Verlag, Berlin (1980)
[38] *Electromagnetic Surface Modes*, edited by A. D. Boardman, Wiley-Interscience, New-York (1982)
[39] *Surface Polaritons*, edited by V.M. Agranovich and A.A. Maradudin, North Holland, Amsterdam (1982)
[40] A. Otto, Z. Physik **216**, 398 (1968)
[41] D. Sarid, Phys. Rev. Lett. **47**, 1927 (1981)
[42] See for example chapter 13 of M. Born and E. Wolf, *Principles of Optics*, Pergamon Press, Oxford (1964)
[43] M.C. Hutley, *Diffraction Gratings*, Academic Press, London (1982)
[44] *Electromagnetic Theory of Gratings*, edited by R. Petit, Springer-Verlag, Berlin (1980)
[45] D. Maystre, in *Progress in Optics XXI*, edited by E. Wolf, Elsevier, Amsterdam (1986)
[46] D. Maystre, M. Nevière, R. Reinisch, J.L. Coutaz, J. Opt. Soc. Am. **B5**, 338 (1988)
[47] M.C. Hutley, D. Maystre, Opt. Commun. **19**, 431 (1976)
[48] See for example F. Abeles, in *Electromagnetic Surface Excitations*, edited by R.F. Wallis and G.I. Stegeman, Springer-Verlag, Berlin (1986)
[49] M.A. Ordal, L. Long, R.J. Bell, S. Bell, R.R. Bell, R.W. Alexander, C.A. Ward, Appl. Opt. **22**, 1099 (1983)
[50] H.J. Simon, R.E. Brenner, J.G. Rako, Opt. Commun. **23**, 245 (1977)
[51] J.G. Rako, J.C. Quail, H.J. Simon, Phys. Rev. **B30**, 5552 (1984)
[52] Z. Chen, W. Wang, G. Wang, Z. Zhang, Kexue Tongbao **29**, 1160 (1984)
[53] F. De Martini, P. Ristori, E. Santamato, A.C.A. Zammit, Phys. Rev. **B23**, 3797 (1980)
[54] J. Quail, J. Rako, H.J. Simon, R. Deck, Phys. Rev. Lett. **50**, 1987 (1983)
[55] J.C. Quail, H.J. Simon, J. Opt. Soc. Am. **B1**, 317 (1984)
[56] J.C. Quail, H.J. Simon, J. Appl. Phys. **56**, 2589 (1984)
[57] M.G. Weber, Phys. Rev. **B33**, 6775 (1986)
[58] G.I. Stegeman, J.J. Burke, D.G. Hall, Appl. Phys. Lett. **41**, 906 (1982)
[59] R.T. Deck, D. Sarid, J. Opt. Soc. Am. **72**, 1613 (1982)
[60] G.I. Stegeman, C. Liao, C. Karaguleff, Opt. Commun. **46**, 253 (1983)

[61] Y.J. Chen, E. Burstein, Nuovo Cimento **39B**, 807 (1977)
[62] M. Fukui, V.C.Y. So, J.E. Sipe, G.I. Stegeman, J. Phys. Chem. Solids **40**, 523 (1979)
[63] C.K. Chen, A.R.B. de Castro,Y.R. Shen, Opt. Lett. **4**, 393 (1979)
[64] K. Sasaki, H. Kawagishi, Y. Ishijima, Appl. Phys. Lett. **47**, 783 (1985)
[65] R. Reinisch, G. Chartier, M. Nevière, M.C. Hutley, G. Clauss, J.P. Galaup, J.F. Eloy, J. Phys. (Paris) **44**, L1007 (1983)
[66] J.L. Coutaz, J. Opt. Soc. Am **B4**, 105 (1987)
[67] J.C. Quail, H.J. Simon, J. Opt. Soc. Am **B5**, 325 (1988)
[68] H.J. Simon, C. Huang, J.C. Quail, Z. Chen, Phys. Rev. **B38**, 7408 (1988)
[69] Z. Chen, H.J. Simon, Opt. Lett. **13**, 1008 (1988)
[70] H.J. Simon, Z. Chen, Phys. Rev. **B39**, 3077 (1989)
[71] Z. Chen, D. Cui, H. Lu, Y. Zhou, Opt. Lett. **8**, 563 (1983)
[72] Z. Chen, D. Cui, Y. Zhou, H. Lu, G. Yang, S. Gu, Appl. Phys. Lett. **51**, 1301 (1987)
[73] D. Cui, Z. Chen, Y. Zhou, H. Lu, G. Yang, S. Gu, Physica Scripta **37**, 746 (1988)
[74] A. Wokaun, J.G. Bergman, J.P. Heritage, A.M. Glass, P.F. Liao, D.H. Olson, Phys. Rev. **B24**, 849 (1981)
[75] G.S. Agarwal, S.S. Jha, Solid State Comm. **41**, 499 (1982)
[76] G.S. Agarwal, S.S. Jha, Phys. Rev. **B26**, 482 (1982)
[77] R. Reinisch, M. Nevière, Phys. Rev. **B28**, 1870 (1983)
[78] M. Nevière, R. Reinisch, D. Maystre, Phys. Rev. **B32**, 3634 (1985)
[79] M. Nevière, P. Vincent, D. Maystre, R. Reinisch, J.L. Coutaz, J. Opt. Soc. Am. **B5**, 330 (1988)
[80] G.A. Farias, A.A. Maradudin, Phys. Rev. **B30**, 3002 (1984)
[81] J.E. Gersten, A. Nitzan, J. Chem. Phys. **73**, 3023 (1980)
[82] *Surface Enhanced Raman Scattering*, edited by R.K. Chang and T.E. Furtak, Plenum Press, New York (1982)
[83] S. Ushioda, A. Aziza, J.B. Valdez, G. Mattei, Phys. Rev. **B19**, 4012 (1979)
[84] A.M. Glass, A. Wokaun, J.P. Heritage, J.G. Bergman, P.F. Liao, D.H. Olson, Phys. Rev. **B24**, 4906 (1981)
[85] J.P. Heritage, J.G. Bergman, Opt. Commun. **35**, 373 (1980)
[86] E.C. Stoner, Philos. Mag. **36**, 803 (1945)
[87] P.F. Liao, A. Wokaun, J. Chem. Phys. **76**, 751 (1982)
[88] C.K. Chen, T. Heinz, D. Ricard, Y.R. Shen, Phys. Rev. **B27**, 1965 (1983)
[89] G.T. Boyd, Th. Rasing, J.R. Leite, Y.R. Shen, Phys. Rev. **B30**, 519 (1984)
[90] See for example R.K. Chang and B.L. Laube, CRC Critical Rev. Sol. State Mat. Sci. **12**, 1 (1984)
[91] K. Arya, Phys. Rev. **B29**, 4451 (1984)
[92] R.T. Deck, R.K. Grygier, Appl. Opt. **23**, 3202 (1984)
[93] A.A. Maradudin, D.L. Mills, Phys. Rev. **B11**, 1392 (1975)
[94] D.V. Murphy, K.U. Von Raben, T.T. Chen, J.F. Owen, R.K. Chang, Surf. Sci. **124**, 529 (1983)
[95] V.E. Albano, S. Daiser, G. Entl, R. Miranda, K. Wandelt, N. Garcia, Phys. Rev. Lett. **51**, 2314 (1983)
[96] H. Seki, T.J. Chuang, Chem. Phys. Lett. **100**, 393 (1983)
[97] O.A. Aktsipetrov, E.M. Dubinina, S.S. Elovikov, D.A. Esikov, E.D. Mishina, N.N. Fominykh, JETP Lett. **44**, 475 (1986)

[98] O.A. Aktsipetrov, S.I. Vasilev, V.Panov, Sov. Phys. JETP **67**, 1010 (1988)
[99] C.K. Chen, T. Heinz, D. Ricard, Y. Shen, Phys. Rev. Lett **46**, 1010 (1981)
[100] O.A. Aktsipetrov, V.Y. Bartenev, E.D. Mishina, A.V. Petukhov, Sov. J. Quantum Electron. **13**, 712 (1983)
[101] G.L. Richmond, Surf. Sci. **147**, 115 (1984)
[102] Y. R. Shen, J. Vac. Sci. Technol. **B3**, 1464 (1985)
[103] J.G. Bergman, D.S. Chemla, P.F. Liao, A.M. Glass, A. Pinczuk, R. M. Hart, D.H. Olson, Opt. Lett. **6**, 33 (1981)
[104] O.A. Aktsipetrov, I.M. Baranova, E.D. Mishina, A.V. Petukhov, JETP Lett. **40**, 1012 (1984)
[105] A.A. Golubtsov, N.F. Pilipetskii, A.N. Sudarkin, V.V. Shelepenko, V.V. Yakimenko, JETP Lett. **43**, 277 (1986)
[106] G.T. Boyd, Y.R. Shen, T.W. Hansch, Opt. Lett. **11**, 97 (1986)
[107] D. Maystre, Opt. Commun. **6**, 50 (1972)

Optical Second-Harmonic Generation from Jellium. The Nonlinear Energy Reflection Coefficient as a Function of the Angle of Incidence

H.R. Jensen and O. Keller

Institute of Physics, University of Aalborg,
Pontoppidanstræde 103, DK-9220 Aalborg Øst, Denmark

1 Introduction

Within the last twenty years optical second-harmonic processes in centrosymmetric metals have been studied intensively [1,2]. Especially, calculations of the nonlinear energy reflection coefficient, a quantity adequate for comparison to experimental data, have attracted a great deal of attention. The large optical response from the surface (selvedge) region stems from the breaking of the inversion symmetry of the crystal lattice and from the pronounced variation of the electron density. For optical frequencies in the vicinity of the plasma edge one expects that the nonlinear response in free-electron-like metals is dominated by the excitations of collective modes.

In the present communication we have, within the framework of the semiclassical infinite-barrier (SCIB) model, studied how the nonlinear energy reflection coefficient, taken as a function of the angle of incidence for the fundamental field, depends on the collective-mode pattern retained in the metal. In our study hydrodynamic response tensors are used. Material data for sodium, which is quite free-electron-like, are adopted. Our numerical results are discussed and compared with those of other models. The present work is closely related to a recent study of the frequency dependence of the reflection coefficient [3].

2 The Collective-Mode Approximation

In this section we consider briefly the form of the result obtained recently [4,5] for the nonlinear reflection coefficient on the basis of the SCIB-model. We limit ourselves to the case where a monochromatic plane wave having the fundamental frequency ω is incident at an angle θ on a semiinfinite metal with a planar surface. In the collective-mode regime a relatively simple expression, which is also easy to interpret physically, is obtained for the nonlinear reflection coefficient [5]. The expression contains the linear and nonlinear nonlocal conductivity response tensors in a model-independent general form. With the use of appropriate boundary conditions for the fields, i.e. the requirement that the tangential components of the electric and magnetic fields have to be continuous at the surface, the nonlinear energy reflection coefficient for the p-polarized second-harmonic light takes the form [4,5]

$$R_2^p(q_{\|}, \omega, \phi) = |A(q_{\|}, \omega)|^2[\sin^4\phi + |B(q_{\|}, \omega)|^2\cos^4\phi$$
$$+2\mathrm{Re}\,\{B(q_{\|}, \omega)\}\sin^2\phi\cos^2\phi] \ . \qquad (1)$$

The quantity $q_{\|} = (\omega/c_0)\sin\theta$ denotes the component of the wave vector of the fundamental field parallel to the surface, c_0 being the vacuum velocity of light. It is assumed in Eq.(1) that the fundamental field is linearly polarized at an angle ϕ with respect to the plane of incidence. The explicit expressions for the complex functions $A(q_{\|}, \omega)$ and $B(q_{\|}, \omega)$ depend on the specific model used for calculation of the linear and nonlinear collective response functions and on the mode structure retained inside the metal. Once the functions $|A|^2$, $|B|^2$, and Re {B} are known, R_2^p can be calculated for various values of the parameters $q_{\|}, \omega$, and ϕ. One should notice that the form of R_2^p in Eq.(1) is unchanged if single-particle excitations are included in the description [5]. The explicit expressions for A and B are, however, more complicated in this case.

3 Numerical Calculation in the Hydrodynamic Approach

Using hydrodynamic response functions [5], we shall, taking sodium as the example, discuss the numerical results obtained for the nonlinear reflection coefficient, R_2^p, as a function of the angle of incidence θ in the polarization configurations $s_1 \rightarrow p_2$ and $p_1 \rightarrow p_2$ ($1 \supset: \omega, 2 \supset: 2\omega$; second-harmonic field p-polarized, fundamental field either s- or p-polarized). Two wavelengths of the incident field are considered, namely λ = 2650Å (= $\lambda_{\mathrm{YAG}}/4$, λ_{YAG} being the YAG-laser wavelength) and λ = 1325Å (= $\lambda_{\mathrm{YAG}}/8$). The material data used for sodium at T = 300K are [5]: Conduction electron density $N_0 = 2.65 \times 10^{28}\mathrm{m}^{-3}$, electron relaxation time $\tau = 3.23 \times 10^{-14}$s, and scalar optical effective mass $m^* = 1.07m$, m being the rest mass of the electron. It is convenient for the subsequent analysis to know the magnitude of the longitudinal plasma frequency (times τ) for sodium, i.e., $\omega_{\mathrm{p}}^{\mathrm{L}}\tau = [N_0e^2/(m^*\epsilon_0)]^{1/2}\tau \simeq 287$, where e and ϵ_0 are the charge of the electron and the permittivity of free space, respectively.

In Fig. 1, numerical calculations of R_2^p as a function of θ for the frequencies $\lambda_{\mathrm{YAG}}/4$ and $\lambda_{\mathrm{YAG}}/8$ are presented. Our results are obtained on the basis of the collective SCIB model retaining all the modes [5]. For comparison is shown also the results obtained in the polariton approximation, i.e., the one used originally by Bloembergen et al. [6], and in the work of Schaich and Liebsch [7]. In the nonretarded hydrodynamic calculation of Schaich and Liebsch only the $p_1 \rightarrow p_2$ configuration was considered. The characteristic edges and peaks in the angular spectra of R_2^p can be identified as follows. From the frequency dependence of R_2^p it turns out [3] that peaks occur in the spectra when $\omega = \omega_{\mathrm{p}}^{\mathrm{T}}/2$ and edges when

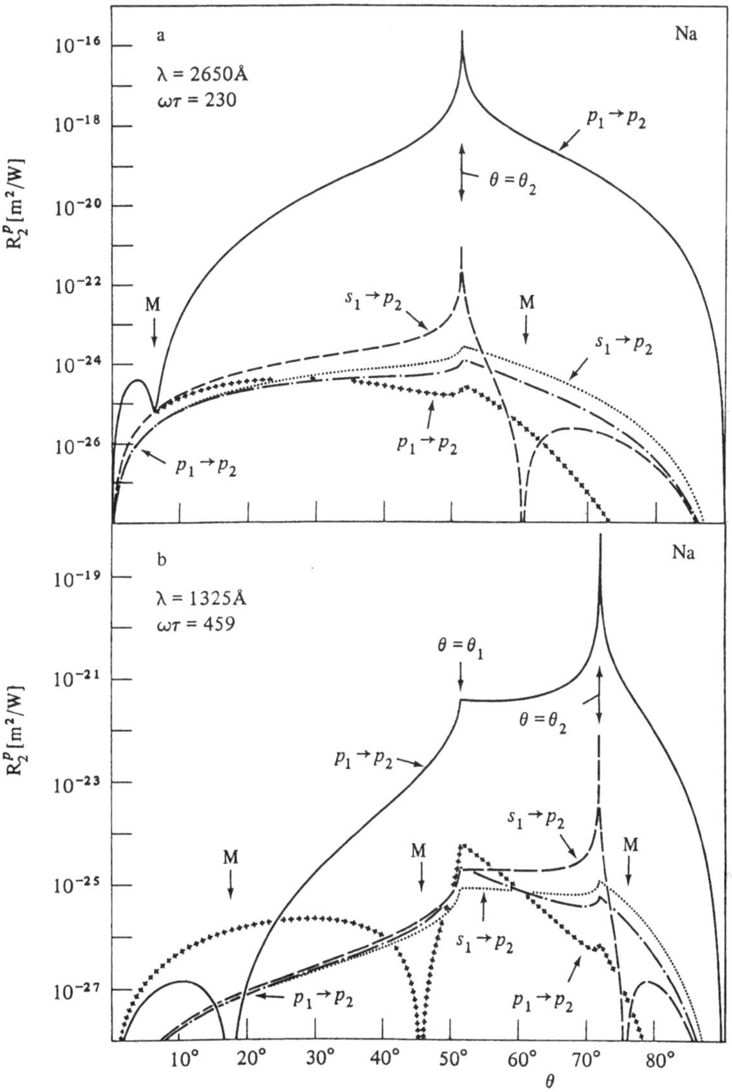

Fig. 1. Nonlinear energy reflection coefficient, R_2^p, as a function of the angle of incidence, θ, for λ = 2650Å (a) and λ =1325Å (b). Curves are plotted for the two scattering configurations $s_1 \rightarrow p_2$ and $p_1 \rightarrow p_2$. The curves ——— and – – –– are the results of our SCIB model (all modes retained). The curves marked with + + + + represent the results of Schaich and Liebsch [7]. The curves – · – · – and · · · · · show the results of Bloembergen et al. [6]. In this model only the modes $2T_1$ and T_2 are incorporated.

$\omega = \omega_p^T$, where ω_p^T is the transverse plasma frequency. Since $\omega_p^T = \omega_p^L/\cos\theta$, it is realized that characteristic edges ($n = 1$) and peaks ($n = 2$) show up in the angular spectrum at angles θ_n determined by

$$\theta_n = \arccos\left(\frac{\omega_p^L}{n\omega}\right) . \tag{2}$$

In Figure 1a, the fundamental frequency fulfils the condition $\omega_p^L/2 < \omega < \omega_p^L$, so that peaks and only peaks are present. In Figure 1b, one has $\omega > \omega_p^L$ so that both peaks and edges are allowed to occur. It is evident from Figure 1 that the calculations of Bloembergen et al. [6], Schaich and Liebsch [7], and those presented by us deviate much mutually.

In order to examine the causes of these differences we compare in Fig. 2 the different models for the $p_1 \rightarrow p_2$ configuration at $\lambda = \lambda_{YAG}/8$. For brevity we classify the mode structure in accordance with the notation of [5]. Second-harmonic reflection coefficients are often expressed in terms of the Rudnick-Stern a and b parameters, and a quantity d giving the strength of the bulk contribution. The basis for the a and b parameter description is the dipole-sheet approximation. Although our theory cannot in principle be reduced to this ap-

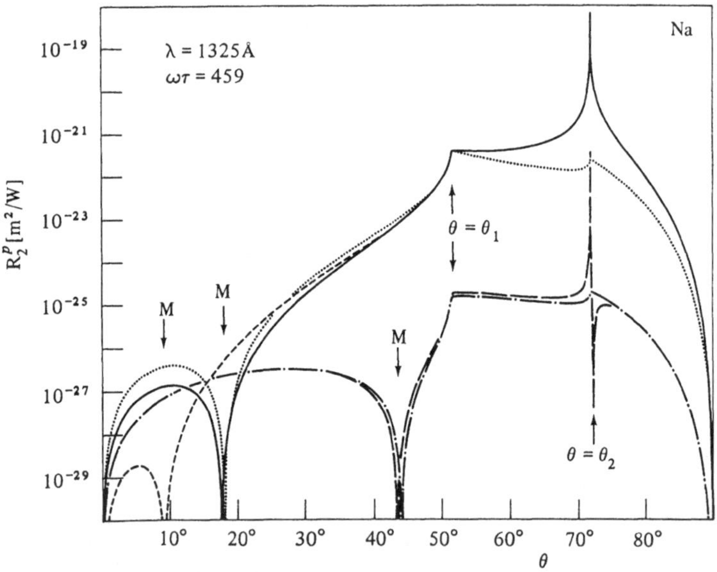

Fig. 2. Nonlinear energy reflection coefficient, R_2^p, as a function of the angle of incidence, θ, for λ = 1325Å in the scattering configuration $p_1 \rightarrow p_2$. The curve ——— represents the result of our SCIB model. Leaving out all the mirror terms from our model one gets the curve — — —. Omitting also the free term $\vec{E}_{2,a}^{\rm free}$ one obtains the curve — · — · —. The curve - - - - - is obtained retaining only modes classified as $2T_1$, T_2, $L_1 - L_1$, and with the free contribution $\vec{E}_{2,a}^{\rm free}$ also omitted. The curve $R_2^p(\theta)$ in [8], setting $a = -23.0$, $b = -1.3$, and $d = 1.0$.

proximation we have tried to make a fit of the dipole-sheet formalism to our result. The dotted $(\cdots\cdots)$ curve shows the (best) fit, which is obtained by setting $a = -23.0$, $b = -1.3$, and $d = 1.0$. We notice that a is an order of magnitude larger than predicted by Schaich and Liebsch [7]. If one leaves out all the so-called mirror terms, i.e. the modes $|L_1 - T_1|$, $L_1 - L_1$, $T_1 - T_1$, and the part $\vec{E}_{2,a}^{\text{free}}$ of the free contribution to the second-harmonic field we obtain the dashed-dotted $(- \cdot - \cdot -)$ curve. This curve gives a result in complete agreement with that of Liebsch and Schaich [8] ($a = -1.0$, $b = -0.3$, and $d = 1.0$). The free contribution $\vec{E}_{2,a}^{\text{free}}$ consists of both "normal" terms and mirror terms, [5]. If one adds to the result of Liebsch and Schaich, the normal terms in $\vec{E}_{2,a}^{\text{free}}$ one obtains the dashed $(- - -)$ curve. Thus, it is realized that the normal terms in $\vec{E}_{2,a}^{\text{free}}$ omitted in [8] give rise to a sharp resonance at $\theta = \theta_2$. It is obvious from our discussion that the mirror terms cause a substantial overall increase in $R_2^p(\theta)$. Not all the mirror terms are of equal importance. Thus, if we retain only the modes $2T_1$, T_2, and $L_1 - L_1$ ($\vec{E}_{2,a}^{\text{free}}$ also zero) we find the dashed (- - - - -) curve. It is seen that this curve gives a rather nice fit to our general model. We notice in Figs. 1 and 2 that $R_2^p(\theta)$ exhibits certain minima (marked with M in the figures). It turns out that these minima, which appear only in the vicinity of the plasma frequency, occur when the second-harmonic energy flow is almost parallel to the surface.

References

[1] J.E. Sipe and G.I. Stegeman, in *Surface Polaritons, Electromagnetic Waves at Surfaces and Interfaces,* eds. V.M. Agranovich and D.L. Mills (North-Holland, Amsterdam 1982) p.661.

[2] G.L. Richmond, J.M.Robinson, and V.L. Shannon, Prog.Surf.Sci. **28,** 1 (1988).

[3] H.R. Jensen and O. Keller, in *Proc. of the Sixth Rochester Conference on Coherence and Quantum Optics,* eds. L. Mandel and E. Wolf (Plenum, New York), (in press).

[4] O. Keller, J.Opt.Soc.Am. **B2,** 367 (1985).

[5] H.R. Jensen, K. Pedersen, and O. Keller, in *Proc. of the International Conference on Nonlinear Optics,* ed. V.J. Corcoran (STS, 1988) p.174.

[6] N. Bloembergen, R.K. Chang, S.S. Jha, and C.H. Lee, Phys.Rev. **174,** 813 (1968).

[7] W.L. Schaich and A. Liebsch, Phys.Rev. **B37,** 6187 (1988).

[8] A. Liebsch and W.L. Schaich, Phys.Rev. (to be published).

On the Nonlocal Response Theory of Optical Rectification and Second-Harmonic Generation in Centrosymmetric Superconductors

O. Keller

Institute of Physics, University of Aalborg,
Pontoppidanstræde 103, DK-9220 Aalborg Øst, Denmark

1 Introduction

The study of the nonlinear interaction between light and centrosymmetric metals has been of interest for a quarter of a century [1–71]. From a fundamental point of view this interest stems from the fact that the process of for instance second-harmonic generation is surface sensitive on the monolayer level and that a microscopic understanding of the basic physics requires a knowledge of exotic aspects of the nonlinear coupling of light to the many-body system of the condensed medium. Since it is very difficult to obtain even a qualitative understanding of the nonlinear interaction in metals where band structure effects are pronounced, much theoretical work has recently been devoted to *ab initio* studies of simple jellium systems [29,32,52–59,61,67–69,71]. To compare microscopic calculations with experiment it has turned out [66] that only by performing the measurements on single crystals of free-electron-like metals kept under UHV-conditions will there be hope for a significant improvement of the agreement between theory and experiment.

In the present work, I shall discuss in the simplest manner recent efforts to extend the nonlinear theory for the normal metallic state to the superconducting state [72,73]. I shall limit myself to BCS-like superconductors [74] and consider only the Cooper-paired jellium. Although the linear electrodynamic properties of superconductors have been extensively investigated for many years [75–78], the idea of studying the nonlinear electrodynamics of superconductors appears to be new. For the old superconductors exhibiting small energy gaps, the far infrared nonlinear properties are especially interesting. In the new so-called high-T_C superconductors [79–81] studies in the entire infrared frequency region are of importance. There are many reasons why Cooper-paired superconductors should be of interest in nonlinear optics. Let me mention just three. Firstly, externally impressed dc-currents and induced screening currents in superconductors are surface currents and second-harmonic generation is surface sensitive. Secondly, even-order nonlinear effects give rise to electromagnetic rectification. This rectification process is extremely important since the dc-conductivity of the medium is infinite in the superconducting state. Thirdly, the superconducting state is an extremely ordered state. This means that one would expect significant differences

Springer Series in Wave Phenomena, Vol. 9 **Nonlinear Optics in Solids**
Editor: O. Keller

in the quantum statistics of the photons in nonlinear experiments performed above and below the transition temperature.

In Sect. 2, a general treatment of the concept of nonlocality in nonlinear optics is given. Since the pair-bound state of the superconductor has a finite size it is important for the understanding of the nonlinear response of the superconductor to establish a nonlocal formulism at the outset. With emphasis on optical second-harmonic generation, I discuss symmetry principles in nonlocal optics and compare the moment and multipole expansion formulisms. In Sect. 3, the nonlinear response of a Cooper-paired jellium is investigated. Using a many-body density matrix formalism, expressions for the response functions associated with optical second-harmonic generation and optical rectification are established. On the basis of these expressions, I discuss the semi-local nonlinear response which for the optical rectification process is related to what I call the nonlinear Meissner effect. Furthermore, I present a discussion of the simultaneous two-photon mechanism and the double-nonlocal nonlinear response. I conclude the section by some aphorisms on the nonlinear response of a Cooper-paired superconductor. In Sect. 4, I outline how the foregoing analyses can be extended to handle the response of a superconductor-vacuum interface. Special emphasis is devoted to discussions of the semiclassical infinite-barrier model and the selvedge response.

2 Nonlocal Response Theory

In this section I shall discuss second-order effects in nonlinear optics within the framework of a nonlocal approach. The main emphasis is devoted to a conceptual study of the optical second-harmonic response tensor. By taking the formal approach as a starting point, symmetry principles in the nonlocal domain are investigated. A recently developed [67] moment expansion formulism is described, and it is shown how the multipole expansion approximation often adopted in studies of the nonlinear current density response of centrosymmetric media can be obtained in a natural fashion via the moment expansion theory. Finally, on the basis of the random-phase-approximation (RPA) description, the structure of the nonlocal nonlinear response tensors are discussed.

2.1 Formal Approach

In linear optics the concept of nonlocality in time is of exceeding importance for our understanding of the interaction between light and matter. A process is nonlocal in time if the light-induced current density $\vec{J}_1(\vec{r}, t)$, at the space point $\vec{r}$ at the time t is determined not only by the electric field $\vec{E}_1(\vec{r}, t)$ present at time t but also by the field prevailing at $\vec{r}$ at earlier times. In linear response theory nonlocality in time is described by a constitutive equation of the form

$$\vec{J}_1(\vec{r}, t) = \int_{-\infty}^{\infty} \overleftrightarrow{\sigma}(\vec{r}, t, t') \cdot \vec{E}_1(\vec{r}, t')dt', \quad (1)$$

where $\overleftrightarrow{\sigma}(\vec{r},t,t')$ is the so-called linear conductivity response tensor. Due to causality $\overleftrightarrow{\sigma}$ will be zero for times $t' > t$. One should emphasize that Eq.(1), considered as a microscopic relation, in principle can be used to describe not only the linear responses of superconductors and metals but also those of dielectrics, semiconductors, fluids, gases, plasmas, molecules, atoms, etc. If the light-unperturbed matter system exhibits time-invariant properties the response tensor can only be a function of the time difference $t-t'$, i.e., $\overleftrightarrow{\sigma}(\vec{r},t,t') \Rightarrow \overleftrightarrow{\sigma}(\vec{r},t-t')$. Time invariance allows one to simplify the analysis of the response. Thus, since the right hand side of Eq.(1) is a simple convolution in this case the Fourier transform of the constitutive equation has the product form

$$\vec{J}_1(\vec{r};\omega) = \overleftrightarrow{\sigma}(\vec{r};\omega) \cdot \vec{E}_1(\vec{r};\omega). \tag{2}$$

When it is written in the form of Eq.(2) one says that the electromagnetic response exhibits frequency (ω) dispersion. As the well known Kramers-Krönig relations derived in linear response theory on the basis of the principle of causality demonstrate, frequency dispersion and damping are quantities which inevitably are connected. In optics one sometimes encounters problems where the concept of spatial nonlocality plays an important role. A response is nonlocal in space if the current density at point $\vec{r}$ depends not only on the electric field at $\vec{r}$ but also on the field at neighbouring points $\vec{r}\,'$. Mathematically, spatial nonlocality can be described by replacing Eq.(2) with the relation

$$\vec{J}_1(\vec{r};\omega) = \int_{-\infty}^{\infty} \overleftrightarrow{\sigma}(\vec{r},\vec{r}\,';\omega) \cdot \vec{E}_1(\vec{r}\,';\omega) d^3r', \tag{3}$$

where $\overleftrightarrow{\sigma}(\vec{r},\vec{r}\,';\omega)$ is the so-called spatially nonlocal linear conductivity tensor. It is of course possible also to describe spatial nonlocality in systems where the time invariance is broken. In such cases one returns to Eq.(1), makes the replacement $\overleftrightarrow{\sigma}(\vec{r},t,t') \Rightarrow \overleftrightarrow{\sigma}(\vec{r},\vec{r}\,',t,t')$, and performs an integration over $\vec{r}\,'$ also. If the system under consideration exhibits infinitesimal translational invariance in space the response can depend only on the difference $\vec{r}-\vec{r}\,'$, i.e., $\overleftrightarrow{\sigma}(\vec{r},\vec{r}\,';\omega) \Rightarrow \overleftrightarrow{\sigma}(\vec{r}-\vec{r}\,';\omega)$. In turn, this enables us to Fourier analyse Eq.(3) in space and thus base our subsequent description on a constitutive equation of the form

$$\vec{J}_1(\vec{q},\omega) = \overleftrightarrow{\sigma}(\vec{q},\omega) \cdot \vec{E}_1(\vec{q},\omega). \tag{4}$$

Spatial ($\vec{q}$) dispersion occurs in many systems. Optical activity is the classical example of a phenomenon which essentially only exists by virtue of spatial dispersion. In excitonic systems spatial dispersion is of decisive importance in certain frequency regions, and for metal optics spatial dispersion plays a crucial role for frequencies near the plasma edge. Also the understanding of local field effects in solid state optics requires that nonlocality in space is considered. To describe surface effects in metal optics one needs the use of $\overleftrightarrow{\sigma}(\vec{r},\vec{r}\,';\omega)$ since the translational symmetry is broken in the surface region, and since the surface region cannot in general be characterized just as a region where the medium is inhomogeneous, cf. Eq.(2). It is worth pointing out that the magnetic effects can be described within the framework of Eq.(3) (or eventually Eq.(4)). As an example one could mention that bulk diamagnetism is related to the difference between the longitudinal and transverse parts of the response function in Eq.(4). From a conceptual point of view the incorporation of spatial nonlocality in optics is important since this allows us to put space and time on the same footing.

Now, when the field strength is increased the linear relation between the field and the induced current density (Eq.(1)) breaks down, and one enters the regime of nonlinear optics. If the nonlinearities are weak it is useful to expand the current density in orders of $\vec{E}_1$. Limiting ourselves to effects associated with the lowest-order nonlinearities, the correction to the spatially local expression in Eq.(1) is given by

$$\vec{J}_2(\vec{r},t) = \int_{-\infty}^{\infty} \overleftrightarrow{\Sigma}(\vec{r},t,t',t'') : \vec{E}_1(\vec{r},t'')\vec{E}_1(\vec{r},t')dt'', dt', \tag{5}$$

where $\overleftrightarrow{\Sigma}(\vec{r},t,t',t'')$ is the (spatially) local, nonlinear conductivity tensor. The quantity $\overleftrightarrow{\Sigma}$ is a third rank tensor, and the notation in Eq.(5) is as follows: $(\overleftrightarrow{\Sigma}: \vec{E}_1\vec{E}_1)_i = \Sigma_{ijk}E_{1,k}E_{1,j}$ with implicit summation over repeated indices. The majority of second-order effects in nonlinear optics are discussed under the assumption that the light-unperturbed system exhibits translational invariance in time, i.e., $\overleftrightarrow{\Sigma}(\vec{r},t,t',t'') \Rightarrow \overleftrightarrow{\Sigma}(\vec{r},t-t',t-t'')$. Time invariance makes it possible to simplify Eq.(5). Thus, in the Fourier domain one has

$$\vec{J}_2(\vec{r};\Omega) = \int_{-\infty}^{\infty} \overleftrightarrow{\Sigma}(\vec{r};\omega,\Omega-\omega) : \vec{E}_1(\vec{r};\omega)\vec{E}_1(\vec{r};\Omega-\omega)d\omega, \tag{6}$$

absorbing a trivial $(2\pi)^{-1}$ factor in $\overleftrightarrow{\Sigma}$. Taking Eq.(6) as a starting point a number of interesting nonlinear phenomena can be studied, e.g., sum- and difference-frequency generation, linear electro-optic effects (one field is quasi-static) and Brillouin and Raman scattering (one field is that associated with the vibrations in the material system). A particularly simple situation occurs if $\vec{E}_1(\vec{r},t)$ consists of only one frequency component, say ω. Thus, in Fourier space only $\vec{E}_1(\vec{r};\pm\omega)$ are present. The essential physics now is contained in a relation

$$\vec{J}_2(\vec{r};2\omega) = \overleftrightarrow{\Sigma}(\vec{r};\omega,\omega) : \vec{E}_1(\vec{r};\omega)\vec{E}_1(\vec{r};\omega), \tag{7}$$

describing optical second-harmonic generation, and one

$$\vec{J}_2(\vec{r};0) = \frac{1}{2}\overleftrightarrow{\Sigma}(\vec{r};\omega,-\omega) : \vec{E}_1(\vec{r};\omega)\vec{E}_1^*(\vec{r};\omega) + \text{c.c.}, \tag{8}$$

describing optical rectification. Since the main theme for our description is spatial nonlocality (in the following just referred to as nonlocality) in nonlinear optics, we shall limit our discussion in the remaining part of this article to optical second-harmonic generation and optical rectification. Furthermore, since the treatment of the 2ω and dc-generation processes in nonlocal response theory follows the same line of deduction we shall focus our attention in the remaining part of this section on second-harmonic generation. Let me emphasize also that the following discussion can easily be extended to comprise sum- and difference-frequency generation.

A beautiful demonstration of the need for a nonlocal theory appears if one wants to understand second-harmonic generation in centrosymmetric media. Thus, adopting the traditional line of reasoning (which in my opinion is not quite transparent, cf. the remarks in Sect. 2.2) it follows from Eq.(7) that an inversion of the field direction ($\vec{E}_1 \Rightarrow -\vec{E}_1$), implying in turn an inversion of the current density ($\vec{J}_2 \Rightarrow -\vec{J}_2$) if the medium exhibits centrosymmetry, leads to the conclusion that $\overleftrightarrow{\Sigma}(\vec{r};\omega,\omega)$ must be zero. The extension of second-harmonic generation studies to the nonlocal domain is achieved by replacing Eq.(7) by the relation

$$\vec{J}_2(\vec{r};2\omega) = \int_{-\infty}^{\infty} \overleftrightarrow{\Sigma}(\vec{r},\vec{r}\,',\vec{r}\,'';\omega,\omega) : \vec{E}_1(\vec{r}\,'';\omega)\vec{E}_1(\vec{r}\,';\omega)d^3r''d^3r'. \tag{9}$$

It is evident from Eq.(9) that the nonlinear current density at the space point $\vec{r}$ has a contribution proportional to the product of the fundamental fields at $\vec{r}\,'$ and $\vec{r}\,''$. The weight factor for this contribution is the nonlinear response tensor $\overleftrightarrow{\Sigma}(\vec{r},\vec{r}\,',\vec{r}\,'';\omega,\omega)$. The total $\vec{J}_2(\vec{r};2\omega)$ is obtained by adding up the contributions from all the possible combinations of the space points $\vec{r}\,'$ and $\vec{r}\,''$. It is worth emphasizing here that all linear as well as nonlinear responses are inherently nonlocal as soon as matter is treated from the basic quantum mechanical (wave function) point of view. It should also be pointed out that nonlocality *a priori* plays a more important role in nonlinear optics than in linear optics due to the fact that the local response, which always dominates in the linear regime, in for instance nonlinear second-order phenomena is eliminated by the symmetry selection rules. If the medium in consideration exhibits translational invariance, i.e., $\overleftrightarrow{\Sigma}(\vec{r},\vec{r}\,',\vec{r}\,'';\omega,\omega) \Rightarrow \overleftrightarrow{\Sigma}(\vec{r}-\vec{r}\,',\vec{r}-\vec{r}\,'';\omega,\omega)$, Fourier transformation in space reduces Eq.(9) to the form (trivial factors being absorbed in $\overleftrightarrow{\Sigma}$)

$$\vec{J}_2(\vec{Q},2\omega) = \int_{-\infty}^{\infty} \overleftrightarrow{\Sigma}(\vec{q},\vec{Q}-\vec{q},\omega,\omega) : \vec{E}_1(\vec{q},\omega)\vec{E}_1(\vec{Q}-\vec{q},\omega)d^3q. \tag{10}$$

In the simplest case, where the fundamental field consists of only one plane-wave component, the constitutive equation essentially takes the form

$$\vec{J}_2(2\vec{q},2\omega) = \overleftrightarrow{\Sigma}(\vec{q},\vec{q},\omega,\omega) : \vec{E}_1(\vec{q},\omega)\vec{E}_1(\vec{q},\omega). \tag{11}$$

In Sect. 3, we shall discuss the structure of $\overleftrightarrow{\Sigma}(\vec{q},\vec{q},\omega,\omega)$ for a Cooper-paired superconduct

2.2 Symmetry Principles

Symmetry considerations are of the outmost importance in nonlinear optics since they allow us to identify vanishing elements in the response tensors and to determine which elements are of equal magnitude. In turn, these considerations reduce the burden of theoretical calculations and provide us with important polarization selection rules for the experiments. Up to now, symmetry principles always seem to have been investigated within the framework of a local approach. In the present subsection, I shall demonstrate, by a few simple examples, how nonlocal response theory enables us to establish the limitations for applying "the local symmetry concept". For simplicity, I shall discuss only the second-harmonic response.

It seems commonly believed that the symmetries of the nonlinear response tensors (local or nonlocal) to be used in second-harmonic generation studies of condensed media are determined exclusively by the bulk and/or surface symmetries of the medium in consideration. In my understanding this is a conceptually wrong view. To realize this, one just needs to consider the simplest model, namely the homogeneous jellium model. Thus, if an electromagnetic wave of wave vector $\vec{\kappa}$ propagates through this jellium, the relevant response tensor is

$$\overleftrightarrow{\Sigma}(\vec{\kappa},\vec{\kappa},\omega,\omega) = \int_{-\infty}^{\infty} \overleftrightarrow{\Sigma}(\vec{r}-\vec{r}\,',\vec{r}-\vec{r}\,'';\omega,\omega) e^{-i\vec{\kappa}\cdot(2\vec{r}-\vec{r}\,'-\vec{r}\,'')} d^3r'' d^3r', \tag{12}$$

cf.Eq.(11), and the discussion preceding this equation. It readily appears from Eq.(12) that the symmetries of $\overleftrightarrow{\Sigma}(\vec{\kappa},\vec{\kappa},\omega,\omega)$ are determined not only by the intrinsic (i.e., light-unperturbed) properties of the medium via $\overleftrightarrow{\Sigma}(\vec{r}-\vec{r}\,',\vec{r}-\vec{r}\,'';\omega,\omega)$ but also by the interaction geometry [here just the direction of wave propagation (via $\exp[-i\vec{\kappa}\cdot(2\vec{r}-\vec{r}\,'-\vec{r}\,'')]$)]. It is obvious from Eq.(12) that the response of the homogeneous jellium is like that of a "crystal" having an "optic axis" along $\vec{\kappa}$ and (i) infinitesimal rotational symmetry around $\vec{\kappa}$ and (ii) mirror symmetry in every plane containing the $\vec{\kappa}$-axis. The resulting symmetry scheme for such a "crystal" is shown in Table 1. It appears from this table that our homogeneous jellium has the same symmetry scheme as the crystal classes 4 mm and 6 mm.

Table 1. Symmetry scheme for the Fourier transformed nonlinear response tensor $\overleftrightarrow{\Sigma}(\vec{\kappa},\vec{\kappa},\omega,\omega)$ of a homogeneous jellium in a coordinate system where the z-axis is parallel to $\vec{\kappa}$.

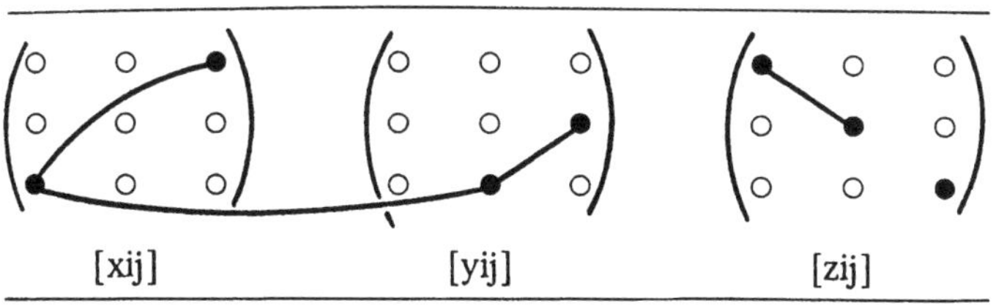

This is due to the fact that a third-rank tensor cannot distinguish between the 4-fold, 6-fold, and ∞-fold rotational symmetries. For a tensor of sufficiently high rank ∞-fold symmetry can of course be distinguished from any other finite-fold symmetry. For an inhomogeneous electron gas exhibiting only density variations perpendicular to a flat surface (the surface normal being parallel to the z-axis) the relevant response tensor is that associated with a mixed (Fourier/direct-space) representation, i.e., $\overleftrightarrow{\Sigma}(z, z', z''; \vec{q}_{\|}, \vec{q}_{\|}, \omega, \omega)$ where $\vec{q}_{\|}$ is the wave-vector component of the fundamental field parallel to the surface. The symmetry class for this response tensor (or a fully Fourier transformed one) is C_{1v}, the mirror plane being the plane of incidence (xz-plane). The symmetry scheme for the inhomogeneous jellium, given in the xyz-coordinate system, is shown in Table 2.

Table 2. Symmetry scheme for the nonlinear response tensor $\overleftrightarrow{\Sigma}(z, z', z''; \vec{q}_{\|}, \vec{q}_{\|}, \omega, \omega)$ of an inhomogeneous jellium exhibiting density variations in the z-direction only. The plane of incidence for the light coincides with the xz-plane.

$$\begin{pmatrix} \bullet & \circ & \bullet \\ \circ & \bullet & \circ \\ \bullet & \circ & \bullet \end{pmatrix} \qquad \begin{pmatrix} \circ & \bullet & \circ \\ \bullet & \circ & \bullet \\ \circ & \bullet & \circ \end{pmatrix} \qquad \begin{pmatrix} \bullet & \circ & \bullet \\ \circ & \bullet & \circ \\ \bullet & \circ & \bullet \end{pmatrix}$$

[xij] [yij] [zij]

In the xyz-system, the symmetry scheme for the homogeneous electron gas has the form given in Table 2 with the addition that it is symmetric about the main diagonal also. In the long-wavelength limit only, the symmetry schemes for the various conductivity response tensors will depend solely on the intrinsic symmetry of the light-unperturbed material system. For the homogeneous jellium $\overleftrightarrow{\Sigma}(\vec{0}, \vec{0}, \omega, \omega) = \overleftrightarrow{0}$ due to the microscopic inversion symmetry, i.e., $\overleftrightarrow{\Sigma}(\vec{r} - \vec{r}\,', \vec{r} - \vec{r}\,''; \omega, \omega) = - \overleftrightarrow{\Sigma}(\vec{r}\,' - \vec{r}, \vec{r}\,'' - \vec{r}; \omega, \omega)$ as one readily verifies from Eq.(12).

2.3 Moment Expansion

To describe the optical second-harmonic generation in centrosymmetric media one has to consider both the local and the nonlocal contributions to the nonlinear conductivity response tensor. In a jellium system, the local contribution stems from inhomogeneities in the electron density. Since these are pronounced in the electron tail, separating the bulk jellium from vacuum, one expects that local second-harmonic generation from free-electron metals is a surface effect. At space point $\vec{r}$, the local (l) contribution, $\vec{J}_2{}^l(\vec{r}; 2\omega)$, to the nonlinear current density is given by

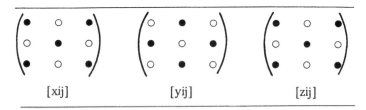

$$\vec{J}_2^{\,l}(\vec{r};2\omega) = \overleftrightarrow{\Sigma}^{l}(\vec{r};\omega,\omega) : \vec{E}_1(\vec{r};\omega)\vec{E}_1(\vec{r};\omega), \tag{13}$$

cf. Eq.(7). In a real metal, the broken symmetry of the ionic arrangement in the vicinity of the surface causes the presence of an ionic surface contribution to the 2ω-generation, also. Furthermore, what is often not appreciated is the fact that the periodic variation in the pseudopotential of bulk metal crystals gives rise to a periodically varying local response tensor. Thus, in the case where, from an optical point of view, the metal can be described in a simple two-band picture, the local response function of the bulk, $\overleftrightarrow{\Sigma}^{l}_{B}$, fulfils the criterion

$$\overleftrightarrow{\Sigma}^{l}_{B}(\vec{r};\omega,\omega) = \overleftrightarrow{\Sigma}^{l}_{B}(\vec{r}+n\vec{d};\omega,\omega), \tag{14}$$

where $d = 2\pi/K$ is the spatial period associated with the relevant reciprocal lattice vector $\vec{K}$, and n is an integer. Although the mean value of $\overleftrightarrow{\Sigma}^{l}_{B}$ over a period $\vec{d}$ is zero, it is not entirely obvious beforehand whether the local bulk contribution, which in principle does give rise to 2ω-generation since the fundamental field decays slightly over the unit cell, *in comparison to* the surface contribution will be small, in general. Qualitatively, the *small* bulk contributions from the different unit cells have to be added over *many* cells (those from a layer of thickness as the penetration depth for the fundamental field) whereas the *large* surface contribution stems from a *few* atomic cell-layers. The nonlocal effects, which are active both at the surface and in the bulk of condensed matter systems, usually are described in lowest order only, i.e., in a scheme where only first-order spatial derivatives of the fundamental field are incorporated. Thus, to lowest order, the nonlocal contribution, $\vec{J}_2^{\,nl}(\vec{r};2\omega)$, to the nonlinear current density at space point $\vec{r}$ is given quite generally by the following so-called near-local (nl) expression:

$$\vec{J}_2^{\,nl}(\vec{r};2\omega) = \overleftrightarrow{\Sigma}^{nl}(\vec{r};\omega,\omega) \vdots \left(\vec{\nabla}\vec{E}_1(\vec{r};\omega)\right) \vec{E}_1(\vec{r};\omega). \tag{15}$$

A description of second-harmonic generation in centrosymmetric media which is based on the expansion indicated in Eqs.(13) and (15) *a priori* works best in dielectrics, where the electronic orbitals are well localized. In metals and superconductors, where transitions between highly delocalized Bloch orbitals are involved, it is usually necessary to return to the general nonlocal constitutive equation in (9). The standard procedure employed to determine $\overleftrightarrow{\Sigma}^{l}$ and $\overleftrightarrow{\Sigma}^{nl}$ is based on a multipole expansion method [82]-[84]. Thus, by expanding the nonlinear current density in a series of multipole terms, i.e.,

$$\vec{J}_2(\vec{r};2\omega) = -2i\omega\vec{P}_2(\vec{r};2\omega) + \vec{\nabla}\times\vec{M}_2(\vec{r};2\omega)$$
$$+2i\omega\vec{\nabla}\cdot\overleftrightarrow{Q}_2(\vec{r};2\omega), \tag{16}$$

where $\vec{P}_2, \vec{M}_2$, and $\overleftrightarrow{Q}_2$ are the electric dipole, magnetic dipole, and electric quadrupole moments, respectively, and by making a well-known multipole expansion of the interaction Hamiltonian involving the fundamental field, time-dependent perturbation theory allows one to obtain explicit expressions for the local and near-local response tensors. There are certain drawbacks attached to the multipole scheme for calculating the series (Eqs.(13), (15), etc.) of nonlinear response tensors. Thus, the tensors involved are of third rank ($\overleftrightarrow{\Sigma}^{l}$), fourth rank ($\overleftrightarrow{\Sigma}^{nl}$), and higher ranks and there appears not to be any simple relationship between the rather lengthy expressions for the various response tensors. Since it is evident from Eq.(9) that *in principle* only a third rank tensor is needed for the description of the nonlocal second-harmonic response it is worth investigating how the local and near-local response tensors $\overleftrightarrow{\Sigma}^{l}(\vec{r};\omega,\omega)$ and $\overleftrightarrow{\Sigma}^{nl}(\vec{r};\omega,\omega)$ can be obtained from a knowledge of the general nonlocal response tensor $\overleftrightarrow{\Sigma}(\vec{r},\vec{r}\,',\vec{r}\,'';\omega,\omega)$.

To accomplish our goal, we make a Taylor series expansion of the fundamental field around the point in space ($\vec{r}$) where the nonlinear current density is to be calculated. Next, by inserting the Taylor expansion into Eq.(9), it is realized that the local response tensor is given by

$$\overleftrightarrow{\Sigma}^{l}(\vec{r};\omega,\omega) = \int_{-\infty}^{\infty} \overleftrightarrow{\Sigma}(\vec{r},\vec{r}\,',\vec{r}\,'';\omega,\omega) d^3r\,'' d^3r\,'. \tag{17}$$

In the context of Eq.(7) it was argued that the local response function would vanish in a medium exhibiting inversion symmetry. This important result cannot be justified alone on the basis of Eq.(7), where the properties of an apparent *local (point) relation* between the field ($\vec{E}_1(\vec{r};\omega)$) and the induced current density ($\vec{J}_2(\vec{r};2\omega)$) are inferred from a *nonlocal argument* (an inversion symmetry argument requires that the characteristics of the medium in the neighbourhood of $\vec{r}$ are known). In the view of Eq.(17), however, it appears that a "local" relation like that of Eq.(7) is inherently nonlocal. If the medium has microscopic inversion symmetry around $\vec{r}$, i.e., $\overleftrightarrow{\Sigma}(\vec{r},\vec{r}\,',\vec{r}\,'';\omega,\omega) = -\overleftrightarrow{\Sigma}(\vec{r},2\vec{r}-\vec{r}\,',2\vec{r}-\vec{r}\,'';\omega,\omega)$, it follows readily from Eq.(17) that $\overleftrightarrow{\Sigma}^{l}(\vec{r};\omega,\omega) = \overleftrightarrow{0}$. Via the Taylor expansion of the fundamental field it follows that the near-local response tensor is obtained from $\overleftrightarrow{\Sigma}(\vec{r},\vec{r}\,',\vec{r}\,'';\omega,\omega)$ via the integral

$$\begin{aligned}\overleftrightarrow{\Sigma}^{nl}(\vec{r};\omega,\omega) = \int_{-\infty}^{\infty} [&\overleftrightarrow{\Sigma}(\vec{r},\vec{r}\,',\vec{r}\,'';\omega,\omega)(\vec{r}\,''-\vec{r}) \\ &+ \tilde{\overleftrightarrow{\Sigma}}(\vec{r},\vec{r}\,',\vec{r}\,'';\omega,\omega)(\vec{r}\,'-\vec{r})]\, d^3r'' d^3r',\end{aligned} \tag{18}$$

where $\sim$ indicates that $\overleftrightarrow{\Sigma}$ is to be transposed in the last two indices, i.e., $\tilde{\Sigma}_{ikj} = \Sigma_{ijk}$. The quantities $\overleftrightarrow{\Sigma}(\vec{r}\,''-\vec{r})$ and $\tilde{\overleftrightarrow{\Sigma}}(\vec{r}\,'-\vec{r})$ are to be considered as tensor products in accordance with the fact that $\overleftrightarrow{\Sigma}^{nl}$ is a fourth rank tensor. It is evident from Eqs.(17) and (18) that the response functions related to the multipole expansion scheme are obtained as tensor moments of the fully nonlocal response function.

2.4 Response Tensor in RPA-Description

In this subsection, I shall consider the tensor-product structure and the nonlocal form of the nonlinear response function responsible for second-harmonic generation. Furthermore, I shall demonstrate how the multipole and moment expansion schemes are related.

We assume that the light-unperturbed electron system can be described on the basis of the effective one-electron time-dependent Schrödinger equation

$$(H^{(0)} - \epsilon_j)\psi_j(\vec{r}) = 0, \tag{19}$$

where $H^{(0)}$ is the Hamiltonian operator for the electron, and $\psi_j(\vec{r})$ is the jth effective one-body wave function belonging to the energy eigenstate of energy ϵ_j. By using the Liouville equation for the one-body density-matrix operator it is possible in a convenient manner to determine $\overleftrightarrow{\Sigma}(\vec{r},\vec{r}\,',\vec{r}\,'';\omega,\omega)$ in terms of the $\psi_j(\vec{r})$'s and the E_j's. As the result of such a derivation one obtains the following tensor-product structure of the response tensor [42,67]:

$$\begin{aligned}\overleftrightarrow{\Sigma}(\vec{r},\vec{r}\,',\vec{r}\,'') &= \delta(\vec{r}\,'-\vec{r}\,'')\vec{\Sigma}_A(\vec{r},\vec{r}\,')\overleftrightarrow{U} + \overleftrightarrow{\Sigma}_B(\vec{r},\vec{r}\,',\vec{r}\,'') \\ &+ \delta(\vec{r}-\vec{r}\,')\,\overleftrightarrow{U}\,\vec{\Sigma}_C(\vec{r},\vec{r}\,''),\end{aligned} \tag{20}$$

where δ is the Dirac delta function, and $\overleftrightarrow{U}$ is the unit dyadics. For brevity, we have left out the ω's from the notation. The term containing $\vec{\Sigma}_A$ stems from simultaneous two-photon processes and thus requires the use of only one nonlocal coordinate, cf. Fig.1. Also the term containing $\vec{\Sigma}_C$ is simple in the sense that nonlocality occurs in one spatial coordinate, only. The excitation from the ground state in the $\vec{\Sigma}_C$-term is of a type where one photon is absorbed locally, the other nonlocally (Fig.1). One should note that the nonlocal, nonlinear problem associated with the terms $\delta(\vec{r}\,'-\vec{r}\,'')\vec{\Sigma}_A(\vec{r},\vec{r}\,')\overleftrightarrow{U}$ and $\delta(\vec{r}-\vec{r}\,')\,\overleftrightarrow{U}\,\vec{\Sigma}_C(\vec{r},\vec{r}\,'')$ is no more difficult to handle than the corresponding nonlocal, linear problem, cf.

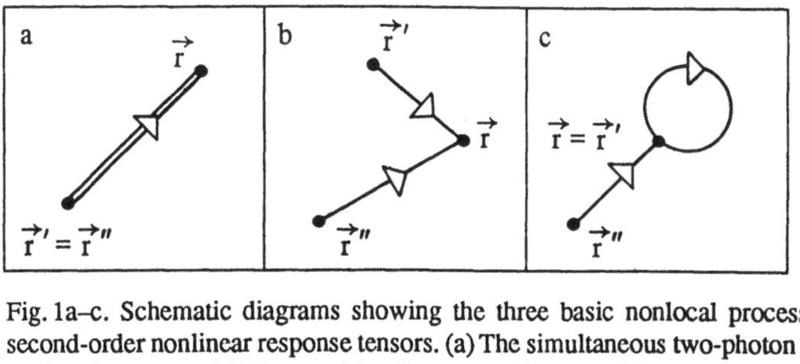

Fig. 1a–c. Schematic diagrams showing the three basic nonlocal processes associated with the second-order nonlinear response tensors. (a) The simultaneous two-photon process. (b) The double-nonlocal process. (c) The semi-local process. In second-harmonic generation from a Cooper-paired superconductor all the three basic processes occur, whereas only the semi-local process contributes in optical rectification, cf. the discussion in Sects. 3.2–3.5.

Eq.(3). The term $\overleftrightarrow{\Sigma}_B$ is the only one with a double nonlocal structure (Fig.1). This term is associated with those excitations between the initial and final states which involve an intermediate quantum state. To examine the structure of the nonlinear response more closely, we need the explicit expressions for $\vec{\Sigma}_A$, $\overleftrightarrow{\Sigma}_B$, and $\vec{\Sigma}_C$. These are

$$\vec{\Sigma}_A(\vec{r},\vec{r}\,') = \sum_{j,j'} \alpha_2(j,j')\vec{\mathcal{J}}_{j\to j'}(\vec{r})\psi_j^*(\vec{r}\,')\psi_{j'}(\vec{r}\,'), \tag{21}$$

$$\overleftrightarrow{\Sigma}_B(\vec{r},\vec{r}\,',\vec{r}\,'') = \sum_{j,j',j''} \beta(j,j',j'')\vec{\mathcal{J}}_{j\to j'}(\vec{r})\vec{\mathcal{J}}_{j'\to j''}(\vec{r}\,')\vec{\mathcal{J}}_{j''\to j}(\vec{r}\,''), \tag{22}$$

$$\vec{\Sigma}_C(\vec{r},\vec{r}\,'') = \sum_{j,j'} \alpha_1(j,j')\psi_{j'}^*(\vec{r})\psi_j(\vec{r})\vec{\mathcal{J}}_{j'\to j}(\vec{r}\,''), \tag{23}$$

where

$$\vec{\mathcal{J}}_{\alpha\to\beta}(\vec{r}) = -\frac{e\hbar}{2im}\left[\psi_\beta^*(\vec{r})\vec{\nabla}\psi_\alpha(\vec{r}) - \psi_\alpha(\vec{r})\vec{\nabla}\psi_\beta^*(\vec{r})\right] \tag{24}$$

is the transition current density between states α and β, $-e$ and m denoting the electron charge and mass, respectively. It is interesting to notice that the individual terms in the sums appearing in Eqs.(21)-(23) are separable in the $\vec{r},\vec{r}\,'$, and $\vec{r}\,''$ coordinates. Furthermore, it is found that $\overleftrightarrow{\Sigma}_B$ consists of a sum of triadic tensor products, $\vec{\mathcal{J}}_{j\to j'}\vec{\mathcal{J}}_{j'\to j''}\vec{\mathcal{J}}_{j''\to j}$, where the individual vectors are functions of one coordinate only. The quantities

$$\alpha_n(j,j') = \frac{e^2}{nm\omega^2}\frac{f(\tilde{\epsilon}_j) - f(\tilde{\epsilon}_{j'})}{\tilde{\epsilon}_{j'} - \tilde{\epsilon}_j + n\hbar\omega}, \qquad n = 1,2, \tag{25}$$

$$\beta(j,j',j'') = \frac{2}{\omega^2}\frac{1}{\tilde{\epsilon}_{j'} - \tilde{\epsilon}_j + 2\hbar\omega}\left[\frac{f(\tilde{\epsilon}_{j''}) - f(\tilde{\epsilon}_j)}{\tilde{\epsilon}_{j''} - \tilde{\epsilon}_j + \hbar\omega} - \frac{f(\tilde{\epsilon}_{j'}) - f(\tilde{\epsilon}_{j''})}{\tilde{\epsilon}_{j'} - \tilde{\epsilon}_{j''} + \hbar\omega}\right] \tag{26}$$

contain the energy resonance factors associated with the different transitions and Fermi-Dirac distribution functions

$$f(\tilde{\epsilon}) = \left[\exp\left(\frac{\tilde{\epsilon}}{k_B T}\right) + 1\right]^{-1}, \qquad \tilde{\epsilon} = \epsilon - \mu, \tag{27}$$

which take into account the occupancy of the eigenstates. In Eq.(27), μ denotes the chemical potential.

In order to establish the link between the multipole and moment expansion formalisms, it is realized from a comparison of Eqs.(17), (18), and Eqs.(20)-(23) that basically one has to calculate the zeroth and first-order moments of the transition current density. An analysis of these moments shows that [67]

$$\int_{-\infty}^{\infty} \vec{\mathcal{J}}_{\alpha\to\beta}(\vec{r})d^3r = i\omega_{\alpha\to\beta}\vec{P}_{\alpha\to\beta}, \tag{28}$$

$$\int_{-\infty}^{\infty} \vec{\mathcal{J}}_{\alpha\to\beta}(\vec{r})d^3r = \mathrm{i}\omega_{\alpha\to\beta} \overleftrightarrow{Q}_{\alpha\to\beta} - \frac{e}{2m} \overleftrightarrow{U} \times \vec{L}_{\alpha\to\beta}, \tag{29}$$

where $\omega_{\alpha\to\beta} = (\epsilon_\beta - \epsilon_\alpha)/\hbar$. Since

$$\vec{P}_{\alpha\to\beta} = -e \int_{-\infty}^{\infty} \psi_\beta^*(\vec{r})\vec{r}\psi(\vec{r})\psi_\alpha(\vec{r})d^3r, \tag{30}$$

$$\overleftrightarrow{Q}_{\alpha\to\beta} = -\frac{e}{2} \int_{-\infty}^{\infty} \psi_\beta^*(\vec{r})\vec{r}\,\vec{r}\psi_\alpha(\vec{r})d^3r, \tag{31}$$

$$\vec{L}_{\alpha\to\beta} = \frac{\hbar}{\mathrm{i}} \int_{-\infty}^{\infty} \psi_\beta^*(\vec{r})(\vec{r} \times \vec{\nabla})\psi_\alpha(\vec{r})d^3r \tag{32}$$

are the matrix elements associated with electric dipole, electric quadrupole, and magnetic dipole (angular momentum) transitions from state α to state β, the link essentially is established. For a detailed comparison of the moment and multipole expansion schemes, the reader is referred to [67].

3 Nonlinear Response of a Cooper-Paired Jellium

In order to give a solid description of the nonlinear response of a Cooper-paired superconductor it is necessary *a priori* to adopt a nonlocal formalism. This necessity stems from the fact that the pair-bound state has a finite size. For a pure superconductor, the effective range of nonlocality is given by the so-called Pippard coherence length. For an impure material, also the mean free path of the electrons plays a role for the effective range of the nonlocal response. In fact, in highly impure specimens this effective range of nonlocality is bounded by the mean free path. For a homogeneous superconductor, the local, nonlinear London response will vanish, cf. the discussion in Sect. 2.2. In this section, I shall in a few words describe the BCS ground state of the superconductor [74], and introduce the quasi-particle concept [75,76]. Based on a many-body density matrix formulation, the quasi-particle formulation is used to derive the nonlinear response of the Cooper-paired jellium. A major part of the chapter is devoted to a discussion of the response tensors for optical rectification and second-harmonic generation. The chapter is terminated by an outlook towards future work in the field.

3.1 BCS Ground State and Quasi-Particle Concept

Let me remind the reader about the pairing approximation and the concept of quasi-particle excitation. Thus, at zero temperature, the normalized ground state $|G\rangle$ of the Cooper-paired superconductor is given by

$$|G\rangle = \left[\prod_{\vec{k}} \left(u_{\vec{k}} + v_{\vec{k}} a^+_{\vec{k}\uparrow} a^+_{-\vec{k}\downarrow}\right)\right] |0\rangle \tag{33}$$

in second quantization. The Fermion vacuum state is denoted by $|0\rangle$, and $a^+_{\vec{k}\uparrow}(a^+_{-\vec{k}\downarrow})$ is the Fermion creation operator for an electron in the single-particle state of wave vector $\vec{k}\,(-\vec{k})$ and spin up ($\uparrow$) (down ($\downarrow$)). The quantities $u_{\vec{k}}$ and $v_{\vec{k}}$, which satisfy the relation $u^2_{\vec{k}} + v^2_{\vec{k}} = 1$, are the probability amplitudes for the pair state $(\vec{k}\uparrow, -\vec{k}\downarrow)$ being empty and full, respectively. At finite temperatures, quasi particles are excited. To describe the quasi-particle excitation one introduces the one-quasi-particle state

$$|\vec{p}\uparrow\rangle = \left[a^+_{\vec{p}\uparrow} \prod_{\vec{k}\neq\vec{p}} (u_{\vec{k}} + v_{\vec{k}} a^+_{\vec{k}\uparrow} a^+_{-\vec{k}\downarrow})\right] |0\rangle. \tag{34}$$

The state $|\vec{p}\uparrow\rangle$ has with certainty an electron in state $\vec{p}\uparrow$ (its mate $-\vec{p}\downarrow$ being empty) and the states $\vec{p}\uparrow$ and $-\vec{p}\downarrow$ are blocked from participating in the pairing interaction. Utilizing that $a^+_{\vec{p}\uparrow}|G\rangle = u_{\vec{p}}|\vec{p}\uparrow\rangle$ and $a_{-\vec{p}\downarrow}|G\rangle = -v_{\vec{p}}|\vec{p}\uparrow\rangle$, one realizes that the quasi-particle operators

$$\gamma^+_{\vec{p}\uparrow} \equiv u_{\vec{p}} a^+_{\vec{p}\uparrow} - v_{\vec{p}} a_{-\vec{p}\downarrow}, \tag{35}$$

$$\gamma_{-\vec{p}\downarrow} \equiv u_{\vec{p}} a_{-\vec{p}\downarrow} + v_{\vec{p}} a^+_{\vec{p}\uparrow}, \tag{36}$$

when applied to the ground state create the excited state $|\vec{p}\uparrow\rangle$ and the null state, respectively, i.e.,

$$\gamma^+_{\vec{p}\uparrow}|G\rangle = |\vec{p}\uparrow\rangle; \qquad \gamma_{-\vec{p}\downarrow}|G\rangle = 0. \tag{37}$$

The relations in Eqs.(35) and (36), and their Hermitian conjugates, define the well-known Bogoliubov-Valatin transformations. At temperature T, the quasi-particle excitation energy is given by

$$E_{\vec{k}} = +\left[\bar{\epsilon}^2_{\vec{k}} + \Delta^2_{\vec{k}}(T)\right]^{1/2} \tag{38}$$

for state $\vec{k}$, where $\Delta_{\vec{k}}(T)$ is the $\vec{k}$-dependent gap parameter determined by the integral equation

$$\Delta_{\vec{k}} = -\sum_{\vec{k}'} W_{\vec{k},\vec{k}'} \frac{\Delta_{\vec{k}'}}{2E_{\vec{k}'}} \tanh\left(\frac{E_{\vec{k}'}}{2k_{\mathrm{B}}T}\right), \tag{39}$$

$W_{\vec{k},\vec{k}'} = \langle\vec{k}\uparrow, -\vec{k}\downarrow|W|\vec{k}'\uparrow, -\vec{k}'\downarrow\rangle$ being the scattering matrix element between the pair states $(\vec{k}'\uparrow, -\vec{k}'\downarrow)$ and $(\vec{k}\uparrow, -\vec{k}\downarrow)$ [W is the two-body potential]. The quasi-particle occupancy in state $\vec{k}$ is given by the Fermi-Dirac distribution function $f(E_{\vec{k}})$ (see Eq.(27)), and the probability amplitudes $u_{\vec{k}}$ and $v_{\vec{k}}$ by

$$u_{\vec{k}} = \frac{1}{\sqrt{2}}\left(1 + \frac{\tilde{\epsilon}_{\vec{k}}}{E_{\vec{k}}}\right)^{1/2}, \tag{40}$$

$$v_{\vec{k}} = \frac{1}{\sqrt{2}}\left(1 - \frac{\tilde{\epsilon}_{\vec{k}}}{E_{\vec{k}}}\right)^{1/2}. \tag{41}$$

3.2 Many-Body Density Matrix Formulation

To determine the nonlinear conductivity response tensors describing optical rectification and second-harmonic generation in a superconductor it is convenient to take as a starting point the Liouville equation for the many-body density matrix operator ρ, i.e.,

$$i\hbar\frac{\partial \rho}{\partial t} = [H,\ \rho], \tag{42}$$

where [...] denotes a commutator. The Hamiltonian operator

$$H = H_0 + \frac{1}{2}\left(H_1^{\vec{p}\cdot\vec{A}}\mathrm{e}^{-i\omega t} + H_2^{\vec{A}\cdot\vec{A}}\mathrm{e}^{-2i\omega t} + \mathrm{h.c.}\right) + H_0^{\vec{A}\cdot\vec{A}} \tag{43}$$

is the sum of the unperturbed Hamiltonian H_0 and the interaction Hamiltonian stemming from the $\vec{p}\cdot\vec{A}$ and $\vec{A}\cdot\vec{A}$ perturbations, $\vec{A}$ being the vector potential associated with the fundamental electromagnetic field. As indicated, the $\vec{A}\cdot\vec{A}$-interaction term gives rise to both a 2ω ($H_2^{\vec{A}\cdot\vec{A}}$) and a dc ($H_0^{\vec{A}\cdot\vec{A}}$) part of H. The $\vec{A}\cdot\vec{A}$ terms *eo ipsi* are nonlinear. Note that since we are dealing with a fully nonlocal description we cannot approximate H by a dipole (or a low-order multipole) Hamiltonian. To solve Eq.(42), one makes the ansatz

$$\rho = \rho_0 + \rho_0^{NL} + \frac{1}{2}\sum_{n=1}^{\infty}(\rho_n \mathrm{e}^{-in\omega t} + \mathrm{h.c.}), \tag{44}$$

where ρ_0 is the density matrix of the light-unperturbed system, and ρ_0^{NL} is the nonlinear correction to the dc-part of the density operator. The Hermitian conjugate to $H_1^{\vec{p}\cdot\vec{A}}$, $H_2^{\vec{A}\cdot\vec{A}}$, and ρ_0 are denoted by $H_{-1}^{\vec{p}\cdot\vec{A}}$, $H_{-2}^{\vec{A}\cdot\vec{A}}$, and ρ_{-n}, respectively. By inserting Eqs.(43) and (44) into Eq.(42), it is realized from an iterative procedure that the 2ω and nonlinear dc-parts of the density matrix obey the Liouville equations

$$2\hbar\omega\rho_2 = [H_0, \rho_2] + \frac{1}{2}[H_1^{\vec{p}\cdot\vec{A}}, \rho_1] + [H_2^{\vec{A}\cdot\vec{A}}, \rho_0], \tag{45}$$

and

$$0 = [H_0, \rho_0^{\mathrm{NL}}] + \frac{1}{4}[H_1^{\vec{p}\cdot\vec{A}}, \rho_{-1}] + \frac{1}{4}[H_{-1}^{\vec{p}\cdot\vec{A}}, \rho_1] + [H_0^{\vec{A}\cdot\vec{A}}, \rho_0], \tag{46}$$

respectively. The operators ρ_1 and ρ_{-1} are obtained from the linear equations of motions for the density matrix. Once ρ_2 and ρ_0^{NL} have been determined, the light-induced current density $\vec{J}$ is obtained from the trace of the product $\rho\vec{j}$, i.e.,

$\vec{J} = \mathrm{Tr}\{\rho\vec{j}\}$. The one-body operator $\vec{j}$ for the current density is composed of an unperturbed part $\vec{j}_0 = [-e/(2m)](\vec{p}\delta(\vec{r}-\vec{r}_e) + \delta(\vec{r}-\vec{r}_e)\vec{p})$ and a part $\vec{j}_{\vec{A}} = \frac{1}{2}(\vec{j}_1 e^{-i\omega t} + \vec{j}_{-1}e^{i\omega t}) = (-e^2/m)\vec{A}\delta(\vec{r}-\vec{r}_e)$ linear in $\vec{A}$. The position vector and momentum operator of the electron are denoted by $\vec{r}_e$ and $\vec{p}$, respectively. On the basis of the general expression for the current density one realizes that the second-harmonic ($\vec{J}_2$) and dc ($\vec{J}_0$) parts of the light-induced current density are given by

$$\vec{J}_2 = \mathrm{Tr}\{\rho_2\vec{j}_0\} + \frac{1}{2}\mathrm{Tr}\{\rho_1\vec{j}_1\}, \tag{47}$$

$$\vec{J}_0 = \mathrm{Tr}\{\rho_0^{NL}\vec{j}_0\} + \frac{1}{4}\mathrm{Tr}\{\rho_1\vec{j}_{-1}\} + \frac{1}{4}\mathrm{Tr}\{\rho_{-1}\vec{j}_1\}. \tag{48}$$

To calculate $\vec{J}_2$ and $\vec{J}_0$ it is convenient to express the interaction Hamiltonian and the current density operator in second quantized form. Furthermore, by limiting ourselves to a homogeneous superconductor with free-electron single-particle wave functions, i.e., $\psi_{\vec{k},\sigma} = V^{-1/2}\exp(i\vec{k}\cdot\vec{r})\eta_\sigma$, ($\sigma = \uparrow$ or $\downarrow$), where $\eta_\uparrow = \binom{1}{0}$ and $\eta_\downarrow = \binom{0}{1}$ are the spin parts of the wave functions for spin up ($\uparrow$) and down ($\downarrow$), and V is the normalization volume, and by assuming that the fundamental field has plane-wave character, i.e., $\vec{A}(\vec{r},t) = \frac{1}{2}\{\vec{A}(\vec{q},\omega)\exp[i(\vec{q}\cdot\vec{r}-\omega t)] + \mathrm{c.c.}\}$ it can be shown that

$$H_1^{\vec{p}\cdot\vec{A}} = \frac{e\hbar}{2m}\vec{A}(\vec{q},\omega)\cdot\left[\sum_{\vec{k},\sigma}(2\vec{k}+\vec{q})a^+_{\vec{k}+\vec{q},\sigma}a_{\vec{k},\sigma}\right], \tag{49}$$

$$H_2^{\vec{A}\cdot\vec{A}} = \frac{e^2}{4m}\vec{A}(\vec{q},\omega)\cdot\vec{A}(\vec{q},\omega)\sum_{\vec{k},\sigma}a^+_{\vec{k}+2\vec{q},\sigma}a_{\vec{k},\sigma}, \tag{50}$$

$$H_0^{\vec{A}\cdot\vec{A}} = \frac{e^2}{4m}\vec{A}(\vec{q},\omega)\cdot\vec{A}^*(\vec{q},\omega)\sum_{\vec{k},\sigma}a^+_{\vec{k},\sigma}a_{\vec{k},\sigma}, \tag{51}$$

and

$$\vec{j}_0 = -\frac{e\hbar}{2mV}\sum_{\vec{k},\vec{k}',\sigma}(\vec{k}+\vec{k}')a^+_{\vec{k}',\sigma}a_{\vec{k},\sigma}e^{i(\vec{k}-\vec{k}')\cdot\vec{r}}, \tag{52}$$

$$\vec{j}_1 = -\frac{e^2}{mV}\vec{A}(\vec{q},\omega)e^{i\vec{q}\cdot\vec{r}}\sum_{\vec{k},\vec{k}',\sigma}a^+_{\vec{k}',\sigma}a_{\vec{k},\sigma}e^{i(\vec{k}-\vec{k}')\cdot\vec{r}}. \tag{53}$$

The operators $H_{-1}^{\vec{p}\cdot\vec{A}}$, $H_{-2}^{\vec{A}\cdot\vec{A}}$, and $\vec{j}_{-1}$ are obtained from Eqs.(49), (50), and (53) by making the replacements $\vec{q} \Rightarrow -\vec{q}$ and $\vec{A} \Rightarrow \vec{A}^*$.

By inserting the inverse Bogoliubov-Valatin transformations (see Eqs.(35) and (36)) into Eqs.(49)-(53), it is possible after a lengthy calculation, where many pitfalls have to be avoided, to obtain explicit expressions for the current densities associated with second-harmonic generation and optical rectification in a homogeneous Cooper-paired jellium. In the four-dimensional Fourier space, the final results are conveniently written in the forms [72,73]

$$\vec{J}_2 = \left(\overleftrightarrow{\Sigma}_2^{\alpha} + \overleftrightarrow{\Sigma}_2^{\beta} + \overleftrightarrow{\Sigma}_2^{\gamma}\right) : \vec{E}_1\vec{E}_1, \tag{54}$$

$$\vec{J}_0 = \frac{1}{2}\overleftrightarrow{\Sigma}_0^{\gamma} : (\vec{E}_1\vec{E}_1^* + \text{c.c.}). \tag{55}$$

In the succeeding Sects. 3.3-3.5, I shall discuss the physical interpretation of the content in the nonlinear conductivity response tensors of Eqs.(54) and (55).

3.3 Semi-Local Nonlinear Response

In Fig.1 we presented a schematic illustration of a semi-local nonlinear process, i.e. a process where spatial nonlocality occurs in one ($\vec{r}'$) coordinate, only. The response tensors $\overleftrightarrow{\Sigma}_2^{\gamma}$ and $\overleftrightarrow{\Sigma}_0^{\gamma}$ are associated with semi-local processes, and the expressions for them are derived from $\frac{1}{2}\text{Tr}\{\rho_1\vec{j}_1\}$, and $\frac{1}{4}\text{Tr}\{\rho_1\vec{j}_{-1}\} + \frac{1}{4}\text{Tr}\{\rho_{-1}\vec{j}_1\}$, cf. Eqs.(47) and (48). To interpret the origin of the semi-local response, let us consider the current density induced in a thermal equilibrium distribution of electrons by the part of the one-body current density operator $\vec{j}_1$, which is linear in the vector potential, i.e.,

$$\text{Tr}\{\rho_0\vec{j}_1\} = \sum_{I,I'}\langle I|\rho_0|I'\rangle\langle I'|\vec{j}_1|I\rangle, \tag{56}$$

where $|I\rangle$ and $|I'\rangle$ are many-body states. Since $\langle I|\rho_0|I'\rangle = P(E_I)\delta_{I,I'}$, where $\delta_{I,I'}$ is the Kronecker delta and $P(E_I)$ is the probability for finding the system in the eigenstate $|I\rangle$ with energy E_I, one finds in Fourier space [with $\exp(\text{i}\vec{q}\cdot\vec{r})$ factor left out]

$$\text{Tr}\{\rho_0\vec{j}_1\} = \sum_I P(E_I)\langle I|\vec{j}_1|I\rangle = -\frac{ne^2}{m}\vec{A}(\vec{q},\omega), \tag{57}$$

where n is the electron density. In linear response theory, $\text{Tr}\{\rho_0\vec{j}_1\}$ represents the well-known *diamagnetic* response. The diamagnetic response involves an inherently local relation between the vector potential and the induced current density. From a photon point of view, the ac-diamagnetic effect is associated with simultaneous absorption and emission of a photon in the mode $(\omega, \vec{q})$. In the linear case, only the mean value $\langle I|\vec{j}_1|I\rangle$ of the diamagnetic current density operator plays a role. In the nonlinear case, where terms

$$\text{Tr}\{\rho_1\vec{j}_{\pm 1}\} = \sum_{I,I'}\langle I|\rho_1|I'\rangle\langle I'|\vec{j}_{\pm 1}|I\rangle \tag{58}$$

are to be calculated, the presence of off-diagonal elements in $\langle I|\rho_1|I'\rangle$ which are different from zero (all diagonal matrix elements vanish) implies that the *nonlinear* diamagnetic response is associated with the transition matrix elements $\langle I'|\vec{j}_{\pm 1}|I\rangle$, $(I' \neq I)$, of the diamagnetic current density operator. In the photon picture, the 2ω and dc-diamagnetic effects are related to a simultaneous absorption of a photon from the mode $(\omega, \vec{q})$ and emission of a photon into the mode $(2\omega, 2\vec{q})$. The absorption of the second photon from $(\omega, \vec{q})$ is associated with the ac-modulation of the populations in the levels of the many-body system.

In explicit triadic form, the "diamagnetic term" to the second-harmonic response tensor is given by

$$\overleftrightarrow{\Sigma}_2^{\gamma}(\vec{q}, \vec{q}, \omega, \omega) = \frac{\hbar e^3}{4m^2\omega^2 V} \overleftrightarrow{U} \vec{M}_0, \tag{59}$$

where

$$\begin{aligned} \vec{M}_0 = &\sum_{\vec{k}} (2\vec{k} + \vec{q}) \\ &\times \{ [(1, +0, -1) + (1, -0, +1)](f_{\vec{k}} - f_{\vec{k}+\vec{q}})(u_{\vec{k}}^2 u_{\vec{k}+\vec{q}}^2 - v_{\vec{k}}^2 v_{\vec{k}+\vec{q}}^2) \\ &+ [(1, -0, -1) + (1, +0, +1)](1 - f_{\vec{k}} - f_{\vec{k}+\vec{q}})(u_{\vec{k}+\vec{q}}^2 v_{\vec{k}}^2 - u_{\vec{k}}^2 v_{\vec{k}+\vec{q}}^2) \}. \end{aligned} \tag{60}$$

For brevity, I have introduced the short-hand notation

$$[l\hbar\omega \pm E_{\vec{k}+m\vec{q}} \pm E_{\vec{k}+n\vec{q}}]^{-1} \equiv (l, \pm m, \pm n) \tag{61}$$

for the energy resonance factors, and denoted here and in the following the Fermi-Dirac distribution by $f_{\vec{\kappa}}$, i.e. $f_{\vec{\kappa}} \equiv f(E_{\vec{\kappa}})$. The Fermi factors and the energy resonance denominators in in Eq.(60) are associated with the ac-modulation of the density matrix. The structure of the response can be discussed on the basis of the three elementary types of quasi-particle processes. As illustrated schematically in Fig.2, the three fundamental processes are: (i) two-quasi-particle absorption, (ii) two-quasi-particle emission, and (iii) quasi-particle scattering, i.e. a process consisting of a quasi-particle emission plus a quasi-particle absorption. The two-quasi-particle absorption and emission processes vanish in the normal state, since one has no pair states to excite and deexcite, respectively. The quasi-particle

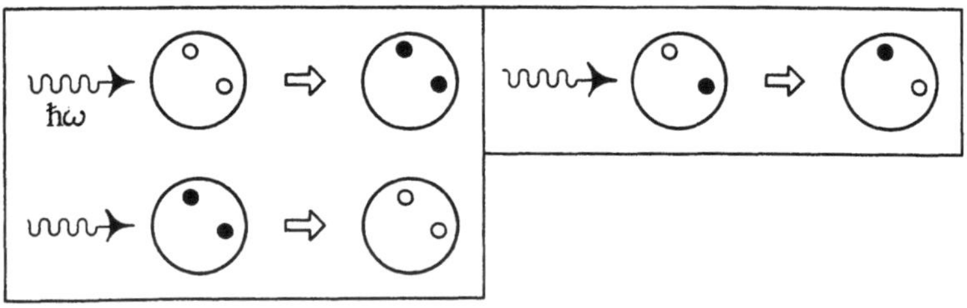

Fig. 2. Schematic diagrams showing the three fundamental photon absorption processes in a Cooper-paired jellium. In the diagram to the right, a photon absorption causes a quasi-particle scattering from state $\vec{k}$ to state $\vec{k} + \vec{q}$ (or vice versa) [•: filled state, ∘ : empty state]. In the diagrams to the left, the photon absorption gives rise to creation and destruction of two quasi particles, respectively. The large circles indicate the Fermi surface. In general quasi particles exist both inside and outside the Fermi surface. Three equivalent diagrams can be drawn for the photon emission process.

scattering processes vanish for $T \to 0$ since we have no quasi particles to scatter. The types of elementary processes giving rise to the two terms in the curly bracket of Eq.(60) can easily be identified via the respective Fermi-Dirac factors. Thus, since $f_{\vec{\kappa}}$ and $1 - f_{\vec{\kappa}}$ are the probabilities for the one-body $\vec{\kappa}$-state being full and empty, respectively, the relations

$$P(E_J) - P(E_I) \Rightarrow$$
$$\pm \left[f_{\vec{k}+\vec{q}} f_{\vec{k}} - (1 - f_{\vec{k}+\vec{q}})(1 - f_{\vec{k}}) \right] = \pm \left[f_{\vec{k}+\vec{q}} + f_{\vec{k}} - 1 \right] \tag{62}$$

demonstrate that the term in Eq.(60) containing the factor $1 - f_{\vec{k}} - f_{\vec{k}+\vec{q}}$ is associated with two-quasi-particle absorption ($-$ sign in Eq.(62)) and emission (+ sign) processes, the many-body system making a transition from state $|I\rangle$ to state $|J\rangle$ in the relation (62). Since

$$P(E_J) - P(E_I) \Rightarrow f_{\vec{k}+\vec{q}}(1 - f_{\vec{k}}) - (1 - f_{\vec{k}+\vec{q}}) f_{\vec{k}} = f_{\vec{k}+\vec{q}} - f_{\vec{k}}, \tag{63}$$

it is evident that the first term in the curly bracket of Eq.(60) stems from quasi-particle scattering processes. In the two-quasi-particle absorption/emission terms of $\vec{M}_0$, the energy resonance factor $[\hbar\omega - E_{\vec{k}} - E_{\vec{k}+\vec{q}}]^{-1}$ corresponds to a process where absorption of a photon results in creation of two quasi particles. The factor $[\hbar\omega + E_{\vec{k}} + E_{\vec{k}+\vec{q}}]^{-1}$, which one omits in a rotating wave approximation scheme, describes simultaneous photon absorption and quasi-particle destruction. The factors $[\hbar\omega + E_{\vec{k}+\vec{q}} - E_{\vec{k}}]^{-1}$ and $[\hbar\omega + E_{\vec{k}} - E_{\vec{k}+\vec{q}}]^{-1}$, appearing in the quasi-particle scattering part of $\vec{M}_0$, stem from resonances associated with processes where a photon absorption causes a scattering of a quasi particle from state $\vec{k}+\vec{q}$ to state $\vec{k}$, and from state $\vec{k}$ to state $\vec{k}+\vec{q}$, respectively. It should be noted that in a physical interaction process with a superconductor quasi particles are *always* involved in pairs, cf. the form of the interaction Hamiltonian and the current density operator in Eqs.(49)-(53). Since, as one easily shows, $\vec{M}_0$ is parallel to $\vec{q}$, it follows from Eq.(59) that $\vec{J}_2 \propto (\vec{M}_0 \cdot \vec{E}_1)\vec{E}_1$, so that (i) the diamagnetic part of the 2ω-current density is parallel to the fundamental field, and (ii) $\vec{J}_2$ will vanish unless $\vec{E}_1$ has a component parallel to the wave vector of the fundamental field. One should notice that the conclusion above has been obtained for an infinite, homogeneous Cooper-paired jellium only. A simple approximate expression for $\vec{M}_0$ can be obtained in the near-local regime. In $\vec{q}$-space, the near-local value for $\vec{M}_0$, denoted by $\vec{M}_0{}^{nl}$, is obtained via a Taylor series expansion of $\vec{M}_0$ to lowest order in $\vec{q}$. Since $\vec{M}_0$ vanishes in the local limit one has

$$\vec{M}_0{}^{nl}(\vec{q}, \omega) = \frac{\partial M_0(0, \omega)}{\partial q} \vec{q}, \tag{64}$$

and consequently after a straightforward calculation

$$\vec{M_0}^{nl}(\vec{q},\omega) = \frac{2\hbar\vec{q}}{m}\sum_{\vec{k}}(\vec{k}\cdot\vec{e}_{\vec{q}})^2$$
$$\times\left\{\frac{2f_{\vec{k}}(1-f_{\vec{k}})}{k_BT\omega}\left(1-\frac{\Delta^2_{\vec{k}}}{E^2_{\vec{k}}}\right)+\frac{\hbar^2\omega\Delta^2_{\vec{k}}}{E^3_{\vec{k}}}\frac{1-2f_{\vec{k}}}{(\hbar\omega)^2-(2E_{\vec{k}})^2}\right\}, \tag{65}$$

where $\vec{e}_{\vec{q}} = \vec{q}/q$. In Eq.(65) the terms stemming from quasi-particle scattering and two-quasi-particle absorption or emission are easily identified. Note also the energy resonance at $\hbar\omega = 2E_{\vec{k}}$. It turns out from the calculation of Eq.(65) , which we shall not present here, that only the coherence factor $u^2_{\vec{k}+\vec{q}}v^2_{\vec{k}} - u^2_{\vec{k}}v^2_{\vec{k}+\vec{q}}$ entering the two-quasi-particle absorption/emission processes plays a role in the near-local regime [it contributes with a factor $\hbar^2\Delta^2_{\vec{k}}(\vec{k}\cdot\vec{e}_{\vec{q}})/(2mE^3_{\vec{k}})$]. From a numerical point of view, the naive $\vec{M_0}^{nl}$ is simple to calculate since, in spherical coordinates, only the integral over $|\vec{k}|$ has to be done numerically. Furthermore, the sum over $\vec{k}$-states is independent of $\vec{q}$. For comparison, the $\vec{k}$-sum in $\vec{M_0}$ does depend on $\vec{q}$, and to evaluate $\vec{M_0}$ numerical integrations over the azimuth angle ($\vec{q}$ is chosen parallel to the polar axis) and $|\vec{k}|$ have to be performed.

The diamagnetic contibution to the nonlinear dc-current density is given by

$$\overleftrightarrow{\Sigma}^{\gamma}_0(\vec{q},\vec{q},\omega,\omega) = -\overleftrightarrow{\Sigma}^{\gamma}_2(\vec{q},\vec{q},\omega,\omega), \tag{66}$$

and thus need not be discussed further. It is important to notice that *only the diamagnetic response contributes to the nonlinear dc-current density,* cf. the discussion of Sects. 3.4 and 3.5.

3.4 Simultaneous Two-Photon Excitation

The response tensor $\overleftrightarrow{\Sigma}^{\alpha}_2$ in Eq.(54) describes the second-harmonic current density generated by simultaneous two-photon absorption. This can be realized by noting that $\overleftrightarrow{\Sigma}^{\alpha}_2$ stems from that part of $\mathrm{Tr}\{\rho_2\vec{j}_0\}$ in Eq.(47) which is associated with $H_2^{\vec{A}\cdot\vec{A}}$, cf. Eq.(45). A response originating in the $\vec{A}\cdot\vec{A}$ part of the interaction Hamiltonian inevitably implies that nonlocality occurs in one ($\vec{r}''$) spatial coordinate, only. A schematic illustration of the nonlocal principle associated with simultaneous two-photon processes is shown in Fig.1. In triadic notation, the explicit expression for $\overleftrightarrow{\Sigma}^{\alpha}_2$ is [72]

$$\overleftrightarrow{\Sigma}^{\alpha}_2(\vec{q},\vec{q},\omega,\omega) = \frac{\hbar e^3}{8m^2\omega^2V}\vec{N_0}\overleftrightarrow{U}, \tag{67}$$

where the quantity $\vec{N_0}$ is obtained from $\vec{M_0}$ making the replacements $\omega \Rightarrow 2\omega$ and $\vec{q} \Rightarrow 2\vec{q}$, i.e.,

$$\vec{N_0} = \vec{M_0}(\vec{q} \Rightarrow 2\vec{q}, \omega \Rightarrow 2\omega). \tag{68}$$

It appears from Eq.(67) that the 2ω-current density associated with the simultaneous two-photon processes has a contribution only from the transverse part of the fundamental field, and that the induced current is parallel to the wave vector of the light. A detailed discussion of the structure of $\overleftrightarrow{\Sigma}_2^{\alpha}$ parallels that for $\overleftrightarrow{\Sigma}_2^{\gamma}$ since $\vec{N}_0$ and $\vec{M}_0$ are so closely related in form.

The contribution to the nonlinear dc-current density arising from the $H_0^{\vec{A}\cdot\vec{A}}$-interaction Hamiltonian is hidden in the term $\mathrm{Tr}\{\rho_0^{NL}\vec{j}_0\}$ of Eq.(48). An explicit derivation [73] shows that this contribution always vanishes. This is to be expected since the two photons absorbed have opposite momentum, cf. the fact that the vector potential enters the calculation in the form $\vec{A}\cdot\vec{A}^*$. Oppositely propagating photons cannot transfer at net momentum to the electron system. The near-local approximation for $\overleftrightarrow{\Sigma}_2^{\alpha}$ is obtained via $\vec{N}_0{}^{nl} = \vec{M}_0{}^{nl}(\vec{q} \Rightarrow 2\vec{q}, \omega \Rightarrow 2\omega)$.

3.5 Double-Nonlocal Nonlinear Response

The nonlinear nonlocality unfolds itself only in the term $\overleftrightarrow{\Sigma}_2^{\beta}$ of Eq.(54). To determine $\overleftrightarrow{\Sigma}_2^{\beta}$, the contribution to ρ_2 stemming from second-order perturbation theory applied to the $\vec{p}\cdot\vec{A}$-interaction Hamiltonian is inserted into the term $\mathrm{Tr}\{\rho_2\vec{j}_0\}$ of Eq.(47). It is apparent from the calculation of ρ_2 that $\overleftrightarrow{\Sigma}_2^{\beta}$ will exhibit the double-nonlocal structure in space which is shown schematically in Fig.1 (the $\overleftrightarrow{\Sigma}_B$-term). A tedious calculation demonstrates that $\overleftrightarrow{\Sigma}_2^{\beta}$ is given by

$$\overleftrightarrow{\Sigma}_2^{\beta}(\vec{q},\vec{q},\omega,\omega) = \frac{\hbar^3 e^3}{8m^3\omega^2 V}\sum_{\vec{k}}\left[(2\vec{k}+2\vec{q})(2\vec{k}+\vec{q})(2\vec{k}+3\vec{q})\sum_{i=1}^{4}\nu_i\right] , \tag{69}$$

where

$$\begin{aligned}\nu_1 = {} & (u_{\vec{k}+2\vec{q}}u_{\vec{k}+\vec{q}} + v_{\vec{k}+2\vec{q}}v_{\vec{k}+\vec{q}})(u_{\vec{k}}v_{\vec{k}+\vec{q}} - u_{\vec{k}+\vec{q}}v_{\vec{k}})(u_{\vec{k}}v_{\vec{k}+2\vec{q}} - v_{\vec{k}}u_{\vec{k}+2\vec{q}})\\ & \times\{(f_{\vec{k}+2\vec{q}} - f_{\vec{k}+\vec{q}})[(2,-0,-2)(1,+1,-2) - (2,+0,+2)(1,-1,+2)]\\ & +(1 - f_{\vec{k}} - f_{\vec{k}+\vec{q}})[(2,-0,-2)(1,-0,-1) - (2,+0,+2)(1,+0,+1)]\} ,\end{aligned} \tag{70}$$

$$\begin{aligned}\nu_2 = {} & (u_{\vec{k}}u_{\vec{k}+\vec{q}} + v_{\vec{k}}v_{\vec{k}+\vec{q}})(u_{\vec{k}+\vec{q}}v_{\vec{k}+2\vec{q}} - v_{\vec{k}+\vec{q}}u_{\vec{k}+2\vec{q}})(u_{\vec{k}}v_{\vec{k}+2\vec{q}} - v_{\vec{k}}u_{\vec{k}+2\vec{q}})\\ & \times\{(f_{\vec{k}+\vec{q}} - f_{\vec{k}})[(2,-0,-2)(1,-0,+1) - (2,+0,+2)(1,+0,-1)]\\ & +(1 - f_{\vec{k}+\vec{q}} - f_{\vec{k}+2\vec{q}})[(2,+0,+2)(1,+1,+2) - (2,-0,-2)(1,-1,-2)]\},\end{aligned} \tag{71}$$

$$\begin{aligned}\nu_3 = {} & (u_{\vec{k}}u_{\vec{k}+\vec{q}} + v_{\vec{k}}v_{\vec{k}+\vec{q}})(u_{\vec{k}+\vec{q}}u_{\vec{k}+2\vec{q}} + v_{\vec{k}+\vec{q}}v_{\vec{k}+2\vec{q}})(u_{\vec{k}}u_{\vec{k}+2\vec{q}} + v_{\vec{k}}v_{\vec{k}+2\vec{q}})\\ & \times\{(f_{\vec{k}+2\vec{q}} - f_{\vec{k}+\vec{q}})[(2,+0,-2)(1,+1,-2) - (2,-0,+2)(1,-1,+2)]\\ & +(f_{\vec{k}+\vec{q}} - f_{\vec{k}})[(2,-0,+2)(1,-0,+1) - (2,+0,-2)(1,+0,-1)]\} ,\end{aligned} \tag{72}$$

$$\begin{aligned}\nu_4 = {} & (u_{\vec{k}+\vec{q}}v_{\vec{k}+2\vec{q}} - u_{\vec{k}+2\vec{q}}v_{\vec{k}+\vec{q}})(u_{\vec{k}+\vec{q}}v_{\vec{k}} - u_{\vec{k}}v_{\vec{k}+\vec{q}})(u_{\vec{k}}u_{\vec{k}+2\vec{q}} + v_{\vec{k}}v_{\vec{k}+2\vec{q}})\\ & \times\{(1 - f_{\vec{k}} - f_{\vec{k}+\vec{q}})[(2,-0,+2)(1,-0,-1) - (2,+0,-2)(1,+0,+1)]\\ & +(1 - f_{\vec{k}+\vec{q}} - f_{\vec{k}+2\vec{q}})[(2,-0,+2)(1,+1,+2) - (2,+0,-2)(1,-1,-2)]\} .\end{aligned} \tag{73}$$

In [72] there is a misprint in ν_4 (in the first coherence factor $u_{\vec{k}}$ should be replaced by $u_{\vec{k}+\vec{q}}$, as is obvious from the general structure of the ν_i's).

A detailed discussion of the structure of the ν_i's is not needed, because the content of the equations above are almost self-explanatory in the view of the discussion in Sect. 3.3. However, I want to emphasize that the four ν_i's describe the four possibilities for two-photon excitation of the superconductor from the initial to the final state via an intermediate state. In each of the two steps either a two-quasi-particle emission or absorption or a quasi-particle scattering is involved, as shown schematically in Fig.3. Note that for $T \to 0$, $\nu_3 \to 0$. It is characteristic that products of three coherence factors are present in the double-nonlocal response, and products of two coherence factors occur in the semi-local and simultaneous two-photon processes. In linear electrodynamics, as

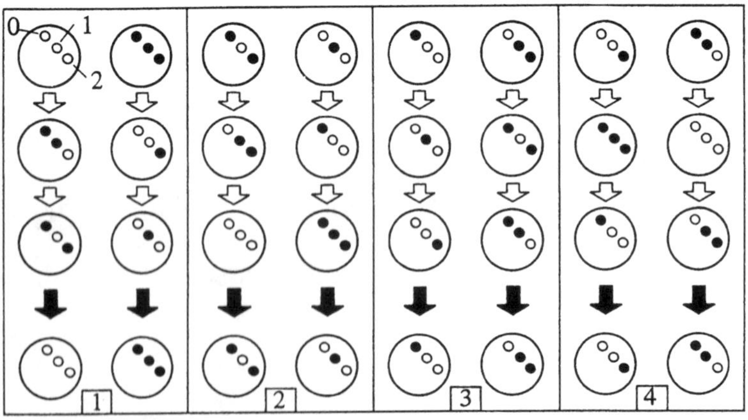

Fig. 3. Schematic diagrams illustrating how the double-nonlocal response tensor is composed of various combinations of the three elementary quasi-particle processes. In the upper row are shown the eight possible combinations for having in total one, two, or three quasi particles in the single-particle states $\vec{k}, \vec{k}+\vec{q}, \vec{k}+2\vec{q}$, denoted by 0, 1, 2 in the figure [•: filled state, o: empty state]. Letting the upper row represent the eight possible initial states, the second row shows the occupancies of the corresponding intermediate excited states. Since the first photon absorption process can only change, by two-quasi-particle emission and absorption or by quasi-particle scattering, the occupancies of the two states $\vec{k}$ and $\vec{k}+\vec{q}$, the intermediate excited states *must* necessarily be occupied as shown in the second row. The absorption of the second photon can only change the quasi-particle occupancies in the states $\vec{k}+\vec{q}$ and $\vec{k}+2\vec{q}$. In consequence, the final excited states are occupied by quasi particles as displayed in the third row. When the 2ω-photon is released, the occupancies of the states $\vec{k}$ and $\vec{k}+2\vec{q}$ are changed in a unique manner in each of the eight channels and the superconductor returns to the ground state shown in the fourth row. As indicated in the figure, the eight channels can be divided into four groups , each belonging to one of the ν_i's, the ν_i's being denoted by 1-4. Note that the two initial states (and hence also the intermediate and final excited states) associated with a given ν_i are those which have opposite quasi-particle occupancy. This principle is in agreement with the fact that (i) the coherence factor product for the two channels belonging to a given ν_i is the same, and (ii) the empty and full pair states interfere.

well as in other interaction methods, only products of two coherence factors can appear, in principle. In the next section a few additional remarks on the coherence factors are given. In the article by J.F. Nielsen and myself in this book, a more detailed study of the coherence factors is undertaken. It can be shown that the double-nonlocal response tensor vanishes in the near-local regime, i.e.,

$$\overleftrightarrow{\Sigma}_2^{\beta,nl}(\vec{q},\vec{q},\omega,\omega) = \overleftrightarrow{0} \ . \tag{74}$$

This means that the structure of the near-local , nonlinear response is as simple as that related to linear phenomena [78]. By realizing that $\sum_i \nu_i$ is an even function of k_x and k_y in a coordinate system where the z-axis is parallel to $\vec{q}$, it is a straightforward matter to obtain the symmetry scheme for $\overleftrightarrow{\Sigma}_2^{\beta}$. For completeness, the symmetry schemes for $\overleftrightarrow{\Sigma}_2^{\alpha}$, $\overleftrightarrow{\Sigma}_2^{\beta}$, and $\overleftrightarrow{\Sigma}_2^{\gamma}(\overleftrightarrow{\Sigma}_0^{\gamma})$ are displayed in Table 3.

Table 3. Symmetry schemes for the nonlinear response tensors $\overleftrightarrow{\Sigma}_2^{\alpha}$, $\overleftrightarrow{\Sigma}_2^{\beta}$, and $\overleftrightarrow{\Sigma}_2^{\gamma}(\overleftrightarrow{\Sigma}_0^{\gamma})$ of a Cooper-paired homogeneous jellium in a Cartesian coordinate system where the z-axis is parallel to the wave vector of the fundamental field. Instead of $\overleftrightarrow{\Sigma}_2^{\beta}$ and $\overleftrightarrow{\Sigma}_{2(0)}^{\gamma}$ new symmetrized response tensors giving the same physical description of the nonlinear response can easily be introduced. Adding these new tensors to $\overleftrightarrow{\Sigma}_2^{\alpha}$ it is realized that the new nonlinear response tensor can be written in the general form shown in Table 1

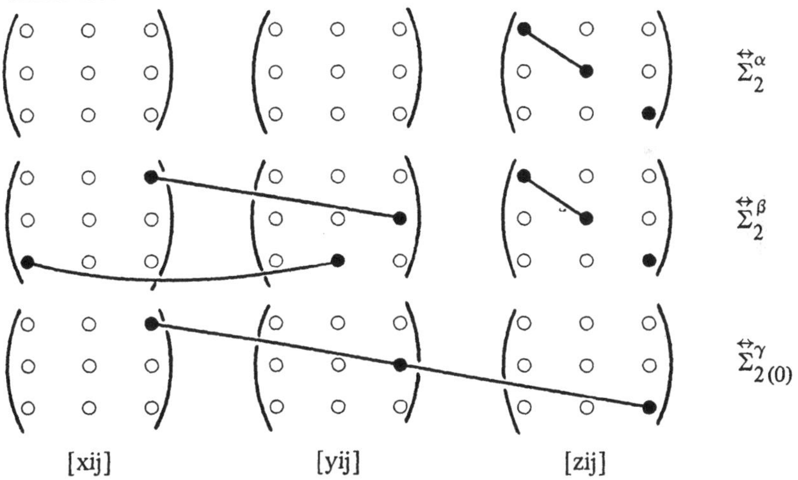

In contrast to what was claimed in [72], it can be shown that the contribution to the nonlinear dc-current density from the double-nonlocal processes vanishes identically. This result is in agreement with the fact that there are no diamagnetic effects associated with the term $\mathrm{Tr}\{\rho_0^{\mathrm{NL}}\vec{j}_0\}$ in Eq.(48). Taking into account the conclusion obtained in Sect. 3.4, it is realized that $\mathrm{Tr}\{\rho_0^{\mathrm{NL}}\vec{j}_0\} = 0$ for a homogeneous Cooper-paired jellium.

3.6 Aphorisms

The coherence factors are unique for the superconducting state, and it is well known that the quantitative outcome of a linear investigation depends on the combination in which the u's and v's occur in the response function. Thus, for electromagnetic experiments the relevant coherence factors are l^2 and p^2, where $l = u_{\vec{k}}u_{\vec{k}'} + v_{\vec{k}}v_{\vec{k}'}$, and $p = u_{\vec{k}}v_{\vec{k}'} - v_{\vec{k}}u_{\vec{k}'}$, electromagnetic absorption being related to p^2. To study the interaction of a superconductor with acoustic waves the coherence factors m^2 and n^2, where $m = u_{\vec{k}}v_{\vec{k}'} + v_{\vec{k}}u_{\vec{k}'}$ and $n = u_{\vec{k}}u_{\vec{k}'} - v_{\vec{k}}v_{\vec{k}'}$, are needed. The acoustic attenuation is described in terms of n^2. In nuclear spin-relaxation experiments l^2 enters. In nonlinear optics new combinations of the coherence factors appear. Hence, for the semi-nonlocal and the simultaneous two-photon processes it is seen from Eqs.(60) and (68) that the actual combinations are ln and pm. In the double-nonlocal response various products of the factors l and p are involved, $\vec{k}'$ being either $\vec{k} + \vec{q}$ or $\vec{k} + 2\vec{q}$. Of course one would also in nonlinear studies other than the electromagnetic ones obtain products of up to three coherence factors.

Since the electron spin plays an important role for the Cooper-pairing it would be of interest to study the nonlinear effects caused by the direct interaction of the electron spin with the fundamental light field. In the simplest approach where spin-orbit coupling is neglected one has to add a contribution

$$H_1^{\text{Spin}} = \frac{e\hbar}{2m}\vec{\sigma} \cdot \vec{\nabla} \times \vec{A} \tag{75}$$

to the interaction Hamiltonian in Eq.(43), and a term

$$\vec{j}_0^{\text{Spin}} = -\frac{ie}{2m}(\delta(\vec{r} - \vec{r}_e)\vec{p} \times \vec{\sigma} - \vec{p} \times \vec{\sigma}\delta(\vec{r} - \vec{r}_e)) \tag{76}$$

to the one-body current density operator. The Cartesian components of the spin operator $\vec{\sigma}$ are the Pauli spin matrices. At the moment work is in progress to determine, on the basis of the Pauli Hamiltonian, the explicit spin contribution to the 2ω and dc-response tensors of a Cooper-paired jellium [73].

The optical rectification process is of special interest in superconductors since the infinite conductivity of a superconductor is known to be a dc-property. From a practical point of view it would be interesting to investigate the possibilities for generating by nonlinear electromagnetic pumping a dc-current in a superconducting ring. From a fundamental point of view such an experiment would also be extremely interesting due to the fact that the magnetic flux (and hence the current) is quantized in such a ring. The understanding of the photon interaction with a superconducting system of finite angular momentum in itself is a theoretical challenge of substantial scope. In the linear regime, instabilities associated with the interaction of electromagnetic surface waves and drifting superconducting electrons of large speed should also be investigated [78].

It seems fruitful to try to understand the optical rectification process from a different point of view, namely as a nonlinear, dynamic Meissner effect. Thus, associated with the fundamental field $\vec{E}_1$ one has a magnetic field $\vec{B}_1$. In turn, this implies that the nonlinear dc part of the electric field, $\vec{E}_1\vec{E}_1^*$, is accompanied by a nonlinear magnetic dc field, $\vec{B}_1\vec{B}_1^*$. This magnetic dc-field has to be screened completely by an induced nonlinear dc-current in the superconductor, cf. the classical Meissner effect. The current density producing the necessary screening is $\vec{J}_0 = \frac{1}{2}\overleftrightarrow{\Sigma}_0^\gamma : (\vec{E}_1\vec{E}_1^* + \text{c.c.})$.

4 Response of a Superconductor–Vacuum Interface

In the previous chapter, I discussed the nonlinear response of an infinitely extended homogeneous superconductor. A solid description must, however, be able to incorporate surface effects. Even for the simplest case, i.e. that of a plane superconductor–vacuum interface, it is a formidable task to tackle the surface problem from *ab initio* considerations, as is well known from many studies on the normal metallic state. Hence, I shall not in this article embark on a major analysis of the surface response theory, but limit myself to a few conceptual considerations and to a discussion of a simple surface model in which response functions for infinite media are used. Furthermore, only optical second-harmonic generation is considered.

4.1 Semiclassical Infinite-Barrier Model

For a plane superconductor–vacuum interface it is possible to reduce the problem to an effectively one-dimensional one. Thus, in a Cartesian coordinate system where the z-axis is perpëndicular to the interface plane, only the coordinates z, z', and z'' appear in the anlysis. Now, if one assumes that the free electrons are specularly scattered from the surface and that the so-called nonclassical quantum interference effects [69] are neglected, it is possible to write the linear conductivity tensor $\overleftrightarrow{\sigma}(z,z')$ in a form $\overleftrightarrow{\sigma}(z,z') = \Theta(z)\Theta(z')[\overleftrightarrow{\sigma}_\infty(z-z') + \xi_j\,\overleftrightarrow{\sigma}_\infty(z+z')]$ where only response functions $\overleftrightarrow{\sigma}_\infty(z \pm z')$ belonging to an infinite (∞), homogeneous medium occur. The quantity ξ_j is given by $\xi_j = 1$ for $j = x, y$ and $\xi_1 = -1$ for $j = z$, writing the tensor elements of $\overleftrightarrow{\sigma}$ as σ_{ij}. In the above-mentioned so-called semiclassical infinite-barrier (SCIB) model, the superconductor–vacuum boundary becomes sharp and the surface barrier seen by the electrons infinitely high, as indicated by the Heaviside unit step function Θ. The natural extension of the SCIB model to the nonlinear case would lead to the formula

$$\overleftrightarrow{\Sigma}_2(z,z',z'') = \Theta(z)\Theta(z')\Theta(z'')\Big[\overleftrightarrow{\Sigma}_{2,\infty}(z-z',z-z'')$$
$$+\xi_\beta^I\,\overleftrightarrow{\Sigma}_{2,\infty}(z+z',z-z'') + \xi_\beta^{II}\,\overleftrightarrow{\Sigma}_{2,\infty}(z-z',z+z'')$$
$$+\xi_\beta^I\xi_\beta^{II}\,\overleftrightarrow{\Sigma}_{2,\infty}(z+z',z+z'')\Big]\,, \tag{77}$$

for the response function, where ξ^I_β, $\xi^{II}_\beta = 1$ for $\beta^I, \beta^{II} = x$ or y and $\xi^I_\beta, \xi^{II}_\beta = -1$ for $\beta^I, \beta^{II} = z$, the beta's being slaved by the response function in the notation $(\overleftrightarrow{\Sigma})_{\alpha\beta^I\beta^{II}}$. If the particle path concept, known from the classical (Boltzmann equation) theory, was correct the form in Eq.(77) would be quite acceptable. From a quantum mechanical point of view it turns out that only the $\overleftrightarrow{\Sigma}{}^\alpha_2$ and $\overleftrightarrow{\Sigma}{}^\gamma_2$ response functions can be written in the form shown in Eq.(77) [69]. For the β-response, the terms $\overleftrightarrow{\Sigma}_{2,\infty}(z+z', z \pm z'')$ are missing within the framework of the SCIB model. The SCIB model has the obvious analytical advantage that it makes use of the relatively simple bulk response functions, only. Furthermore, *if* the form in Eq.(77) is used for the nonlinear response function despite its shortcomings in presenting, properly, the $\overleftrightarrow{\Sigma}{}^\beta_2$ response within the common scheme, the calculation of the second-harmonic field becomes tractable. The response formula in Eq.(77) enables us to study, in a naive approach, bulk and surface effects and the coupling between them. Thus, the term $\overleftrightarrow{\Sigma}_{2,\infty}(z-z', z-z'')$ represents a truncated bulk response. The response is truncated in the sense that only the superconducting halfspace contributes to the nonlinear current density. The term $\overleftrightarrow{\Sigma}_{2,\infty}(z+z', z+z'')$ is a pure surface term since only processes involving electron reflection at the boundary are present in this term. The tensors $\overleftrightarrow{\Sigma}_{2,\infty}(z-z', z+z'')$ and $\overleftrightarrow{\Sigma}_{2,\infty}(z+z', z-z'')$ describe the bulk-surface coupling since only combined direct and indirect electron propagation from the source points (z', z'') to the observation point (z) is involved.

It is a straightforward matter to generalize the response formalism in chapter 3 to obtain the response in Eq.(77). What is needed is to calculate the nonlinear response of the superconductor in the case where the fundamental field consists of two plane waves $(\vec{q}, \omega)$ and $(\vec{Q}-\vec{q}, \omega)$, say. As the outcome of the calculation one obtains an expression for $\overleftrightarrow{\Sigma}_{2,\infty}(\vec{q}, \vec{Q}-\vec{q}, \omega, \omega)$, cf. Eq.(10). An inverse Fourier transformation in turn gives us $\overleftrightarrow{\Sigma}_{2,\infty}(\vec{r}\mp\vec{r}', \vec{r}\mp\vec{r}''; \omega, \omega)$ and hence the response functions on the right hand side of Eq.(77). Once the SCIB model has been established, it is possible to construct hydrodynamic theories for the nonlinear response of the superconductor. In a hydrodynamic approach the Cooper-paired superconductor is treated as an electron fluid and only collective polariton and plasmon excitations are allowed in the system.

4.2 Selvedge Response

It is known from studies on the normal metallic state that it is important to be able to go beyond the theoretical level of the sharp-boundary models. For a Cooper-paired superconductor work is in progress to determine the nonlinear response functions associated with optical second-harmonic generation and optical rectification [73]. With a knowledge of the linear, $\overleftrightarrow{\sigma}_2(\vec{r}, \vec{r}'; 2\omega)$, and nonlinear, $\overleftrightarrow{\Sigma}_2(\vec{r}, \vec{r}', \vec{r}''; \omega, \omega)$, response functions, the second-harmonic electric field

$\vec{E}_2(\vec{r};2\omega)$ can in principle be determined from the integro-differential equation

$$\overleftrightarrow{\mathcal{L}}\,[A\vec{\nabla};2\omega)\cdot\vec{E}_2(\vec{r};2\omega) = 2i\mu_0\omega\left[\int_{-\infty}^{\infty}\overleftrightarrow{\sigma}_2 A(\vec{r},\vec{r}';2\omega)\cdot\vec{E}_2(\vec{r}';2\omega)d^3r'\right.$$

$$\left.+\int_{-\infty}^{\infty}\overleftrightarrow{\Sigma}_2(\vec{r},\vec{r}',\vec{r}'';\omega,\omega):\vec{E}_1(\vec{r}'';\omega)\vec{E}_1(\vec{r}';\omega)d^3r''d^3r'\right], \tag{78}$$

where the tensorial operator $\overleftrightarrow{\mathcal{L}}$ is given by

$$\overleftrightarrow{\mathcal{L}}(\vec{\nabla};2\omega) = \vec{\nabla}\vec{\nabla} - \overleftrightarrow{U}\left[\nabla^2 + \left(\frac{2\omega}{c_0}\right)^2\right]. \tag{79}$$

To make progress in the study of the selvedge field problem it is convenient to divide the response functions into selvedge (SE) and background ((0)) parts, i.e. $\overleftrightarrow{\sigma}_2=\overleftrightarrow{\sigma}_0^{(0)}+\overleftrightarrow{\sigma}_2^{SE}$ and $\overleftrightarrow{\Sigma}_2=\overleftrightarrow{\Sigma}_2^{(0)}+\overleftrightarrow{\Sigma}_2^{SE}$, and introduce a Green's function propagator $\overleftrightarrow{G}_2(\vec{r},\vec{r}';2\omega)$ for the background problem via the integro-differential equation

$$\overleftrightarrow{\mathcal{L}}(\vec{\nabla}';2\omega)\cdot\overleftrightarrow{G}_2(\vec{r}',\vec{r};2\omega)$$

$$-2i\mu_0\omega\int_{-\infty}^{\infty}\overleftrightarrow{\sigma}_2^{(0)}(\vec{r}',\vec{r}'';2\omega)\cdot\overleftrightarrow{G}_2(\vec{r}'',\vec{r};2\omega)d^3r'' = \delta(\vec{r}'-\vec{r})\,\overleftrightarrow{U}\;. \tag{80}$$

By means of the electromagnetic propagator formulism one can establish an integral equation for the field inside the selvedge. Thus, if we limit ourselves to the flat surface geometry one obtains, leaving out ω from the notation

$$\vec{E}_2(z) = \vec{E}_2^{(0)}(z) + \int_{SE}\overleftrightarrow{K}_2(z,z')\cdot\vec{E}_2(z')dz', \tag{81}$$

where the kernel $\overleftrightarrow{K}(z,z')$ is given by

$$\overleftrightarrow{K}_2(z,z') = -2i\mu_0\omega\int_{SE}\overleftrightarrow{G}_2(z,z'')\cdot\overleftrightarrow{\sigma}_2^{SE}(z'',z')dz''\;. \tag{82}$$

Once the field inside the selvedge is determined by solving the integral equation (81) for a known background field $\vec{E}_2^{(0)}(z)$, the nonlinear field outside the selvedge can be obtained by a simple integration, cf. Eq.(81). It should be emphasized here that the extension of the selvedge depends on the choice one makes for the background response functions. Only the sum of the background field, $\vec{E}_2^{(0)}(z)$, and the selvedge field, $\vec{E}_2(z)-\vec{E}_2^{(0)}(z)$, is a physically meaningful quantity. From a conceptual point of view substantial progress can be obtained in the study of the field inside the selvedge by choosing the SCIB model as the background model [59]. The propagator associated with the SCIB model can be obtained in explicit form, and it can be shown that it has the tensor-product structure [56,57]

$$\overleftrightarrow{G}_2(z,z') = \sum\!\!\!\!\!\!\int_n^{R(z,z')}\vec{\Gamma}_O^{(n)}(z)\vec{\Gamma}_S^{(n)}(z'), \tag{83}$$

where we have introduced a generalized summation symbol $\sum\!\!\!\!\!\int_n$ to indicates that the tensor-product superposition contains both a summation $\sum_n$ over discrete n

values and an integration $\int dn$ over a continuous n spectrum. The superscript $R(z, z')$ added to $\sum\!\!\!\!\int$ indicate that z and z' in the individual tensor products are restricted (R) to specific intervals. It is the $R(z, z')$ restrictions which prevent us from solving the integral equation in (81) exactly. The nonseparable part of the equation stems from the so-called rotational-free direct contribution to the propagator and from the divergence-free and rotational-free self-field contributions.

Knowing the field inside the selvedge, the selvedge current density $\vec{J}_2^{\mathrm{SE}}(z)$ is determined by the expression

$$\vec{J}_2^{\mathrm{SE}}(z) = \int_{\mathrm{SE}} \Big[\delta(z'' - z)\, \overleftrightarrow{\sigma}^{\,\mathrm{SE}}(z, z') \cdot \vec{E}_2(z') + \overleftrightarrow{\Sigma}^{\,\mathrm{SE}}(z, z', z'') : \vec{E}_1(z'')\vec{E}_1(z')\Big]\, dz'' dz' \,. \tag{84}$$

From a phenomenological point of view, the selvedge current density is of substantial interest since, by integration over the selvedge, it allows us to obtain the so-called Rudnick-Stern a and b parameters [13] for the superconducting phase. The importance of a and b for the normal metallic phase is well known, cf. the article written by J.L. Coutaz in this book.

References

[1] F. Brown, R.E. Parks, and A.M. Sleeper, Phys.Rev.Lett. **14,** 1029 (1965).
[2] F. Brown and R.E. Parks, Phys.Rev.Lett. **16,** 507 (1966).
[3] S.S. Jha, Phys.Rev. **140,** A2020 (1965).
[4] S.S. Jha, Phys.Rev. **145,** 500 (1966).
[5] N. Bloembergen and Y.R. Shen, Phys.Rev. **141,** 298 (1966).
[6] S.S. Jha and C.S. Warke, Phys.Rev. **153,** 751 (1967).
[7] N. Bloembergen, R.K. Chang, and C.H. Lee, Phys.Rev.Lett. **16,** 986 (1966).
[8] H. Sonnenberg and H. Heffner, J.Opt.Soc.Am. **58,** 209 (1968).
[9] G.V. Krivoshchekov and V.I. Stroganov, Sov.Phys.–Solid State **11,** 89 (1969).
[10] G.V. Krivoshchekov and V.I. Stroganov, Sov.Phys.–Solid State **11,** 2151 (1970).
[11] C.S. Wang, J.M. Chen, and J.R. Bower, Opt.Commun. **8,** 275 (1973).
[12] N. Bloembergen, R.K. Chang, S.S. Jha, and C.H. Lee, Phys.Rev. **174,** 813 (1968).
[13] J. Rudnick and E.A. Stern, Phys.Rev. **B4,** 4274 (1971).
[14] J. Rudnick and E.A. Stern, in *Polaritons,* p.329, ed. E. Burstein and F.DeMartini (Plenum, New York, 1974).
[15] J.R. Bower, Phys.Rev. **B14,** 2427 (1976).
[16] H.J. Simon, D.E. Mitchell, and J.G. Watson, Phys.Rev.Lett. **33,** 1531 (1974).

[17] H.J. Simon, D.E. Mitchell, and J.G. Watson, Opt.Commun. **13,** 294 (1975).
[18] H.J. Simon, R.E. Benner, and J.G. Rako, Opt.Commun. **23,** 245 (1977).
[19] D.L. Mills, Solid State Commun. **24,** 669 (1977).
[20] Y.J. Chen and E. Burstein, Il Nuovo Cimento, **39B,** 807 (1977).
[21] M. Fukui and G.I. Stegeman, Solid State Commun. **26,** 239 (1978).
[22] M. Fukui, V.C.Y. So, J.E. Sipe, and G.I. Stegeman, J.Phys.Chem.Solids, **40,** 523 (1979).
[23] J.E. Sipe, V.C.Y. So, M. Fukui, and G.I. Stegeman, Phys.Rev. **B21,** 4389 (1980).
[24] J.E. Sipe, V.C.Y. So, M. Fukui, and G.I. Stegeman, Solid State Commun. **34,** 523 (1980).
[25] J.C. Quail and H.J. Simon, Phys.Rev. **B31,** 4900 (1985).
[26] J.E. Sipe and G.I. Stegeman, in *Surface Polaritons,* p.661, Vol.1 of *Modern Problems in Condensed Matter Sciences,* eds. V.M. Agranovich and D.L. Mills (North-Holland, Amsterdam, 1982).
[27] C.K. Chen, A.R.B. de Castro, and Y.R. Shen, Phys.Rev.Lett. **46,** 145 (1981).
[28] A. Wokaun, J.G. Bergman, J.P. Heritage, A.M. Glass, P.F. Liao, and D.H. Olson, Phys.Rev. **B24,** 849 (1981).
[29] G.T. Boyd, Th. Rasing, J.R.R. Leite, and Y.R. Shen, Phys.Rev. **B30,** 519 (1984).
[30] C.K. Chen, T.F. Heinz, D. Ricard, and Y.R. Shen, Phys.Rev. **B27,** 1965 (1983).
[31] G.S. Agarwal and S.S. Jha, Phys.Rev. **B26,** 482 (1982).
[32] G.A. Farias and A.A. Maradudin, Phys.Rev. **B30,** 3002 (1984).
[33] R. Reinisch and M. Nevière, Phys.Rev. **B28,** 1870 (1983).
[34] J.L. Coutaz, M. Nevière, E. Pic, and R. Reinisch, Phys.Rev. **B32,** 2227 (1985).
[35] R. Reinisch and M. Nevière, in *Electromagnetic Surface Excitations,* p.232, Vol.3 of Springer Series on Wave Phenomena, eds. R.F. Wallis and G.I. Stegeman (Springer, Heidelberg, 1986).
[36] H.W.K. Tom and G.D. Aumiller, Phys.Rev. **B33,** 8818 (1986).
[37] O. Keller and K. Pedersen, Proc. ECO1, SPIE **1029,** 149 (1989).
[38] O. Keller and K. Pedersen, J.Opt.Soc.Am. **B,** (in press).
[39] K.C. Rustagi, Il Nuovo Cimento LIIIB, 1178 (1968).
[40] O. Keller, Phys.Rev. **B31,** 5028 (1985).
[41] P. Apell, Physica Scripta **27,** 211 (1983).
[42] O. Keller, Phys.Rev. **B33,** 990 (1986).
[43] M. Corvi and W.L. Schaich, Phys.Rev. **B33,** 3688 (1986).
[44] W.L. Schaich and A. Liebsch, Phys. Rev. **B37,** 6187 (1988).
[45] D. Maystre and M. Nevi ere, Appl.Phys. **A39,** 1 (1986).

[46] M. Weber and A. Liebsch, Phys.Rev. **B35,** 7411 (1987).
[47] M. Weber and A. Liebsch, Phys.Rev. **B37,** 1019 (1988).
[48] M.G. Weber and A. Liebsch, Phys.Rev. **B36,** 6411 (1987).
[49] A. Chizmeshya and E. Zaremba, Phys.Rev. **B37,** 2805 (1988).
[50] V. Mizrahi and J.E. Sipe, Phys.Rev. **B34,** 3700 (1986).
[51] J.E. Sipe, V. Mizrahi, and G.I. Stegeman, Phys.Rev. **B35,** 9091 (1987).
[52] P. Guyot-Sionnest and Y.R. Shen, Phys.Rev. **B35,** 4420 (1987).
[53] P. Guyot-Sionnest and Y.R. Shen, Phys. Rev. **B38,** 7985 (1988).
[54] J.E. Sipe, J.Opt.Soc.Am. **B4,** 481 (1987).
[55] O. Keller, Optica Acta **33,** 673 (1986).
[56] O. Keller, Phys.Rev. **B34,** 3883 (1986).
[57] O. Keller, Phys.Rev. **B37,** 10588 (1988).
[58] V.Mizrahi and J.E. Sipe, J.Opt.Soc.Amer. **B5,** 660 (1988).
[59] O. Keller, Phys.Rev. **B38,** 8041 (1988).
[60] P. Guyot-Sionnest, W. Chen, and Y.R. Shen, Phys.Rev. **B33,** 8254 (1986).
[61] J.E. Sipe, D.J. Moss, and H.M. van Driel, Phys.Rev. **B35,** 1129 (1987).
[62] C.-C. Tzeng and J.T. Lue, Surf.Sci. **192,** 491 (1987).
[63] G.L. Richmond, J.M. Robinson, and V.L. Shannon, Progr.Surf.Sci. **28,** 1 (1988).
[64] H.R. Jensen, K. Pedersen, and O. Keller, in *Proc. International Conference on Nonlinear Optics,* p.174, ed. V.J. Corcoran, (STS, McLean, 1988).
[65] A. Liebsch and W.L. Schaich, Phys.Rev. **B,** (to be published).
[66] R. Murphy, M. Yeganeh, K.J.Song, and E.W. Plummer, Phys.Rev.Lett. **63,** 318 (1989).
[67] O. Keller, phys.stat.sol. (b) **157,** 459 (1990).
[68] H.R. Jensen and O. Keller, in *Coherence and Quantum Optics VI,* eds. L. Mandel and E. Wolf (Plenum, New York), (in press).
[69] O. Keller, Materials Science and Engineering **B5,** 183 (1990).
[70] J.L. Coutaz, in *Nonlinear Optics in Solids,* ed. O. Keller (Springer-Verlag, Berlin, 1990).
[71] N.N. Akhmediev, I.V. Mel'nikov, and L.J. Kobur, General Physics Institute, Acad. of Sci. of the U.S.S.R., Preprint N10 (1989).
[72] O. Keller, in *Coherence and Quantum Optics VI,* eds. L. Mandel and E. Wolf (Plenum, New York), (in press).
[73] O. Keller (to be published).
[74] J. Bardeen, L.N. Cooper, and J.R. Schrieffer, Phys.Rev. **108,** 1175 (1957).
[75] J.R. Schrieffer, *Theory of Superconductivity* (Benjamin/Cummings, New York, 1983).
[76] G.Rickayzen, *Theory of Superconductivity* (Interscience, New York, 1964).
[77] M.Tinkham, in *Optical Properties and Electronic Structure of Metals and* Alloys, ed. A. Abelèles, p.431 (North-Holland, Amsterdam, 1966).

[78] O. Keller, J.Opt.Soc.Amer. **B** (to be published).
[79] J.G. Bednorz and K.A. Müller, Z.Phys. **B64,** 189 (1986).
[80] C.W. Chu *et al.,* Phys.Rev.Lett. **58,** 405 (1987).
[81] J. Woods Halley, ed. *Theories of High Temperature Superconductivity* (Addison-Wesley, New York, 1988).
[82] P.S. Pershan, Phys.Rev. **130,** 919 (1963).
[83] E. Adler, Phys.Rev. **134,** A728 (1964).
[84] J. Ducuing and C.Flytzanis, in *Optical Properties of Solids,* p.859, ed. F.Abèles, (North-Holland, Amsterdam, 1972).

Coherence Factors in the Optical Second-Harmonic Response Tensor of a Cooper-Paired Superconductor

J.F. Nielsen and O. Keller

Institute of Physics, University of Aalborg,
Pontoppidanstræde 103, DK-9220 Aalborg Øst, Denmark

1 Introduction

In recent years there has been a rapid increase in the number of studies on superconductors due to the discovery of high-T_C materials. From an optical point of view these new materials possess many interesting features. For instance, since the low temperature energy gap is as large as 30-40 meV, one expects that linear optical experiments in the mid-infrared frequency region will exhibit substantial changes in the optical response when the magnitude of the energy gap is changed via a change of the temperature of the sample.

Second-harmonic generation is expected to be even more interesting [1], since e.g. the superconducting current is located just inside the surface, where also most of the optical second-harmonic response from centrosymmetric materials is generated.

There is still no final explanation of how the superconducting state of the new high-T_C materials can be described, but from a lot of experiments it turns out that pairs of electrons are formed [2]. The existence of these so-called Cooper pairs is very important from an optical point of view, since a lot of the dynamical behaviour can then probably be described along the same lines as for the old BCS superconductors.

2 Structure of the Coherence Factors

The dynamical properties of a superconductor are hidden to a substantial extent in the various combinations of the Bogoliubov amplitudes. These combinations are called coherence factors. The Bogoliubov amplitudes are the probability amplitudes of occupied ($v_{\vec{k}}$) and unoccupied ($u_{\vec{k}}$) quasi-particle states in the initial ($u_{\vec{k}}, v_{\vec{k}}$) or final ($u_{\vec{k}'}, v_{\vec{k}'}$) state. The amplitudes which are related via $u_{\vec{k}}^2 + v_{\vec{k}}^2 = 1$ due to the normalization requirement are given by $u_{\vec{k}}^2 = \frac{1}{2}(1 + \tilde{\epsilon}_{\vec{k}}/E_{\vec{k}})$ and $v_{\vec{k}}^2 = \frac{1}{2}(1 - \tilde{\epsilon}_{\vec{k}}/E_{\vec{k}})$, where $\tilde{\epsilon}_{\vec{k}}$ is the difference between the single-particle energy $\epsilon_{\vec{k}} = \hbar^2 k^2/(2m)$ and the chemical potential μ, and $E_{\vec{k}} = +(\tilde{\epsilon}_{\vec{k}}^2 + \Delta^2(T))^{1/2}$ is the quasi-particle energy, $\Delta(T)$ being the temperature-dependent gap-parameter. As a rough estimate one can use $\Delta(T) = \Delta(0)[1 - (T/T_C)^4]^{1/2}$ for $T < T_C$, where T_C is the superconducting transition temperature.

Springer Series in Wave Phenomena, Vol. 9 **Nonlinear Optics in Solids**
Editor: O. Keller

For $T > T_C$, $u_{\vec{k}}$ and $v_{\vec{k}}$ are step functions, i.e. $v_{\vec{k}} = 1 - u_{\vec{k}} = 1$ and 0 for $\epsilon_{\vec{k}} < \epsilon_F$ and $\epsilon_{\vec{k}} > \epsilon_F$, respectively, assuming the chemical potential to equal the Fermi energy ϵ_F.

There are in general four different combinations of the Bogoliubov amplitudes. These are

$$l(\vec{k},\vec{k}') = (u_{\vec{k}}u_{\vec{k}'} + v_{\vec{k}}v_{\vec{k}'}) \ , \tag{1}$$

$$m(\vec{k},\vec{k}') = (u_{\vec{k}}v_{\vec{k}'} + u_{\vec{k}'}v_{\vec{k}}) \ , \tag{2}$$

$$n(\vec{k},\vec{k}') = (u_{\vec{k}}u_{\vec{k}'} - v_{\vec{k}}v_{\vec{k}'}) \ , \tag{3}$$

$$p(\vec{k},\vec{k}') = (u_{\vec{k}}v_{\vec{k}'} - u_{\vec{k}'}v_{\vec{k}}) \ . \tag{4}$$

For linear investigations only quadratic terms of l, m, n, and p occur [3,4]. If the superconductor interacts with an acoustic wave the relevant coherence factors are m^2 and n^2. In electromagnetic interactions l^2 and p^2 take part, and here absorption is related to p^2. Nuclear spin-relaxation experiments involve l^2.

3 Coherence Factors in Second-Harmonic Generation

In a recent paper by Keller [5] a model for optical second-harmonic generation from a Cooper-paired superconductor based on the quasi-particle formulation is presented. Since the explicit expression for the second-harmonic current density is rather lengthy we only write down the coherence factors appearing in it. For a general discussion and an identification of the different quantities the reader is referred to [1].

$$M_0 \sim \beta_1 = n(\vec{k},\vec{k}')l(\vec{k},\vec{k}') \ , \quad \text{and} \quad \beta_2 = m(\vec{k},\vec{k}')p(\vec{k},\vec{k}') \ , \tag{5}$$

$$\nu_1 \sim \eta_1 = p(\vec{k},\vec{k}')l(\vec{k}',\vec{k}'')p(\vec{k},\vec{k}'') \ , \tag{6}$$

$$\nu_2 \sim \eta_2 = l(\vec{k},\vec{k}')p(\vec{k}',\vec{k}'')p(\vec{k},\vec{k}'') \ , \tag{7}$$

$$\nu_3 \sim \eta_3 = l(\vec{k},\vec{k}')l(\vec{k}',\vec{k}'')l(\vec{k},\vec{k}'') \ , \tag{8}$$

$$\nu_4 \sim \eta_4 = p(\vec{k},\vec{k}')p(\vec{k}',\vec{k}'')l(\vec{k},\vec{k}'') \ , \tag{9}$$

where $\vec{k}' = \vec{k} + \vec{q}$ and $\vec{k}'' = \vec{k} + 2\vec{q}$, $\vec{q}$ being the wave vector of the fundamental light wave. It can be seen from relation (5) that M_0 involves two products of two coherence factors. These products are not quadratic like in the linear case (l^2, m^2, n^2, and p^2), but combinations of coherence factors known from linear acoustic (m, n) and linear electromagnetic (l, p) interactions. It is especially in-

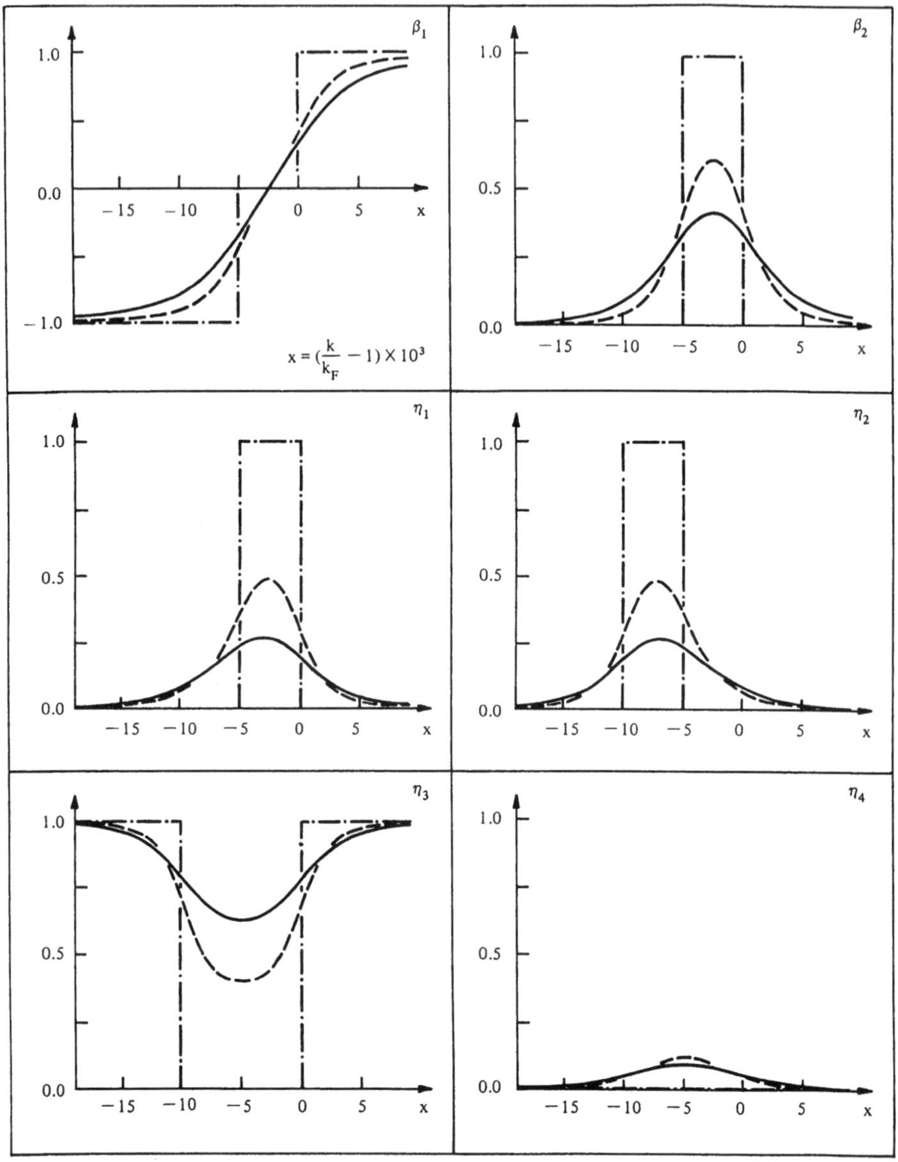

Fig. 1. Coherence factors as a function of the magnitude of the electron wave vector in the vicinity of the Fermi surface for three temperatures: —— $T = 0$, – – – $T = 0.9\, T_C$, and –·–·– $T = T_C$. The plots are drawn for $\vec{k} \parallel \vec{q}$ with $|\vec{q}| = 3.0 \times 10^7 \mathrm{m}^{-1}$. The material data used are typical for $YBa_2Cu_3O_7$, namely $T_C = 93\mathrm{K}$ and $n = 7 \times 10^{27} \mathrm{m}^{-3}$.

teresting to have a look at the η_i's, because here a product of three coherence factors occurs. All the factors in the η_i's are known from linear electromagnetic interactions. The first factor in the η_i's is associated with a process where a scattering of a quasi-particle ($l(\vec{k}, \vec{k}')$) or a two-quasi-particle creation or destruction ($p(\vec{k}, \vec{k}')$) excites the superconductor from the initial ($\vec{k}$) to an intermediate ($\vec{k}'$) state. The second factor involves the same quasi-particle processes, but here the excitations are from the intermediate state ($\vec{k}'$) to the final state ($\vec{k}''$) . The third factor accounts for the deexcitation of the system to the initial state. One has to notice that p occurs either two times or nowhere in the η_i's due to the constant number of quasi-particles before and after the excitation plus deexcitation cycle. The plots in Fig. 1 all show that the coherence factors vary over an interval length of the order q around k_F. Only $\vec{k}$-states in this interval contribute to the difference between the nonlinear electromagnetic properties in the normal and superconducting phase. It is also seen that a change in temperature from a little below T_C, here 0.9 T_C, to T_C results in a larger change of the response than if the temperature is changed from 0.9 T_C to zero. In the plot of β_1 the curves fall below zero for $|\vec{k}| < |\vec{k} - \vec{q}|$, a result which never occurs in linear analyses. A summation of the coherence factors for β_1 and β_2 or the η_i's at a fixed temperature gives ± 1 for all values of $\vec{k}$. In the plot of η_1 one observes a small asymmetry around $k = k_F - q/2$. This asymmetry is due to the fact that the large $p(\vec{k}, \vec{k}')$-function is multiplied by the small $l(\vec{k}', \vec{k}'')$-function which is symmetric about a displaced point, namely $k = k_F - 3q/2$. An analogue asymmetry is observed in η_2.

There are no experimental data yet, neither from the old metallic superconductors nor from the new high-T_C materials.

References

[1] O. Keller, in these proceedings

[2] J.C. Phillips, *Physics of High-T_C Superconductors* (Academic, New York 1989)

[3] G. Rickayzen, *Theory of Superconductivity* (Interscience, New York 1964)

[4] D.R. Tilley and J. Tilley, *Superfluidity and Superconductivity*, (Adam Hilger, Bristol 1986)

[5] O. Keller, in *Coherence and Quantum Optics VI*, eds. P. Mandel and E. Wolf (Plenum, New York) (in press).

[6] Z. Schlesinger et al., Phys.Rev.Lett. **59**, 1958 (1987).

Physical Picture of Parametric Phenomena and Ponderomotive Effects in Solids

P. Mulser

Theoretische Quantenelektronik, Institut für Angewandte Physik,
TH Darmstadt, Hochschulstr. 4A, D-6100 Darmstadt, Fed. Rep. of Germany

Abstract. In this paper a physical picture is given of the important class of three wave parametric decay processes occurring in nonlinear optics of solids. Starting from the radiation pressure on a single particle a general expression for the ponderomotive force in dense matter at rest and in motion is derived. In a second step the formulae are applied to parametric processes of nonlinear optics and it is shown that in terms of wave pressure the quantitative analysis is much more concise, systematic and intuitive.

1. Introduction

1.1 Stimulated Processes

Nonlinear optical phenomena in matter can be classified according to the number of modes involved in a process under consideration: two-wave, three-wave, four-wave interaction, etc. The two-wave process only occurs in mode conversion, i.e. when light incident on matter is transformed into a wave of different nature. For this to happen momentum and energy balance must be fulfilled

$$\hbar\vec{k}_1 = \hbar\vec{k}_2, \quad \hbar\omega_1 = \hbar\omega_2, \tag{1}$$

where $\vec{k}_1, \omega_1$ and $\vec{k}_2, \omega_2$ are the wave vectors and frequencies of light and the excited wave in the medium. This process is possible in the linear as well as in the nonlinear regime. An example of it is the conversion of photons into plasmons, a process which is also known under the name of resonance absorption [1],[2]. For the phenomenon to be stimulated both modes must fulfill a dispersion relation $D_1(\omega_1, \vec{k}_1) = 0$ and $D_2(\omega_2, \vec{k}_2) = 0$.

A whole variety of optical phenomena is summarized as three-wave interactions, e.g. three wave mixing in which two waves generate a third one (for example the second harmonic, 2ω) or three-wave decay in which one wave decays into two other waves (for example Brillouin or Raman scattering, two-plasmon decay). For such processes to occur effectively three wave vectors and frequencies must match,

$$\hbar\vec{k}_1 \pm \hbar\vec{k}_2 = \hbar\vec{k}_3, \quad \hbar\omega_1 \pm \hbar\omega_2 = \hbar\omega_3. \tag{2}$$

Thereby mixing is characterized by the + sign; the − sign stands for the decay process. Again, for a three-wave process to be stimulated each of the three waves must represent a freely propagating volume or surface mode, i.e. three dispersion relations $D_i(\omega_i, \vec{k}_i) = 0$, $i = 1, 2, 3$, must be satisfied. In four and more wave processes the situation is completely analogous. In contrast to two-wave interaction they cannot occur in a linear medium. If one or more of the frequencies involved are zero we classify the process as non-resonant. Examples of such phenomena are light beam self-focusing and the modulational instability (phase self-modulation) of the beam.

Springer Series in Wave Phenomena, Vol. 9 **Nonlinear Optics in Solids**
Editor: O. Keller

1.2 Formal (Conventional) View of Multimode Interaction

Nonlinear optics is governed by (i) Maxwell's wave equation

$$\nabla \times \nabla \times \vec{E} + \frac{1}{c^2}\frac{\partial^2}{\partial t^2}\vec{E} = -\frac{1}{c^2\epsilon_o}\frac{\partial \vec{j}}{\partial t}, \tag{3}$$

and (ii) an equation of motion for the polarisation $\vec{P}$ (magnetisation may be set zero, $\vec{M} = 0$). The connection between the current density $\vec{j}$ and $\vec{P}$ is

$$\vec{j} = \vec{j}_{free} + \vec{j}_{displacement} = \frac{\partial}{\partial t}\vec{P}. \tag{4}$$

When a monochromatic plane wave

$$\vec{E} = \hat{E}_o e^{i\vec{k}\vec{x} - i\omega t}$$

is incident onto a medium a polarisation consisting of a linear and a nonlinear term is induced,

$$\vec{P} = \vec{P}_L + \vec{P}_{NL}; \;\; \vec{P}_L = \epsilon_o \chi_\omega \vec{E}, \;\; \frac{\partial \vec{P}_L}{\partial t} = j_L = \sigma_\omega \vec{E}. \tag{5}$$

Refractive index η and conductivity σ are related to each other by

$$\eta^2 = 1 + \chi = 1 + i\,\frac{\sigma}{\epsilon_o \omega}. \tag{6}$$

$\vec{P}_L$ provides for dispersion of the modes but leaves them uncoupled. Coupling is accomplished by $\vec{P}_{NL}$. In the simplest case of a three-wave process, in presence of a second wave $(\vec{k}_2, \omega_2)$, $\vec{P}_{NL}$ generates a third one,

$$\vec{E}_3 = \hat{E}_3 e^{i\vec{k}_3\vec{x} - i\omega_3 t},$$

however, this is effectively produced only if it satisfies its own wave equation

$$\nabla^2 \vec{E}_3 + \left(\frac{\omega_3}{c}\right)^2 \eta_3^2(\omega_3)\vec{E}_3 \simeq 0,$$

from which $D_3(\omega_3, \vec{k}_3) = \omega_3^2 - c^2\vec{k}_3^2/\eta_3^2 \simeq 0$ is recovered. $\vec{P}_{NL}$, when inserted in eq.(3), in the slowly varying amplitude approximation (SVA) leads to the amplitude equation for each of the three modes $\hat{A}_l(\omega_l, \vec{k}_l)$

$$\frac{d}{dz_l}\hat{A}_l = \gamma_l \hat{A}_j \hat{A}_k^{(*)} e^{i(\Delta\vec{k}\vec{x} - \Delta\omega t)}; \;\; j \neq k \neq l, \;\; z_l = z \parallel \vec{x}. \tag{7}$$

$d/dz_l = \partial/\partial z + \partial/\partial(v_{g,l}t)$ is the convective derivative following the $\hat{A}_l$-mode with group velocity $v_{g,l}$. γ_l is a coupling constant and $\Delta\vec{k}$ represents a possible phase mismatch. The amplitudes are indicated by $\hat{A}_j, \hat{A}_k, \hat{A}_l$ since they are not necessarily of electric nature (e.g. acoustic wave in Brillouin scattering). The asterisk in brackets stands for the complex conjugate of amplitude $\hat{A}_k$ when ω_l is related to ω_j and ω_k by $\omega_l = \omega_j - \omega_k$ (compare with eq.(2)). The derivation of eq.(7) from the general wave eq.(3) is standard and can be found in textbooks of nonlinear optics. This equation makes clear why the matching conditions from eq.(2) must be fulfilled for effective generation of the l-mode, at least approximately (small mismatch $\Delta\vec{k}$, $\Delta\omega$).

1.3 Unified Treatment of Decay Processes

The analysis sketched so far is the standard picture of three- and more-wave interactions. It allows one to calculate temporal and spatial amplification of modes and to determine the corresponding growth rates. Growth occurs due to secular forces, i.e. such ones acting in the same direction over many oscillations of the lowest frequency mode involved. The disadvantage of the standard analysis is twofold:

(i) the analysis does not tell us what kind these forces are and does not give an intuitive physical picture of the phenomenon;

(ii) as a consequence, no quick qualitative access is possible to whether under given conditions a stimulated, i.e. unstable, process evolves and how fast it will develop.

There is, however, for the class of parametric decay processes, i.e. $\hbar\omega_1 = \hbar\omega_2 + \hbar\omega_3$ (− sign in eq.(2)), the possibility of giving a unified picture in physical terms and of calculating growth rates in a far more immediate way. This is accomplished by making use of the concept of light pressure.

In solids the following stimulated decay processes may occur:

Stimulated Brillouin scattering (SBS): photon$\Longrightarrow$ acoustic phonon + photon

Stimulated Raman scattering (SRS): photon $\Longrightarrow$ optical phonon or plasmon + photon

Two-plasmon decay (TPD): photon $\Longrightarrow$plasmon + plasmon

Parametric decay instability (PDI): photon $\Longrightarrow$ phonon + plasmon

Oscillating two-stream instability (OTSI): photon $\Longrightarrow$ phonon ($\omega = 0$) + plasmon.

In addition, there are the already mentioned non-resonant phenomena of self-focussing and self-phase modulation. In what follows we treat first radiation pressure on single particles and derive then the radiation pressure for dense matter. In the next section we apply this to the parametric decay processes and give a quantitative analysis in terms of light pressure. The presentation of the whole subject is self-contained.

2. Light Pressure on Single Particles and Collective Ponderomotive Force

In contrast to the oscillating force induced by an electric wave in a system of charges, e.g. electric dipoles, light pressure is a secular force, i.e. a zero frequency force for which the synonymous expressions "radiation pressure", "wave pressure" and "ponderomotive force (density)" are used also. Conceptually, force, pressure and force density are different from each other; however, since there is no risk of ambiguity, in the following we use one expression for the others and vice versa. As we shall see, classically the ponderomotive force originates form the nonlinearity of the Lorentz force on a charged particle.

2.1. Historical Remarks

In 1861 James Clerk Maxwell found out that a light beam of flux density I, when impinging normally upon a surface of reflectivity R in vacuo, exerts the pressure [3]

$$p_L = (1 + R)\frac{I}{c}. \tag{8}$$

Soon afterwards, Maxwell derived the electromagnetic stress tensor, which was given the correct form of a conservation law by Lorentz not much before 1895. In 1884 Boltzman used eq.(8) to deduce Stefan's T^4 law from the first and second principles of thermodynamics in a very elegant way. A new interpretation of p_L was brought in with Einstein's concept of the photon (1905, 1909, 1917). The first experimental proof of formula eq.(8) was given in 1901 by Lebedev with a precision of 20% [4]. Two years later Nichols and Hull performed another experiment [5]. Only in 1923 was eq.(8) confirmed within a 2% precision by Gerlach and Golsen [6].

Nearly all important applications of radiation pressure before the invention of the laser were believed to be related to astrophysics (equilibrium of massive stars, planetary nebulae dynamics, evolution of the early universe) and there was tacit agreement on the assertion "In the laboratory light pressure is small". As nowadays, however, lasers emit flux densities up to $10^{19}\ W/cm^2$, p_L-values up to 10 $Gbars$, i.e. several percent of the gas pressure at the center of the sun, are possible. Theoretically, only in 1958 was a more advanced formula than eq.(8) presented by the expression for the ponderomotive force $\vec{f}_p$ on a single charged particle. It was shown that a monochromatic wave of the form

$$\vec{E}(\vec{x},t) = \hat{E}(\vec{x})e^{-i\omega t} \tag{9}$$

exerts the force on the particle of charge q and mass m

$$\vec{f}_p = -\frac{q^2}{4m\omega^2}\nabla(\hat{E}\hat{E}^*), \tag{10}$$

[7], [8]. Since then the same formula has been independently derived in at least 4 different places [9], [10]. It is interesting to note that the first author to publish this formula in another context was Landau in the first volume of his famous textbook on theoretical physics (before 1957, after an idea of Kapitza in 1951; [11]) but its relevance to radiation pressure was not recognized. $\vec{f}_p$ is a gradient force and hence the quantity

$$\phi_p = \frac{q^2}{4m\omega^2}\,\hat{E}\hat{E}^* \tag{11}$$

represents a potential and is referred to as the ponderomotive potential of a charged particle. Meanwhile ϕ_p has been recognized as an entity of central importance in physical phenomena such as the free electron laser [10], particle acceleration [12], magnetized plasmas [13], trapping and cooling of atoms [14], [15], dressed atom model [16], and multiphoton ionization [17]. It is of particular relevance for the dynamics and nonlinear optics it induces in laser plasmas [18].

2.2 Ponderomotive Force on Single Particles

The traditional derivation of eq.(10) starts from the Lorentz equation of a charged particle in an electromagnetic field of the form given by eq.(9),

$$m\frac{d\vec{v}}{dt} = q(\vec{E} + \vec{v}\times\vec{B}).$$

Position $\vec{x}$ and velocity $\vec{v}$ are decomposed into the oscillatory components $\vec{\xi}$ and $\vec{w}$ plus the components of the oscillation center $\vec{x}_o$ and $\vec{v}_o$,

$$\vec{x}(t) = \vec{x}_o(t) + \vec{\xi}(t), \quad \vec{v}(t) = \vec{v}_o(t) + \vec{w}(t). \tag{12}$$

$\vec{x}_o(t)$ and $\vec{v}_o(t)$ are smooth quantities slowly varying with time [19]. Under the restriction that the oscillation amplitude $\hat{\xi}$ is much smaller than the local wavelength $\lambda(\vec{x}_o)$, i.e. $\hat{\xi} \ll \lambda(\vec{x}_o)$, the Lorentz equation may be linearized in $\vec{\xi}$ and solved separately,

$$m\frac{d\vec{v}}{dt} \simeq m\left[\frac{\partial\vec{w}}{\partial t} + (\vec{v}_o\nabla)\vec{w}\right] = q\vec{E}(\vec{x}_o). \tag{13}$$

In general $\hat{E}(\vec{x})$ is the superposition of a whole variety of modes,

$$\hat{E}(\vec{x}) = \sum_{\{\vec{k}\}} \hat{E}_k e^{i\vec{k}\vec{x}}.$$

For a single mode eq.(13) yields

$$\vec{w}_k(t) = i\frac{q}{m\omega_k}\vec{E}_k, \quad \omega_k = \omega - \vec{v}_o\vec{k}.$$

The Doppler shift $\Delta\omega_k = \vec{v}_o\vec{k}$ originates from the convective derivative $(\vec{v}_o\nabla)\vec{w}$. Generally, at least in solid state physics, all $\Delta\omega_k$ are small so that all Fourier components $\vec{w}_k$ superpose to yield

$$\vec{w}(t) = i\frac{q}{m\omega}\hat{E}(\vec{x}_o)e^{-i\omega t}. \tag{14}$$

In the next order from the Lorentz equation $\vec{v}_o(t)$ is determined,

$$\vec{f}_p = m\frac{d\vec{v}_o}{dt} = q\{\vec{E}(\vec{x}_o + \vec{\xi}(t)) + \vec{w}(t) \times \vec{B}(\vec{x}_o)\}_o.$$

The bracket contains terms varying with 2ω and zero frequency. Therefore the subscript "o" has been added to indicate that in determining the secular force $\vec{f}_p$ the 2ω-term has to be suppressed. Taylor expansion of $\vec{E}$ in $\vec{\xi} = i\vec{w}/\omega$, substituting $\vec{B}$ from Faraday's law, $i\omega\vec{B} = \nabla \times \vec{E}$, and observing that physical quantities have to be real leads to

$$\{\vec{E} + \vec{w} \times \vec{B}\}_o = -\frac{q^2}{4m\omega^2}\{(\hat{E}^*\nabla)\hat{E} + \hat{E}^* \times \nabla \times \hat{E} + cc\} = -\nabla\phi_p, \tag{15}$$

in agreement with expressions (10) and (11). In a transverse plane wave $(\vec{w}\nabla)\vec{E}$ is zero; if $\vec{E}$ is longitudinal (e.g. plasmons) $\vec{B}$ is zero and $\vec{f}_p$ entirely stems from $(\vec{w}\nabla)\vec{E}$. In a wave of constant amplitude $\vec{f}_p$ is zero.

In terms of the momentum equation the physical interpretation of $\vec{f}_p$ according to eq. (10) is as follows. Let us first assume a pure electrostatic wave in the x-direction of the form $\hat{E}(x)e^{-i\omega t}$ (Fig.1). A free electron is shifted by the E-field from its original position x_o up to x_1. From there it is accelerated to the right until it has passed x_o. Then it is decelerated by the reversed E-field and is stopped at position x_2. If x'_o designates the position in which the field is reversed ($x' > x_o$), the deceleration interval $x_2 - x'_o$ is larger than that of acceleration since on the RHS of x_o the E-field is weaker and therefore a longer distance is needed to take away the energy gained in the former quarter period of oscillation. On its way back the electron is stopped in the region of higher amplitude; the turning point is shifted from x_1 to x_3 into the direction of decreasing wave amplitude $\hat{E}(x)$. In an (inhomogeneous) medium this drift produces, by charge separation, a static E-field which transmits the force to the ions. Since this consideration is based on energy and work only it follows that the drift (or force) is independent of the sign of charge. In an electromagnetic wave the drift is caused by the $\vec{v} \times \vec{B}$ force which also acts along the x-direction.

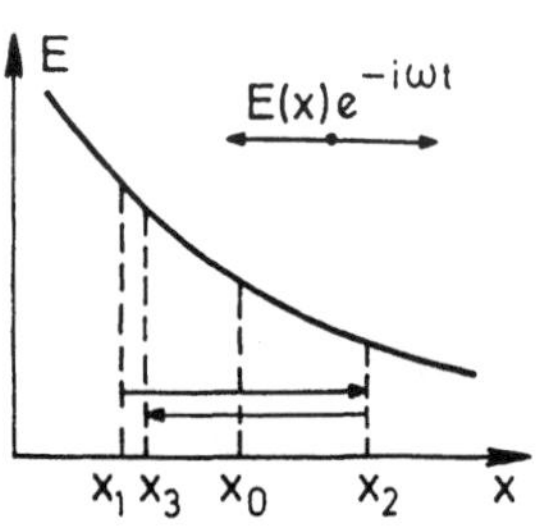

Fig.1: A free electron executes a drift in the direction of decreasing wave amplitude. x_1, x_2, x_3 turning points.

Derivations of $\vec{f}_p$ from a momentum equation are clumsy. Now we show, by basing our derivation on energy considerations, how this result can be extended without much formal effort to arbitrarily strong and anharmonic fields for particles without and with internal degrees of freedom. To this end we consider first a particle whose Hamiltonian is

$$H = \frac{\vec{p}^2}{2m} + V(\vec{x}), \tag{16}$$

with a quasi-periodic potential $V(\vec{x})$ in which the motion has an oscillation center. By decomposing this according to eq.(12) and integrating H over one period τ we find

$$H\tau = \frac{1}{2}\int_t^{t+\tau} m(\vec{v}_o + \vec{w})^2 dt + \int_t^{t+\tau} V(\vec{x}(t))dt = (\frac{1}{2}m\vec{v}_o^2 + W + < V >)\tau,$$

since

$$\int_t^{t+\tau} \vec{v}_o^2 dt \simeq \vec{v}_o^2\tau, \quad \int_t^{t+\tau} \vec{v}_o\vec{w}dt \simeq \vec{v}_o \int \vec{w}dt = 0, \quad W = \frac{1}{\tau}\int_t^{t+\tau} \frac{1}{2}m\vec{w}^2 dt,$$

$$< V > = \frac{1}{\tau}\int_t^{t+\tau} V(\vec{x}(t))dt = \frac{1}{\tau}\int_t^{t+\tau} V(\vec{x})\frac{d\vec{x}}{\vec{v}}.$$

Thus, the oscillation center motion is governed by

$$H_o(p_o, \vec{x}_o) = \frac{\vec{p}_o^2}{2m} + \phi_p, \quad \vec{p}_o = m\vec{v}_o, \quad \phi_p = W + < V >, \quad H_o = H = E = const. \tag{17}$$

ϕ_p plays the role of an effective potential for the motion of $\vec{v}_o$. W is the cycle-averaged oscillation energy. ϕ_p depends on $\vec{x}_o$ and $\vec{v}_o$. Eq.(17) may be considered as an implicit constraint between $\vec{v}_o$ and $\vec{x}_o$. $< V >$ is a temporal average, or a spatial average of V times the weight function $1/v(\vec{x})$. Regions of low velocity contribute more to $< V >$ than potential regions which are crossed at high speed by the particle. As the ponderomotive force is the force on the fictitious oscillation center of mass m it is given by

$$\vec{f}_p = m\frac{d\vec{v}_o}{dt} = -\nabla(W + < V >). \tag{18}$$

It is also relativistically correct as long as only W is relativistic.

In most relevant applications V represents an electric wave and is either of the form $V = V(\vec{x}, t, \eta)$ when $\nabla \times \vec{B} = 0$ holds or, as for an electromagnetic wave, no scalar potential exists at all. In $V(\vec{x}, t, \eta)$ t indicates a fast time variation whereas η stands for a set of parameters slowly varying in space and time. For an electric wave it is

$$\eta = \{\hat{E}(\vec{x}, t), \ \vec{v}_o(\vec{x}, t), \ \vec{k}(\vec{x}, t)\}. \tag{19}$$

The oscillation center speed $\vec{v}_o$ must be considered here because of the Doppler effect. When η is kept constant, i.e. no space and time dependence, the orbit in the oscillation center frame $\vec{\xi}(t)$ is closed. For the case of a quasi-periodic particle orbit it is shown now (i) that any rapidly time-dependent Hamiltonian $H(t)$ can be replaced by an effective Hamiltonian of the form

$$H_{eff} = \frac{\vec{p}^2}{2m} + V(\vec{x}, t, \eta) \tag{20}$$

in which $\vec{p}$ is the *mechanical* momentum $\vec{p} = m\vec{v} = \vec{p}_o + m\vec{w}$ and (ii) that from averaging H_{eff} over one period τ one is lead back to eq.(17) if η does not depend on time.

Proof. In an inertial reference system co-moving with $\vec{v}_o$ over the oscillation interval τ, the orbit $\xi(t)$ is nearly closed and $\vec{w}^2 = (d\vec{\xi}/dt)^2$ is, except at a finite number of points, a unique function of position $\vec{\xi}$. Hence, $\vec{p}_1^2 = m^2\vec{w}^2(\vec{\xi},\eta)$ and it can be shown to be piecewise invertible, $\vec{\xi} = \vec{\xi}(\vec{p}_1,\eta)$. The averaged oscillatory energy W is the sum of the contributions W_i in the three coordinate directions ξ_1, ξ_2, ξ_3 of $\vec{\xi}$, i.e. $W = W_1 + W_2 + W_3$. By setting

$$V_i(\xi_i,\eta) = W_i - \frac{1}{2m}(p_1^i(\xi_i,\eta))^2, \quad V_{eff}(\vec{\xi},\eta) = \sum_{i=1}^{3} V_i(\xi_i,\eta)$$

the Hamiltonian $\mathcal{H}_{eff} = \vec{p}_1^2/2m + V_{eff}$ in the co-moving inertial system is constructed. It is a slowly varying quantity in time owing to $\mathcal{H}_{eff} = W + < V_{eff} >= E_1(\eta)$ and it describes the true motion $\dot{\vec{p}}_1 = -\partial\mathcal{H}_{eff}/\partial\vec{\xi}$, since

$$\dot{w}_i = -\frac{dw_i}{d\xi_i}\frac{d\xi_i}{dt} = w_i\frac{\partial w_i}{\partial\xi_i} = \frac{1}{2m^2}\frac{\partial}{\partial\xi_i}(p_1^i)^2 = -\frac{1}{m}\frac{\partial V_{eff}}{\partial\xi_i} = -\frac{1}{m}\frac{\partial\mathcal{H}_{eff}}{\partial\xi_i}.$$

The transformation to the lab frame is canonical and is obtained by observing that $\dot{\vec{p}} = \dot{\vec{p}}_1$ holds and by setting $V(\vec{x},t,\eta) = V_{eff}(\vec{\xi},\eta) = V(\vec{x} - \int\vec{v}_o dt,\eta)$; hence

$$\dot{\vec{p}} = -\frac{\partial V(\vec{x},t,\eta(\vec{x},t))}{\partial\vec{x}},$$

from which expression (20) follows immediately (i). From now on we assume that η does not explicitly depend on time and hence

$$\hat{E} = \hat{E}(\vec{x}), \quad \vec{k} = \vec{k}(\vec{x}), \quad \vec{v}_o = \vec{v}_o(\vec{x}), \quad \frac{d}{dt}\mathcal{H}_{eff} = \frac{\partial}{\partial t}\mathcal{H}_{eff} = 0.$$

Now cycle-averaging of H_{eff} yields

$$< H_{eff} >= \frac{\vec{p}_o^2}{2m} + \frac{\vec{p}_1^2}{2m} + < V(\vec{x},t,\eta) >= \frac{\vec{p}_o^2}{2m} + E_1(\eta(\vec{x}_o)),$$

$$\frac{d}{dt} < H_{eff} >= \frac{\partial}{\partial t} < H_{eff} >= 0.$$

Thus $H_o(\vec{p}_o,\vec{x}) =< H_{eff} >$ is the correct Hamiltonian governing the oscillation center motion and satisfying $H_o = E = const$. Equality strictly holds for constant $\vec{v}_o$. With $\dot{\vec{v}}_o \neq 0$, as is the case here, it is immediate to show that the relative variation of $\mathcal{H}_{eff}$ is at most of the order of $(2\pi/kL)^2$, with L being the characteristic scale length of the variation of η. This proofs part (ii) of the assertion and it shows that $H_o(\vec{p}_o,\vec{x})$ is the desired adiabatic invariant •

As a consequence, eq.(17) applies and the ponderomotive force on a single particle without internal degrees of freedom is given by eq.(18) with $< V >$ replaced by $< V_{eff} >$. Let now $\vec{E}$ of an electromagnetic wave be antisymmetric in space and time with respect to the center of the oscillation interval. Then $\vec{w},\xi$ and V_{eff} are also antisymmetric with respect to this point and V_{eff} reduces to zero. In an electromagnetic wave such symmetry holds if τ is a finite and well-defined quantity. In this very important case for applications ponderomotive potential ϕ_p and force $\vec{f}_p$ are given by

$$\phi_p = W, \quad f_p = -\nabla W = -\nabla\phi_p. \tag{21}$$

These relations have an immediate interpretation. $W = f(|\hat{E}|^2)$ is a unique function of $\vec{x}_o$ (see Fig.2). When the particle is slowly shifted from $\vec{x}_o$ to $\vec{x}_o'$, W changes. It is evident that the energy change can only be accomplished by external work of a dc force (in the quantum picture: no photons can be absorbed or emitted by a free particle) and hence eq.(21) must hold. As a special case, for harmonic particle motion, $W = q^2\hat{E}\hat{E}^*/4m\omega^2$ results in accordance with eq.(11). However, $\Phi_p = W$ is valid also for anharmonic motion (i), large

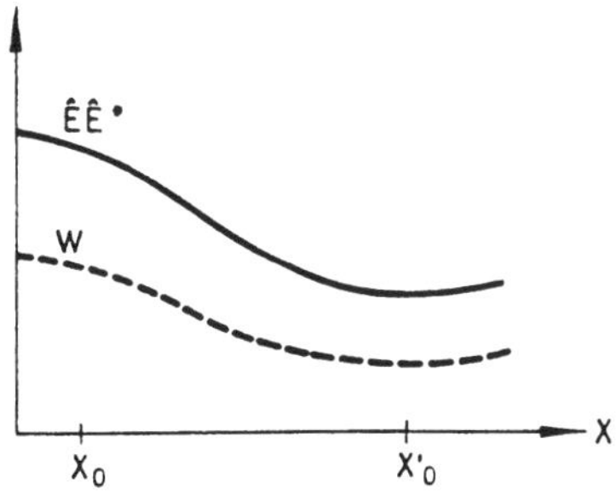

Fig.2: The time averaged oscillation energy W of a particle (dashed line) in an electromagnetic wave is a unique function of position as long as in passing from x_o to x'_o it undergoes several oscillations per wavelength. W is identical with the ponderomotive potential ϕ_p.

oscillations no longer restricted to $\hat{\xi} \ll \lambda$ (ii) and fields of constant $\hat{E}$, but undergoing adiabatic changes of their shape or wavelength (iii). In a very simple way eq.(21) follows from directly averaging the electromagnetic Hamiltonian $H_{em} = (\vec{p} - q\vec{A})^2/2m = m\vec{v}^2/2$ once the interchangeability of d/dt and $<>$ is accepted:

$$\frac{d}{dt} < H_{em} > = < \frac{d}{dt} H_{em} > = < \frac{\partial}{\partial t} H_{em} > = \frac{\partial}{\partial t} < H_{em} > = \frac{\partial}{\partial t} < \frac{m}{2} \vec{v}^2 > = 0,$$

i.e. $< H_{em} > = 1/2m\vec{v}_o^2 + W = const$ results. However, in a longitudinal electric wave $< V_{eff} > \neq 0$ may hold and $\vec{f}_p$ may differ strongly from expression (10) although the standard derivation of f_p from the momentum equation above for a harmonic wave also in this case leads to eq.(10). It follows from the definition of $< V >$ that $< V > \simeq 0$ holds only for $|\vec{v}_o| \ll v_{Phase}$ (see example given below). In the following we study a few relevant cases and give an extension of $\vec{f}_p$ to a point particle with internal degrees of freedom.

I. For an arbitrarily strong monochromatic electromagnetic wave we have

$$W = mc^2\{(1 + \frac{q^2}{\alpha m^2 c^2 \omega^2} \hat{E}\hat{E}^*)^{1/2} - 1\}, \quad \alpha = 1, 2, \tag{22}$$

for the oscillation energies corresponding to plane ($\alpha = 2$) and circular polarisation [20]. For moderate fields Taylor expansion of this expression leads back to Φ_p of eq.(11) for which V_{eff} is easily determined to be $V_{eff}(y) = m\omega^2(y - y_o)^2/2 - q^2\hat{E}\hat{E}^*/4m\omega^2$.

II. Next we consider a point particle with an internal degree of freedom that is a harmonic oscillator in dipole approximation,

$$\ddot{\vec{\delta}} + \omega_o^2 \vec{\delta} = \frac{q}{\mu} \hat{E}(\vec{x}) e^{-i\omega t}; \quad \mu = m_1 m_2/(m_1 + m_2). \tag{23}$$

The averaged oscillation energy is

$$W = < \frac{1}{2} \mu \dot{\vec{\delta}}^2 > = \frac{q}{4\mu} \frac{\omega^2 + \omega_o^2}{(\omega^2 - \omega_o^2)^2} \hat{E}\hat{E}^*.$$

In determining ϕ_p one has to keep in mind that when the field is switched on the oscillator gains internal energy $E_{in} = E_{pot,max}$ also and that the forces producing this energy are internal forces and as such, by definition, cancel each other. Thus ϕ_p is given by

$$\phi_p = W - < E_{in} > = W - \frac{q^2}{4\mu} \frac{2\omega_o^2}{(\omega^2 - \omega_o^2)^2} \hat{E}\hat{E}^* = \frac{q^2}{4\mu(\omega^2 - \omega_o^2)} \hat{E}\hat{E}^*. \tag{24}$$

The same result is obtained by solving eq.(23) up to the first order and determining its secular component. However, the procedure used here is shorter and more general, e.g. a force may result from constant $\hat{E}$ but changing wavelength. $\phi_p = W - < E_{in} >$ also holds for nonharmonic $\vec{E}$ fields or when the Lorentz force is included. For $\omega_o < \omega$ the oscillator behaves like a free particle, i.e. $\vec{f}_p$ tries to drive it into a region of decreasing field amplitude. For $\omega_o > \omega$ it moves in the opposite direction. By introducing a damping term it is shown that at exact resonance $\vec{f}_p$ reduces to zero. ϕ_p is identical with what in quantum mechanics is called the expectation value of interaction energy E_{int}. With the help of the dipole moment $\vec{p} = q\vec{\delta}$ expression (24) can be written as

$$E_{int} =< -\frac{1}{2}\vec{p}\vec{E} > . \tag{25}$$

At $\omega_o > \omega$, $\vec{p}$ is parallel to $\vec{E}$ and the oscillator moves in the direction of increasing field; at $\omega_o < \omega$ $\vec{p}$ is antiparallel to $\vec{E}$ and at resonance the phase shift is $\pi/2$.

III. It is both important for applications and instructive to generalize $\vec{f}_p$ to include a charged particle in a static magnetic field $\vec{B}_o$. With $\vec{E}(x,t)$ in the y-direction and $\vec{B}_o$ parallel to z one has to set

$$W = \frac{1}{2}m < v_x^2 + v_y^2 >, \quad < V >= \mu B_o.$$

$\mu B_o = QIB_o$ (μ magnetic moment, Q cross section of the closed orbit) is the "inner" energy of the orbiting particle configuration which has to be subtracted since, by definition, internal forces do not contribute to $\vec{f}_p$. Specializing for $\vec{E} = E_x \sim e^{-i\omega t}$ yields in perfect analogy to eqs.(23) and (24)

$$\ddot{v}_x + \omega_c^2 v_x = -i\frac{q}{m}E_x, \quad v_y = -i\frac{\omega_c}{\omega}v_x, \quad \omega_c = \frac{q}{m}B_o;$$

$$\begin{aligned} \phi_p &= W - < V >= \frac{q^2}{4m}\frac{\omega^2 + \omega_c^2}{(\omega^2 - \omega_c^2)^2}\hat{E}\hat{E}^* - \frac{q^2}{4m}\frac{2\omega_c^2}{(\omega^2 - \omega_c^2)^2}\hat{E}\hat{E}^* \\ &= \frac{q^2}{4m(\omega^2 - \omega_c^2)}\hat{E}\hat{E}^*. \end{aligned} \tag{26}$$

It is in agreement with results obtained from the momentum equation with weak fields [21]. The cyclotron frequency ω_c takes the place of ω_o.

IV. As a last example we consider an electron entering a longitudinal travelling wave of the form $E(x,t) = \hat{E}(x)sin(kx - \omega t)$ with $k, \omega = const$ and $\hat{E}(x)$ adiabatically increasing from zero to infinity along x. It is convenient to normalize its zeroth order Hamiltonian $H = p^2/2m - e\hat{E}cos(kx - \omega t)/k$ to the kinetic energy $m_c v_\varphi^2/2$ of an electron travelling at phase speed $v_\varphi = \omega/k$. To calculate V_{eff} and its average $< V_{eff} >$ the orbit $\xi(t)$ must be known. Instead, and this is the simpler way here, a procedure developed by Kruskal [22] can be used to construct an adiabatic invariant J from the constancy of which the motion of the oscillation center is determined.

With the normalized kinetic and potential energies $w = 2T/m_e v_\varphi^2$, $v = 2e\hat{E}/km_e v_\varphi^2$ and the ratio $\kappa = 2v/(v+w)$ in the wave frame moving at v_φ the result is given in terms of the elliptic integral of the second kind $E(\kappa)$ [23],

$$J = (v+w)^{1/2}E(\kappa) \mp \frac{\pi}{4}w = const. \tag{27}$$

With $v_o = \int pdt/m_c\tau$ and an initial electron energy $w_i = u^2/v_\varphi^2$ in the lab frame $4J$ becomes $1 + w \mp 4(v+w)^{1/2}E(\kappa)/\pi = w_i$ and

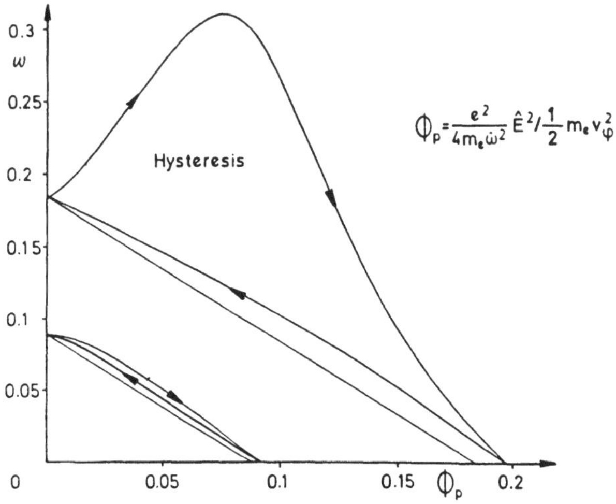

Fig.3: Normalized translational electron energy w as a function of the classical ponderomotive potential ϕ_p. When entering a wave of increasing amplitude and travelling in the same direction with initial energy w_i the particle is reflected (see direction of arrows) if $w_i < 1 - 8/\pi^2 = 0.19$ holds. The motion, instead of following the straight line $w = w_i - \phi_p$ shows a hysteresis in phase space but no energy dissipation.

$$\frac{v_o}{v_\varphi} = 1 \mp \frac{\pi}{2K(\kappa)}(v + w)^{1/2}. \tag{28}$$

$K(\kappa)$ is the elliptic integral of the first kind. The evaluation of v_o in terms of $w_o = v_o^2/v_\varphi^2$ is reported in Fig.3 for $w_i = 0.09$ and $w_i = 0.185$ as a function of the classical ponderomotive potential Φ_p from eq.(11). The electron enters the wave with these energies and, according to eq.(11), it should lose translational energy continuously until being reflected at the point at which w_i equals $2\Phi_p/m_e v_\varphi^2$, all the way forward (– sign in eqs.(27), (28)) and back (+ sign) following the straight lines indicated in the figure. In reality the ponderomotive action shows a hysteresis in phase space which is more pronounced for higher initial energy. Instead of being always repulsive, for initial energies $w_i > 1/9$, Φ_p from eq.(17) even becomes attractive over a finite space region. Finally, however, the electron is reflected without dissipation of energy as long as w_i does not exceed the trapping limit $w_i = 1 - 8/\pi^2$ at $\kappa = 1$ [23]. This drag of the particle or pre-trapping effect observed in Fig.3 is mainly a consequence of $< V > \neq 0$ and is best understood if the sinusoidal potential is changed into a rectangular one. Then, in its rest frame a particle of initial velocity $v_o = u < 0$ oscillates between the speeds $v_\pm = (u^2 \mp 2qV/m)^{1/2}$ to the left. Thereby it spends more time at v_- than v_+ velocities, and hence, v_o is increased. The situation reported here is relevant for wave-particle interaction in plasmas, e.g. resonance absorption in laser plasmas. This example clearly shows that ϕ_p from eq.(11) is only of asymptotic validity (weak fields or $v_o \ll v_\varphi$) whereas eqs.(17) and (18) are far more general.

2.3 Ponderomotive Force in Dense Matter

When starting to calculate the ponderomotive force density, designated by $\vec{\pi}$ in this paper, which a stationary wave field (i.e. time-independent amplitude) exerts on the unit volume of matter of particle density $n(\vec{x})$, one could naively argue that the desired expression is

$\vec{\pi} = -n\nabla\phi_p$. ϕ_p is the single particle ponderomotive potential. In the general nonlinear case it is given by $\phi_p = W + < V > - < E_{in} >$. In the linear case for particles (atoms, molecules) of one resonance frequency ω_o reduces to eq.(24) multiplied by its oscillator strength f when driven at frequency $\omega \neq \omega_o$ by a transverse wave. In the presence of several resonances the dipole moment $\vec{p}$ and ϕ_p are sums of such terms times the appropriate oscillator strengths. Remembering the relations between polarisation $\vec{P}$, electric dipole moment $\vec{p}$ and dielectric susceptibility χ_ω

$$\vec{P} = n\vec{p} = \epsilon_o\chi_\omega\vec{E}, \tag{29}$$

in the linear case one would deduce for this latter $\vec{\pi} = \epsilon_o n\nabla(\chi_\omega\hat{E}\hat{E}^*/4n)$. However, this expression is incomplete. To see this we have to consider what happens when an arbitrary volume V filled with matter is displaced by an infinitesimal amount in a steady state wave field. Due to $\vec{f}_p$ on the single particle the individual volume elements $d\tau$ are shifted by the amount $\delta\vec{r}(\vec{x})$ each until due to "mechanical" stress and strain $\overleftrightarrow{\Pi}$ (or pressure p in a gas) a new equilibrium is reached. Thereby V deforms into the displaced volume V'. The work done by the ponderomotive force in such a shift of matter from V to V' is $\int \vec{\pi}_o\delta\vec{r}d\tau$. On the other hand this must be equal to the negative change of the interaction (=ponderomotive) potential $\int \phi_p n d\tau$, thus

$$\int_V \vec{\pi}_o\delta\vec{r}d\tau = -\{\int_{V'} \phi_p(\vec{x}')n(\vec{x}')d\tau - \int_V \phi_p(\vec{x})n(\vec{x})d\tau\}, \tag{30}$$

where $\vec{x}' = \vec{x} + \delta\vec{r}(\vec{x})$. By keeping in mind that $n(\vec{x}')d\tau' = n(\vec{x})d\tau$ holds since the number of nuclei in a volume element is unaffected by the displacement the RHS can be expressed as

$$\int_{V'} \phi_p(\vec{x}')n(\vec{x}')d\tau' - \int_V \phi_p(\vec{x})n(\vec{x})d\tau = \int_V [\phi_p(\vec{x} + \delta\vec{r}) - \phi_p(\vec{x})]nd\tau.$$

During the displacement $\delta\vec{r}$, $\phi_p = \phi_p(\hat{E})$ changes in space owing to $\hat{E} = \hat{E}(\vec{x})$, and in time because in general the shift of a macroscopic mass causes a redistribution of the wave field and matter at each point. With $\delta\vec{r} = \vec{v}\delta t$ this implies

$$\phi_p(\vec{x} + \delta\vec{r}) - \phi_p(\vec{x}) = \delta\vec{r}\nabla\phi_p + \left(\frac{\partial\phi_p}{\partial t}\right)_{\vec{x}}\delta t = \{\nabla\phi_p + \frac{\vec{v}}{v^2}\frac{\partial\phi_p}{\partial t}\}\delta\vec{r},$$

and eq.(30) becomes

$$\int_V \vec{\pi}_o\delta\vec{r}d\tau = -\int_V n(\nabla\phi_p + \frac{\vec{v}}{v^2}\frac{\partial\phi_p}{\partial t})\delta\vec{r}d\tau.$$

V is arbitrary and hence

$$\vec{\pi}_o = -n(\nabla\phi_p + \frac{\vec{v}}{v^2}\frac{\partial\phi_p}{\partial t})$$

must hold. In the linear case ϕ_p reduces to $\phi_p = \epsilon_o\chi\hat{E}\hat{E}^*/4n$ (the subscript ω is suppressed in χ_ω). In addition, there is another term which originates from a periodic space charge $\rho_{el} = -\nabla\vec{P}$ induced by the $\hbar f$ field $\vec{E}$ so that the total ponderomotive force density $\vec{\pi}$ reads

$$\vec{\pi} = \vec{\pi}_o - < \vec{E}\nabla\vec{P} > . \tag{31}$$

In the linearized version this is

$$\vec{\pi} = \frac{\epsilon_o}{4}n(\nabla + \frac{\vec{v}}{v^2}\frac{\partial}{\partial t})\frac{\chi}{n}\hat{E}\hat{E}^* - < \vec{E}\nabla\vec{P} > . \tag{32}$$

The time derivative stands for the variation of ϕ_p at a fixed position $\vec{x}$ during the displacement $\delta\vec{r}$. There are important applications (e.g. parametric effects in nonlinear optics) in which this term cancels exactly or is very small. Then we obtain

$$\vec{\pi} = \vec{\pi}_o - < \vec{E}\nabla\vec{P} >, \quad \vec{\pi}_o = \frac{\epsilon_o}{4} n \nabla \left(\frac{\chi}{n} \hat{E}\hat{E}^* \right). \tag{33}$$

In deriving eqs.(30) - (33) it is assumed that (i) the wave field is stationary, (ii) the medium is non-absorbing (χ real), and (iii) all variables are continuous. Discontinuities in χ have to be treated separately. In the non-stationary case additional terms appear. However, they may be disregarded here since their magnitude is smaller by the ratio τ/T, $T = |\hat{E}|/|\partial_t \hat{E}|$ being the characteristic rise time. A typical value is $\tau/T < 10^{-3}$.

In general there is a contribution to $\vec{\pi}_o$ from the gradient of χ/n in eq.(33) which means that in dense matter χ is no longer proportional to the density of atoms or molecules. Why this is so is best understood from the Clausius-Mosotti fluid model (see for instance [24], p.35 and 412) of harmonic oscillators of the type of eq.(23). The driving local field in dense matter is no longer the local wave field $\vec{E}$ but an $\vec{E}_{eff} = \vec{E} + (\vec{P}/3\epsilon_o)$. Thus

$$\ddot{\vec{\delta}} + \omega_o^2 \vec{\delta} = -\frac{e}{\mu}(1 + \frac{\chi}{3})\vec{E}$$

and

$$\vec{P} = -ne\vec{\delta} = \frac{ne^2}{\mu(\omega_o^2 - \omega^2)}\vec{E}(1 + \frac{\chi}{3}) = \epsilon_o \chi \vec{E},$$

from which

$$\chi = \frac{\omega_p^2}{(\omega_o^2 - \omega_p^2/3) - \omega^2} = \frac{\omega_p^2}{\omega_1^2 - \omega^2}; \quad \omega_p^2 = \frac{ne^2}{\mu\epsilon_o}$$

follows. The polarisation of the surrounding medium results in a density-dependent reduction of the eigenfrequency from ω_o to ω_1,

$$\omega_1^2 = \omega_o^2 - \omega_p^2/3. \tag{34}$$

ω_p is the plasma frequency.

One should expect that $\vec{\pi}$ from eq.(32) would become equal to the static ponderomotive force density for ω reducing to zero and the factor 1/4 being substituted by 1/2. In ref.[24] for this case

$$\vec{\pi}_s = \frac{\epsilon_o}{2}\nabla[(n\frac{\partial\chi}{\partial n} - \chi)\hat{E}^2] + \frac{\epsilon_o}{2}\chi\nabla\hat{E}^2 + \rho_{el}\hat{E} \tag{35}$$

is derived. It was obtained first by Helmholtz. The leading ponderomotive force term would read accordingly

$$\vec{\pi}_{so} = \frac{\epsilon_o}{4}\nabla[(n\frac{\partial\chi}{\partial n} - \chi)\hat{E}\hat{E}^*] + \frac{\epsilon_o}{4}\chi\nabla\hat{E}\hat{E}^*.$$

For a Clausius-Mosotti fluid the difference between expression (33) and this formula is

$$\vec{\pi}_{so} - \vec{\pi}_o = \frac{2}{9} \cdot \frac{\epsilon_o}{4}\hat{E}\hat{E}^*\chi^3\nabla n/n,$$

which in solids generally is a very small quantity. In nonlinear optics the typical change of $|\hat{E}|^2$ occurs over an extension of a wavelength λ, i.e. $\nabla|\hat{E}|^2 \simeq |\hat{E}|^2/\lambda$, whereas $\nabla n \simeq n/L$ with L of the order of $10^3\lambda$ or higher holds; hence

$$\left|\frac{\pi_o - \pi_{so}}{\frac{\varepsilon_o}{4}\chi\nabla|\hat{E}|^2}\right| < \frac{2}{9}\times 10^{-3},$$

so that the difference is perfectly negligible. In other applications, however, and from a fundamental point of view it may be important.

Expression (35) for static fields was reconfirmed by Landau and Lifshitz [25], Becker and Sauter [26] and by Stratton [27]. There exist two main ways of deriving $\vec{\pi}_s$: one starts from a more or less intuitive expression for the Helmholtz free energy density [25], [28]; the other one calculates the electrostatic energy variation in a virtual displacement of a solid [24], as done in this paper. The first method applies locally, but it is not easy to recognize how the partition is made between mechanical stress and strain $\overleftrightarrow{\Pi}$ and the electric force density. The second method needs some partial integrations over large volumes in order to eliminate some undesired terms. Eq.(35) is then obtained by equating the integrands of two integral expressions of equal value. However, this is conclusive only if either V or $\delta\vec{r}$ can be chosen arbitrarily. Neither is the case in [24]. In our derivation the splitting between $\nabla\overleftrightarrow{\Pi}$ and $\vec{\pi}$ is done in such a way that $\overleftrightarrow{\Pi}$ contains all terms which do not depend explicitly on $\hat{E}$. A very detailed discussion of the problem (but unfortunately, no final answer) may be found in ref. [28]. In jellium (i.e. fully ionized Lorentz plasma of solid state density) owing to $\omega_o = 0$ χ is particularly simple in the limiting case, $\chi = -\omega_p^2/\omega^2$, and hence the difference $n\partial\chi/\partial n - \chi$ is zero. We are convinced that $\vec{\pi}_{so}$ cannot be of general validity and we doubt that eq.(32) can be simplified further without specifying particular boundary conditions.

3. Light Pressure Effects

In this section we apply the ponderomotive force concept to explain the parametric 3-wave decay processes. The simplest situation occurs when no eigenmode is excited in matter. Therefore we treat this case first.

3.1 Nonresonant Effects

They are characterized by the fact that the density modulations induced by the wave pressure have zero frerquency and, as a consequence, no three wave matching condition has to be fulfilled. The best known phenomena of this kind are light beam self-focusing and filamentation. Self-focusing is sketched in Fig.4 for a conductor and a dielectric medium. The intensity of a regular light beam is highest on the axis and decreases outwards. This causes a ponderomotive force as indicated by the arrows. In a conductor in which the influence of the free electrons on the refractive index prevails on the bound electrons, the force is directed outwards and a density depression is induced along the beam axis. This in turn leads to a refractive index increase there and to a corresponding reduction in phase velocity $v_\varphi = c/\eta$, which causes the phase plane distortion as indicated. In a simple dielectric with a susceptibility proportional to the polarizability α of the single atom or molecule and $\alpha = const$ it is always such that the beam would collapse on the axis if diffraction did not limit this process. The direction of $\vec{\pi}$ depends on whether $\omega_o > \omega$ (lower case of Fig.4) or $\omega_o < \omega$ holds (upper case). In general the situation is more complicated. According to eq.(33), the inequality

$$(n + \delta n)(\alpha + \delta\alpha) > or < n\alpha$$

determines whether self-focusing or defocusing occurs. Quantitatively, self-focusing in an

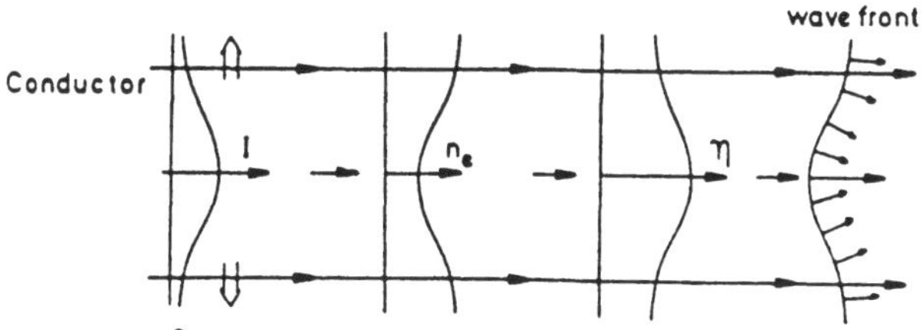

$$\chi = -\frac{\omega_p^2}{\omega^2} = -\frac{n_e}{n_c}, \quad \eta = (1 - \frac{n_e}{n_c})^{1/2} < 1, \quad c_\varphi = \frac{c}{\eta}$$

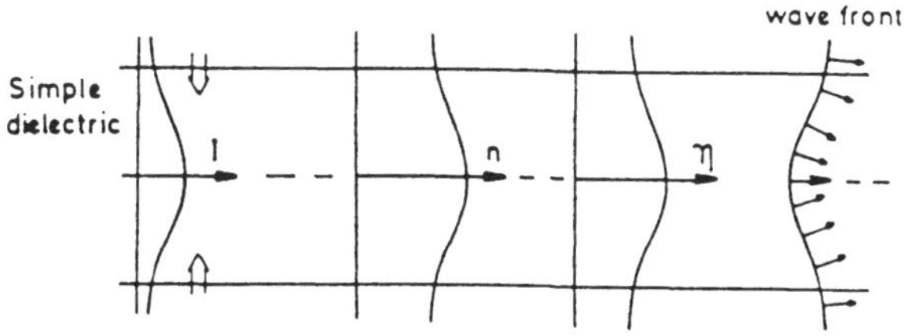

$$\chi = n\alpha, \quad \alpha = const, \quad \alpha \sim \frac{1}{\omega_o^2 - \omega^2}, \quad \omega_o^2 > \omega^2, \quad \chi > 0.$$

Fig.4: Self-focusing in a conductor and a simple dielectric medium. Distribution of intensity I, particle density n, refractive index η across the beam. The phase front distortion in all cases is such that beam focusing occurs.

isotropic solid $\overleftrightarrow{\Pi} = (\overleftrightarrow{I}\,\Pi)$ is described by the equations

$$\nabla\Pi + n\nabla(\frac{\chi}{n}\hat{E}^2) = 0, \quad \nabla^2\hat{E} + \frac{\omega^2}{c^2}\eta^2(\vec{x})\hat{E} = 0; \quad \vec{E} = \hat{E}(\vec{x})e^{-i\omega t}. \tag{36}$$

Other variants of self-focusing are filamentation (filamentary instability) and striation (Fig.5): The light beam breaks up into single beamlets or layers if the intensity is modulated in space as indicated. When $\vec{E}$ can be linearized it is convenient to set

$$n(\vec{x}) = n_o + n_1 e^{i\vec{q}\vec{x}}, \quad \hat{E} = \hat{E}_o e^{i\vec{k}_o\vec{x}} + \hat{E}_1 e^{i\vec{k}_1\vec{x}}, \quad \vec{k}_1 = \vec{k}_o + \vec{q},$$

from which it becomes apparent that self-focusing in its linear (weak) version is a degenerate 3-wave process (acoustic frequency $\omega_a = 0$).

In order to show the usefulness of the ponderomotive concept we consider a density perturbation $n_1(\vec{x})$ along the direction of the beam and ask whether this can be unstable under the condition that the refractive index is given by $\eta^2 = 1 + (n_o + n_1\alpha)$, $n_o = const$,

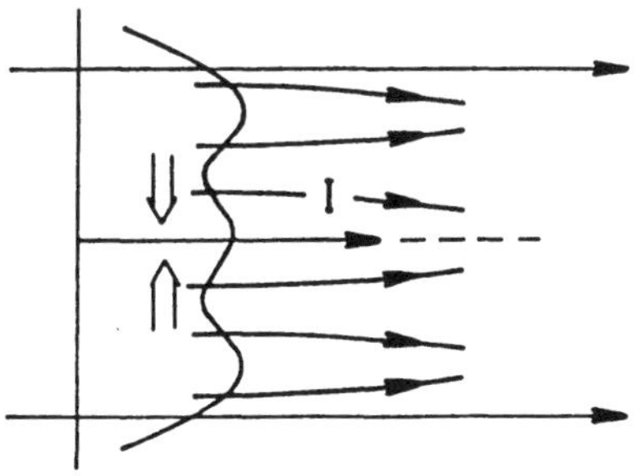

Fig.5: Filamentation and striation are induced by a transverse beam intensity modulation or a modulation of η .

$\alpha = const.$ If the modulation of n_1 has periodicity of at least several wavelengths WKB approximation applies to the amplitude variation of $\hat{E}(\vec{x})$,

$$\hat{E}(\vec{x}) = \frac{\hat{E}_o}{\sqrt{\eta}}; \quad \hat{E}_o = const.$$

Thus, the density modulation induces the ponderomotive force

$$\vec{\pi}_o = \frac{\epsilon_o}{4}\hat{E}_o^2\nabla\frac{1}{\eta} = -\frac{\epsilon_o}{8}\hat{E}_o^2\alpha^2\frac{\nabla n_1}{1+\alpha n_o} \sim -\nabla n_1.$$

Owing to the - sign $\vec{\pi}_o$ is opposite to the gradient of n_1 and exerts therefore a stabilizing effect on a spontaneously (e.g. thermally) excited density fluctuation.

3.2 Stimulated Decay Processes

Ponderomotive force density $\vec{\pi}$ is a gradient force and as such it is highest when two counterpropagating waves or waves propagating obliquely to each other are superposed. In fact, let the two waves be

$$\vec{E}_o = \hat{E}_o e^{i(\vec{k}_o\vec{x}-\omega t)}, \quad \vec{E}_1 = \hat{E}_1 e^{i(\vec{k}_1\vec{x}-\omega_1 t)}.$$

Although the single amplitudes $\hat{E}_o$ and $\hat{E}_1$ may be constant the amplitude of the total wave field $\vec{E}$ is strongly modulated,

$$|\vec{E}| = \{\hat{E}_o\hat{E}_o^* + \hat{E}_1\hat{E}_1^* + \hat{E}_o\hat{E}_1^* e^{i(\vec{k}_o-\vec{k}_1)\vec{x}} + \hat{E}_o^*\hat{E}_1 e^{-i(\vec{k}_o-\vec{k}_1)\vec{x}}\}^{1/2}. \tag{37}$$

In the case of exactly counterpropagating waves $\vec{k}_1 = -\vec{k}_o$ holds and the amplitude oscillates between $\{|\hat{E}_o|^2+|\hat{E}_1|^2-2|\hat{E}_o\hat{E}_1|cos(2\vec{k}_o\vec{x}+\varphi)\}^{1/2}$ and $\{|\hat{E}_o|^2+|\hat{E}_1|^2+2|\hat{E}_o\hat{E}_1|cos(2\vec{k}_o\vec{x}+\varphi)\}^{1/2}$ over half a wavelength. When two waves of different frequencies ω_o and ω_1 are superposed the amplitude modulation propagates at a velocity $v = |\omega_o - \omega_1|/|\vec{k}_o - \vec{k}_1|$. By changing to a reference system moving with v in the $\vec{k}_2 = (\vec{k}_o - \vec{k}_1)$ direction the amplitude modulation again becomes static as in eq.(37) since the Doppler-shifted frequencies ω_o' and ω_1' become equal,

$$\omega_o' = \gamma(\omega_o - \vec{k}_o\vec{v}),\ \omega_1' = \gamma(\omega_1 - \vec{k}_1\vec{v}),\ \omega_o' - \omega_1' = \gamma[\omega_o - \omega_1 - (\vec{k}_o - \vec{k}_1)\vec{v}] = 0. \tag{38a}$$

γ is the Lorentz factor. The $\vec{k}$'s transform as follows:

$$\vec{k}' = \vec{k} + \frac{\gamma-1}{v^2}(\vec{v}\vec{k})\vec{v} - \gamma\vec{v}\frac{\omega}{c^2}. \tag{38b}$$

A standing wave structure in a solid causes a density perturbation and, as a consequence, a modulation of the refractive index η of the same periodicity. In an isotropic solid the equation of state can be described by a pressure p which is a function of particle density n and temperature T. In sufficiently rapid phenomena heat flow may be neglected and n and T are connected by an adiabatic law. Hence p becomes a function of n alone, $p = p(n)$, and the sound velocity s is given by

$$s^2 = \frac{dp}{dn} = \left(\frac{\partial p(n,T)}{\partial n}\right)_\sigma,$$

where the partial derivative is taken at constant entropy density σ. A static amplitude modulation causes a ponderomotive density perturbation n_1 as follows:

$$-\nabla p + \vec{\pi} = -s^2 \nabla n + \frac{\epsilon_o}{4} \chi \nabla |\hat{E}|^2 = 0; \Rightarrow n_1 = +\frac{\epsilon_o}{2} \frac{\chi}{s^2} |\hat{E}_o \hat{E}_1| cos[(\vec{k}_o - \vec{k}_1)\vec{x} + \varphi].$$

The maxima and minima of n_1 and $|\hat{E}|^2$ coincide for $\chi > 0$.

Wave pressure in moving media. In the new reference system matter is moving at velocity $\vec{v}$. To see what now the ponderomotive force density $\vec{\pi}'_o$ is imagine a volume element at rest in this frame of reference. Then eq.(33) applies,

$$\hat{\pi}_o = \frac{\epsilon_o}{4} n' \nabla' (\frac{\chi \omega'_o}{n'} \hat{E}' \hat{E}'^*).$$

Keeping in mind that the quantity $\vec{E}'_s = \hat{\pi}_o / n'q$, q being the effective electronic charge of a particle, is a static electric field it becomes clear that $\vec{\pi}'_o$ is given by the Lorentz formula

$$\vec{\pi}'_o = n'q(\vec{E}'_s + \vec{v} \times \vec{B}'_s) = n'q\vec{E}'_s = \hat{\pi}_o.$$

The last two steps follow from the absence of a static magnetic field $\vec{B}'_s$. To obtain $\vec{\pi}_o$ in the lab frame one has merely to transform n' and $\vec{E}'_s$ according to the standard formulae given in any textbook of electrodynamics. n' transforms like the inverse of a volume and q is a Lorentz invariant. Since $\vec{E}'_s$ depends on $\vec{x}'$, $\vec{\pi}_o$ becomes space- *and* time-dependent. In most applications of non-linear optics in solids γ may be set to unity and hence $n' = n$. Then, if the amplitude $|\vec{E}'|$ is the result of two superposed plane waves $\vec{E}_o(\vec{k}'_o, \omega'_o)$ and $\vec{E}_1(\vec{k}'_1, \omega'_1 = \omega'_o)$, owing to the Lorentz invariance of the phases, $\vec{\pi}_o$ becomes in the lab frame

$$\vec{\pi}_o = \frac{\epsilon_o}{4} n \nabla [\frac{\chi \omega'_o}{n} (\hat{E}_o \hat{E}_1^* e^{i(\vec{k}_o - \vec{k}_1)\vec{x} - i(\omega_o - \omega_1)t} + cc)], \tag{39}$$

i.e. $\vec{\pi}_o$ is a spatially and temporally oscillating force which, when a matching condition like eq.(2) holds, can drive a material wave unstable. In a conductor χ' from eq.(39) is $\chi' = -\omega_p^2 / \omega_o'^2$. ω_p is a relativistic invariant and ω'_o reads in the lab frame, under our condition of $\gamma = 1$, $\omega_o'^2 = (\omega_o - \vec{k}_o \vec{v})(\omega_1 - \vec{k}_1 \vec{v})$, and hence

$$\vec{\pi} = -\frac{\epsilon_o}{4} n \nabla \frac{\omega_p^2 / n}{(\omega_o - \vec{k}_o \vec{v})(\omega_1 - \vec{k}_1 \vec{v})} (\hat{E}_o \hat{E}_1^* e^{i\psi} + cc). \tag{40}$$

If $\vec{k}_1 \simeq -\vec{k}_o$ (e.g. Brillouin scattering) ω'^2 may be replaced by $\omega_o \omega_1$.

In the case the electric field amplitude varies over a length L much larger than the local wavelength γ-values appreciably less than unity may occur in $\vec{\pi}$ when the matter moves sufficiently fast. However, as soon as $\vec{E}$ is the superposition of two or more modes $L \simeq \lambda$ holds and $\vec{\pi}$ loses its meaning as soon as γ deviates considerably from unity. This does not imply that any modulation of $|\vec{E}|$ has to move at speed $v \ll c$. In the section on Raman scattering it will be shown that $v > c$ and simultaneously $\gamma \simeq 1$ may occur since, in matter the velocity to be used in γ is generally different from v in the Doppler formulae in eqs.(38a,b).

Now we are in a position to study ponderomotive modulation when the amplitude structure is moving at speed $\vec{v}$. Particle and momentum conservation require for the local velocity $u(x)$

$$nu = n_o v = const, \quad nu\frac{\partial u}{\partial x} = -s^2 \frac{\partial n}{\partial x} + \frac{\epsilon_o}{4m} \chi \frac{\partial}{\partial x} |\vec{E}_o + \vec{E}_1|^2.$$

With the help of the Mach number $M = u/s$, $M_o = v/s$, and by eliminating n the equations combine to

$$(\frac{1}{M} - M)\frac{\partial M}{\partial x} = -\frac{\epsilon_o \chi}{4mns^2} \frac{\partial}{\partial x} |\vec{E}_o + \vec{E}_1|^2. \tag{41}$$

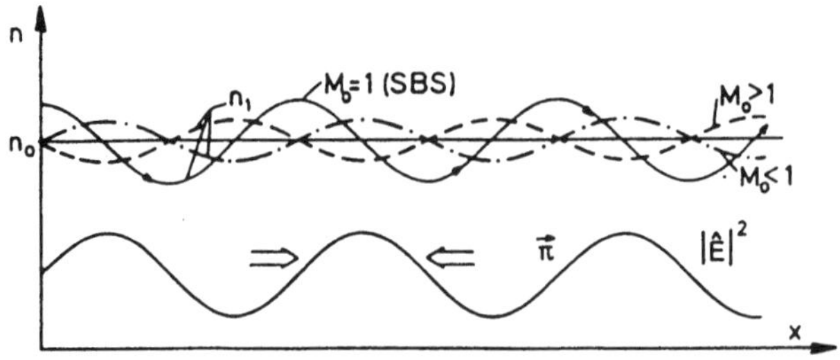

Fig.6: The density n in a solid periodically modulated by the pressure of a partially standing wave. $\chi > 0$, $\hat{E}$, electric field amplitude, n_o average solid density, $n = n_o + n_1(x)$ local solid density, $M_o = v/s$ Mach number. $M_o < 1$: maxima of n_1 in phase with maxima of $\hat{E}^2$; $M_o > 1$: n dephased by π; $M_o = 1$: n dephased by $\pi/2$ (resonance); lower arrows: direction of ponderomotive force for positive χ.

The solutions of this equation are characterized as follows for positive χ (Fig.6). If M is less than unity everywhere a wave field of periodic amplitude modulation produces a stationary density variation the maxima of which coincide with the maxima of $|\vec{E}|^2$ and the modulation amplitude increases with the Mach number increasing. If $\vec{v}$ is supersonic everywhere the density maxima are in phase with the minima of $|\vec{E}|^2$ and decrease with the Mach number increasing. At $M_o = 1$ an infinitesimal modulation of $|\vec{E}|$ (e.g. infinitesimal reflected wave $\vec{E}_r$) is already capable of producing a density modulation for which in the linear regime no steady state exists since $1/M - M = 0$. This is the case of stimulated Brillouin scattering. The dephasing between $|\vec{E}|^2$ and n for $M_o < 1, M_o = 1$, and $M_o > 1$ is the same as that of a linear pendulum driven at $\omega < \omega_o$, $\omega = \omega_o$, and $\omega > \omega_o$.

Stimulated Brillouin scattering. In the lab frame the density modulation n_1 at $M_o = 1$ is the resonantly excited sound wave $(\vec{k}_a, \omega_a)$ which introduces a refractive index modulation. From this inhomogeneity a part of the incident light is scattered. In the frame co-moving with the acoustic wave the acoustic frequency ω_a' is zero; thus, incident and scattered wave must have the same frequency because once the matching conditions eq.(2) are fulfilled in one inertial frame,

$$\vec{k}_o = \vec{k}_a + \vec{k}_1, \quad \omega_o = \omega_a + \omega_1,$$

they hold in any inertial system of reference,

$$\vec{k}_o' = \vec{k}_a' + \vec{k}_1', \quad \omega_o' = \omega_a' + \omega_1'; \quad \omega_a' = 0.$$

The scattered wave is nothing but the Bragg-like reflected wave from the optical inhomogeneity (Fig.7). Since, however, in the co-moving frame matter is moving at $v = -s$ an anisotropy is introduced and, as a consequence, angle of incidence and angle of reflection are equal only in special cases. Normally, $s \ll c$ holds and the angles are nearly equal even in the lab frame. In contrast, in SRS the situation may be different. The equations governing SBS in the SVA approximation are easily deduced. By setting $\vec{E} = \hat{E}_o(\vec{x},t)exp(-i\omega_o t) + \hat{E}_l(\vec{x},t)exp(-i\omega_1 t)$ we find that for the single mode $i = 0, 1$

$$\vec{P} = \sum_i \epsilon_o \chi_{oi} \vec{E}_i + \sum_i \epsilon_o \frac{\partial \chi_{oi}}{\partial n} n_1 \vec{E}_i = \vec{P}_L + \vec{P}_{NL}.$$

Then, after separating according to frequencies and wave vectors (normal mode analysis) eq.(3) splits into

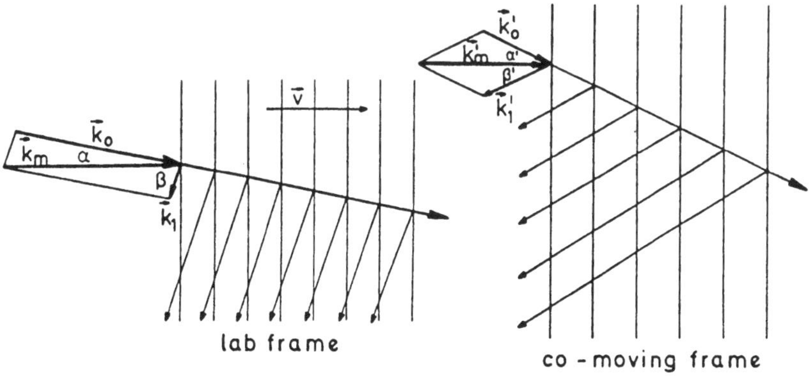

Fig. 7: SBS and SRS in the lab frame and the frame co-moving with the resonantly excited wave in the solid (acoustic phonons and optical phonons or plasmons). In the latter reference system scattered light appears as a wave which is reflected without frequency shift from the inhomogeneity grid induced by light pressure. In SBS α and β are nearly equal.

$$\begin{aligned}\nabla^2 \hat{E}_o + \frac{\omega_o^2}{c^2}(1+\chi_{oo})\hat{E}_o + i\frac{\omega_o}{c^2}\chi_{oo}\frac{\partial \hat{E}_o}{\partial t} &= -\frac{1}{2}\frac{\omega_1^2}{c^2}\frac{\partial \chi_{oo}}{\partial n}\hat{n}_1\hat{E}_1, \\ \nabla^2 \hat{E}_1 + \frac{\omega_1^2}{c^2}(1+\chi_o)\hat{E}_1 + i\frac{\omega_l}{c^2}\chi_{o1}\frac{\partial \hat{E}_1}{\partial t} &= -\frac{1}{2}\frac{\omega_o^2}{c^2}\frac{\partial \chi_{oo}}{\partial n}\hat{n}_1^*\hat{E}_o,\end{aligned} \tag{42}$$

where we have set $n_1 = (\hat{n}_1 exp - i\omega_a t + \hat{n}_1 exp\ i\omega_a t)/2$. The equations governing n_1 are

$$\frac{\partial n_1}{\partial t} + n_o \nabla \vec{u} = 0, \quad n_o \frac{\partial \vec{u}}{\partial t} = -s^2 \nabla n_1 + \frac{\epsilon_o}{4m}\chi_o \nabla |\hat{E}|^2.$$

Eliminating $\vec{u}$ and proceeding as above,

$$\nabla^2 \hat{n}_1 + \frac{\omega_a^2}{s^2}\hat{n}_1 + i\frac{\omega_a}{s^2}\frac{\partial \hat{n}_1}{\partial t} = \frac{\epsilon_o}{8ms^2}\chi_o \nabla^2 \hat{E}_o \hat{E}_1^* \tag{43}$$

follows. The nonlinearity providing for coupling is produced by the light pressure. In the frame co-moving with the acoustic wave it is a secular force.

It should be mentioned that there are 3 situations in which reflection can become large: abrupt change in refractive index at interfaces (i), regions of $\eta = 0$ (may occur in semidonductors or inhomogeneous jellium) (ii), and periodic modulations of η such that the reflected partial waves interfere constructively (iii). SBS and SRS are of the latter type. As soon as a single transition from one η-value to another is smeared out over more than $\lambda/5$ reflection drops drastically [3].

Stimulated Stokes Raman scattering is very similar to SBS, the main difference consisting in the excitation of a mode of generally much higher frequency than ω_a (optical phonon or plasmon). The situation is simplest in a conductor where, besides the phononic normal modes, longitudinal electron plasma waves can be excited. In the simplest case (jellium) their dispersion is given by the Bohm-Gross relation in terms of electron plasma frequency ω_p and electron sound speed s_e,

$$\omega_e^2 = \omega_p^2 + s_e^2 \vec{k}_e^2. \tag{44}$$

The ions are not involved in the collective motion. It is interesting to note that the electromagnetic dispersion relation in such a conductor is obtained by replacing s_e by the vacuum speed of light c in eq.(44). The product of group and phase velocities is $v_g v_\varphi = s_e^2$ and $v_g v_\varphi = c^2$, respectively. Now, frequency downshifted Raman scattering works in exactly the same way as Brillouin scattering: The incident and Stokes waves produce a ponderomotive force travelling with the plasma wave and amplifying it. This wave represents a periodic refractive index modulation

$$\eta = (1+\chi)^{1/2} = (1 - \frac{\omega_e^2}{\omega^2})^{1/2} \simeq \eta_o - \frac{e^2}{2\epsilon_o m_e \eta_o} n_1,$$

where n_1 this time represents the electron density disturbance. With this Figs.6 and 7 hold again if we set $M = u_e/v_{\varphi e}$ and $|\hat{E}|^2$ is shifted by $kx = \pi$ in Fig.6. Again, when travelling at the phase speed of the electron plasma wave ω_e' is zero and incident and Stokes waves assume the same frequency. In this frame also Raman scattering, generally classified as inelastic, is nothing but elastic scattering like classical reflection of light from a mirror. There is no physics in the distinction between elastic and inelastic scattering in an infinitely extended homogeneous medium; the difference is of kinematic nature only. The eqs. governing SRS in jellium are given again by eqs.(42) and (43) if s, ω_a^2, ω_a, m and χ_o are replaced by s_e, $\omega_o^2 - \omega_p^2$, ω_e, m_e and by $-\omega_p^2/\omega_o'^2$.

In SRS in jellium ω_o can be nearly as low as $\omega_o = 2\omega_p$. Then $\vec{k}_e$ of the plasma wave is nearly zero and $v_{\varphi e} = s_e^2/(\omega_e/k_e)$ may exceed c. In applying the relativistic Doppler formulae (38a,b) in a medium the question arises which value of v has to be chosen in $\gamma = (1 - v^2/c^2)^{-1}$ when changing to the rest frame of the plasma wave. A little bit of concentration leads to the correct answer. In the co-moving system ω_o' and ω_1' have to be equal, $\omega_o' = \omega_1'$. When the density of the solid is thought to reduce gradually to zero the relation $\omega_o' = \gamma(v)[\omega_o - (\vec{k}_o/\eta)(v_\varphi \eta)] = const$ holds everywhere including the vacuum. Hence $v = v_{\varphi e}\eta$, since eqs.(38a,b) are defined in the vacuum.

It results from our treatment that in SBS and SRS coupling of the electromagnetic waves to the corresponding material mode in isotropic media occurs in the directions of $\vec{k}_a$ and $\vec{k}_e$, respectively. This however does not exclude transverse phonons from being Brillouin or Raman active since, in general, χ is a tensor and $\vec{\pi}$ does not point in the $\vec{k}_a$ or $\vec{k}_e$ direction. In the anisotropic medium it is therefore necessary or at least convenient to expand $\overleftrightarrow{\chi}$ in normal displacement coordinates and not in n. Finally, one may ask what the effect would be of a dephasing angle $\vartheta = \pm\pi/2$ between $\vec{\pi}$ and n_1 of the material mode instead of $\vartheta = 0$ for positive feedback (see Fig.6). To answer the question we look at the (linearized) momentum equation with $\vartheta = \pm\pi/2$,

$$n_o \frac{\partial \vec{u}}{\partial t} = -s^2 \nabla n_1 \mp \frac{\epsilon_o}{4} \frac{\chi_o}{m} \nabla (\frac{|\vec{E}|^2}{n_1} n_1) \simeq -(s^2 \pm \frac{\epsilon_o}{2} \frac{\chi_o}{m} \frac{\hat{E}_o \hat{E}_1^*}{n_1}) \nabla n_1 = -s_1^2 \nabla n_{\prime}.$$

Hence, in such a case the ponderomotive force contributes to accelerate or decelerate the wave instead of increasing its amplitude. As a consequence, SBS and SRS active modes saturate when the phase mismatch between $\vec{\pi}$ and n_1 approaches $\vartheta = \pm\pi/2$.

PDI and OTSI: a 20 year old paradox. Both processes occur in conductors (jellium) and plasmas and consist in the stimulated decay process

photon $\Longrightarrow$ phonon + plasmon,

$$\hbar\omega_o = \hbar\omega_a + \hbar\omega_e.$$

When all frequencies are positive the process is named parametric decay instability (PDI);

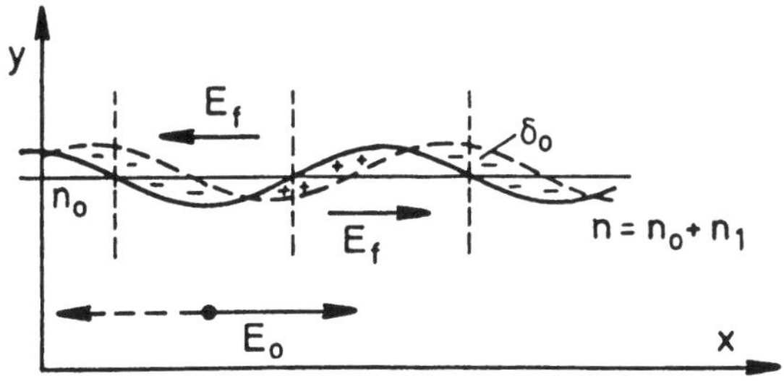

Fig.8: Oscillating two-stream instability. The laser beam is propagating in the z-direction with $\vec{E}_o$ pointing along x. In the presence of an ion density modulation n an electric field $\vec{E}_f$ is produced by charge separation of magnitude δ_o. The amplitude $|\vec{E}_o + \vec{E}_f|$ is in phase with n and drives it unstable if $\omega_o < \omega_e$.

it was the first of its type [29]. Three years later a variant of it was found and named the oscillating two-stream instability (OTSI). It occurs when ω_e is larger than ω_o and the real part of ω_a is zero. For the OTSI no perfect frequency matching is possible. The $\vec{k}$-vectors fulfill the condition $\vec{k}_o = \vec{k}_a + \vec{k}_e$. Since in general $\vec{k}_o$ is much less in magnitude than $\vec{k}_a$ and $\vec{k}_e$ it can be set equal to zero. It was exactly for these two processes that F.F. Chen proposed an explanation in terms of light pressure [30]. The geometry for PDI and OTSI to occur is seen from Fig.8. An electromagnetic wave $\vec{E}_o$ is impinging in the z-direction. When an acoustic wave $n_1(\omega_a, \vec{k}_a)$ (PDI case) or simply a static ion modulation $n_1(\omega_a = 0, \vec{k}_a)$ is present with $\vec{k}_a$ along $\vec{E}_o$ an electron plasma wave $n_{e1}(\omega_e, \vec{k}_e)$ is excited of the same wave number as k_a and grows unstable. In the frame in which n_1 is at rest, which for OTSI is also the lab frame, a space charge is introduced by the presence of the pump wave $\vec{E}_o$ since the electron density, which in the absence of $\vec{E}_o$ is $n_{e1} = n_1$, is periodically shifted in $\pm x$-direction by the amount

$$\delta_o = \frac{e}{m_e \omega_o'^2} \hat{E}_o e^{-i\omega_o' t}.$$

This disturbance gives rise to a forced electric field $\vec{E}_f$ the strength of which is governed by Poisson's equation

$$\frac{\partial E_f}{\partial x} = -\frac{e}{\epsilon_o}[n_1(x - \delta_o) - n_1(x)] \simeq \frac{e}{\epsilon_o}\delta_o \frac{\partial n_1}{\partial x};$$

thus

$$\vec{E}_f = \frac{e}{\epsilon_o}\vec{\delta}_o n_1. \tag{45}$$

$\vec{E}_f$ is directed as in the figure and the product $\hat{E}_o \hat{E}_f^*$ is in phase with the acoustic disturbance n_1. At this point the paradox arises when $\vec{\pi}$ for free electrons is calculated since

$$\vec{\pi} = -\frac{\epsilon_o}{4}\frac{\omega_p^2}{\omega_o^2}\nabla|\vec{E}_o + \vec{E}_f|^2 = -\frac{\epsilon_o}{4}\frac{\omega_p^2}{\omega_o^2}\nabla(\hat{E}_o \hat{E}_f^* + cc) \tag{46}$$

holds and hence $\pi \sim -\nabla n_1$, instead of amplifying n_1, stabilizes it [31], whereas the correct theory predicts instability for the given situation [32]. In addition, a correct theory has to explain the following features [33]:

(i) in the case of PDI ($\omega_o > \omega_e$), n_1 is shifted by $\vartheta = \pi/2$ with respect to $\hat{E}_o \hat{E}_f^*$;
(ii) growth rate γ in the linear regime is zero at exact resonance $\omega_o = \omega_e$;
(iii) γ is the same for n_1 as well as for n_{e1}.

Point (ii) is particularly surprising on the basis of ponderomotive action on free electrons. However, the electrons are not free, and this is the key to the solution of the paradox! To see this the reader may consider the motion of a single electron in an electron plasma wave of wave number k_e. Since the density perturbation is governed by

$$\frac{\partial^2}{\partial t^2} n_{e1} + (\omega_p^2 - s_e^2 \frac{\partial^2}{\partial x^2}) n_{e1} = 0$$

and the particle velocity is $u_o = \dot{\delta}_e = v_\varphi n_{e1}/n_o$, $v_\varphi = \omega_e/k_e$, the electron represents a harmonic oscillator with resonance frequency ω_e given by eq.(44),

$$\delta_e + \omega_e^2 \delta_e = 0.$$

The product $\delta_o n_1$ from eq.(45) acts as a driver of spatial and temporal periodicities k_a and ω_o on the wave (k_a' is very close or equal to k_a) ,

$$\ddot{\delta}_e + \omega_e^2 \delta_e = -\frac{e}{m_e} E_f, \tag{47}$$

as is easily derived from linearized particle conservation, momentum and Poisson's equations. At the mismatch $\Delta' = \omega_o - \omega_e' \neq 0$ $\delta_e(x,t)$ exhibits a steady state solution at frequency ω_o, wavenumber $k_e = k_a$ and with an amplitude $\hat{\delta}_e$ proportional to n_1. Hence, the growth rate γ is necessarily the same for n_1 and n_{e1} (iii); however, their amplitudes $\hat{n}_1$ and $\hat{n}_{e1}$ differ by the ratio

$$\frac{\hat{n}_{e1}}{\hat{n}_1} = \frac{\omega_p^2}{|\Delta|(\omega_o + \omega_e')} k_e \hat{\delta}_o.$$

With regard to eqs.(47) and (23), due to the presence of the electron plasma wave, the ponderomotive force is given by

$$\vec{\pi} = -i\frac{\epsilon_o}{4} \frac{\omega_p^2}{\omega_o^2 - \omega_e'^2} \vec{k}_e (\hat{E}_o \hat{E}_f^* - cc). \tag{48}$$

For $\omega_e' > \omega_o$ it predicts instability (the sign has changed relative to expression (47)), in agreement with ref. [32], whereas stability follows for $\omega_o > \omega_e'$. Remembering that at exact resonance $\omega_o = \omega_e'$ $\vec{\pi}$ is zero, point (ii) is also clarified in the most immediate way. Inserting $\vec{\pi}$ from eq.(48) in the ion momentum equation and using mass conservation to eliminate u straightforwardly leads to

$$\frac{\partial^2}{\partial t^2} n_1 = \left\{ s^2 - \frac{\epsilon_o}{2} \frac{\omega_p^2}{\omega_o^2} \frac{|\hat{E}_o|^2}{m_e m_i (\omega_e'^2 - \omega_o^2)} \right\} \frac{\partial^2}{\partial x^2} n_1. \tag{49}$$

The interesting unstable domain for ω_o in the OTSI case is close to $\omega_e' = \omega_e$, which in turn, owing to Landau damping, is close to ω_p. With these approximations the dispersion relation

$$\frac{\gamma^2}{k_a^2} = \frac{e^2}{4 m_e m_i \omega_o (\omega_e - \omega_o)} |\hat{E}_o|^2 - s^2, \quad (n_1 \sim e^{\gamma t}) \tag{50}$$

is deduced from eq.(49), in perfect agreement with the standard theory [32], if there the damping coefficients are suppressed in the denominator.

At first glance the PDI seems to contradict our assertion since it occurs at $\Delta > 0$ as well as at $\Delta < 0$ but not at exact resonance. However, one has to keep in mind that the concept of light pressure as a secular force applies to a reference system in which the perturbation is static and that in the PDI n_1 is an acoustic mode of frequency $\omega_a = s k_a$ in the lab frame. If one transforms to a system co-moving with the acoustic mode at sound speed, $\partial^2/\partial t^2$ of eq.(49) must be replaced by $d^2/dt^2 = (\partial/\partial t + sk_a)^2$. With this the correct dispersion equation for PDI follows again. The physical reason for such a behaviour is the following. If $0 < \Delta < \omega_a$ holds in the lab frame the observer co-moving with the ion disturbance sees an upshift in frequency from ω_{es} to $\omega_{es} + \omega_a$ and $\Delta' = \omega - \omega'_{es} < 0$, when he looks at the electron plasma wave moving opposite to the sound wave. Thus, in the co-moving system the situation of OTSI is restored for the counterpropagating plasma mode. However, in contrast to the OTSI case the electrostic mode moving to the right is not amplified. Furthermore, we know that in an unstable static structure $|\vec{E}|^2$ must be in phase with $n_1(x)$ when the eigenfrequency of the oscillator is higher than that of the driver, in agreement with the statements above. In the situation of of PDI a travelling structure is amplified which means that now π must be in phase with $n_1(x)$, as indicated by Fig.6 for $M = 1$. With this point (i) is explained, too. Comparing the analysis given here in terms of $\vec{\pi}$ with the standard treatment, e.g. ref.[32], its advantage becomes clear immediately. We also observe that there is a continuous transition from OTSI to filamentation with decreasing wave number k_a. When $2\pi/k_a$ becomes larger than approximately $\lambda_o = 2\pi/k_o$ plasmon excitation becomes ineffective and the only transverse instability we are left with is filamentation.

Finally, the two plasmon decay (TPD) process works in the same way as SBS and SRS. The first of the two electron plasma waves, when superposed on the pump wave, generates a modulation of $|\vec{E}_o + \vec{E}_1|^2$ which, if eqs.(2) are fulfilled, propagates with the phase speed of the second plasma wave into $\vec{k}_{e2}$-direction and drives it unstable. Since there is no preference between the two plasma waves the 2nd one combines with the pump to excite the 1st wave in exactly the some way. The analysis in terms of wave pressure is straightforward [34] and shows that growth is fastest when the two plasma waves emanate under angles of approximately 45 and 135 degrees with respect to $\vec{k}_o$ (Fig.9).

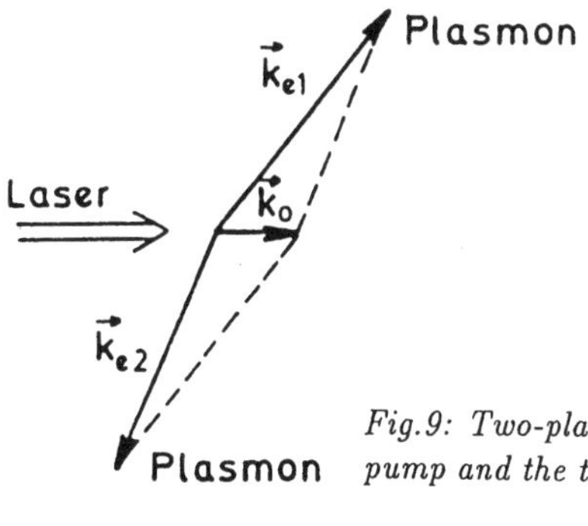

Fig.9: Two-plasmon decay. $\vec{k}_o$, $\vec{k}_{e1}$ and $\vec{k}_{e2}$ wave vectors of the pump and the two unstable plasma waves; $|\vec{k}_{e1}| \simeq |\vec{k}_{e2}| \gg \vec{k}_o$.

4. Conclusion

The analysis of nonlinear optical effects presented here tries to show that a unified picture of a whole class of parametric phenomena can be given on the basis of radiation and wave pressure. Not only is the analysis more concise and shorter (i), it is also much more intuitive (ii) and yields a qualitative basis for predicting stable or unstable behaviour on simple considerations (iii). Finally, it permits an approximate description of unstable optical phenomena when the waves become nonlinear (iv).

References

[1] R.P. Godwin, Phys. Rev. Letters 28, 85 (1972).

[2] J.P. Freidberg, R.W. Mitchell, R.L. Morse, and L.I. Rudsinski, Phys. Rev. Letters 28, 795 (1972).

[3] P. Mulser, J. Opt. Soc. Am. B 2, 1814 (1985).

[4] P.N. Lebedev, Ann. Phys. 6, 433 (1901).

[5] E. Nichols and G.F. Hull, Ann. Phys. 12, 225 (1903).

[6] W. Gerlach and A. Golsen, Z. Phys. 15, 1 (1923).

[7] H.A.H. Boot, S.A. Self, and R.B.R. Shersby-Harvie, J. Electron. Control 4, 434 (1958).

[8] A.V. Gapunov and M.A. Miller, Sov. Phys. JETP 7, 168 (1958).

[9] T.W.B. Kibble, Phys. Review 150, 1060 (1966).
H. Hora, D. Pfirsch, and A. Schlüter, Z. Naturforsch. 22a, 278 (1967).

[10] F.A. Hopf, P. Meystre, M.O. Scully, and W.H. Louisell, Opt. Comm. 18, 413 (1976); Phys. Rev. Letters 37, 1342 (1976).
R.W. Müller, Averaged Effect of Strong-Focusing Fields, in Annual Report GSI-88-17 (ISSN 0171-4546), Darmstadt, 1988, p.26 .

[11] L.D. Landau and E.M. Lifshitz, *Mechanics*, Pergamon, Oxford, 1976; p. 93.

[12] J. Kupersztych, Phys. Rev. Letters 54, 1385 (1985). R.D. Brooks and Z.A. Pietrzyk, Phys. Fluids 30, 3600 (1987).

[13] D.A. D'Ippolito and J.R. Myra, Phys. Fluids 28, 1895 (1985). B.I. Cohen and Th.D. Rognlien, Phys. Fluids 28, 2793 (1985). Ph.L. Similon, A.N. Kaufman, and D.D. Holm, Phys. Fluids 29, 1900 (1986).

[14] See: The Mechanical Effects of Light, in J. Opt. Soc. Am. B2, Nr. 11 (1985), p. 1751, 1776.

[15] E.L. Raab, M. Prentiss, A. Cable, S. Chu, and D.E. Pritchard, Phys. Rev. Letters 59, 2631 (1977). F. Diedrich, E.Peik, J.M. Chen, W. Quint, and H. Walther, Phys. Rev. Letters 59, 2931 (1987).

[16] J. Dalibard and C. Cohen-Tanoudji, J. Opt. Soc. Am. B2, 1707 (1985).

[17] B.W. Boreham and B. Luther-Davies, J. Appl. Phys. 50, 2533 (1979). W. Becker, R.R. Schlicher, M.O. Scully, and K. Wodkiewicz, J. Opt. Soc. Am. B4, 743 (1987).

[18] K. Lee, D.W. Forslund, J.M. Kindel, and E.L. Lindmann Phys. Fluids 20, 51 (1977). C. Max and C. McKee, Phys. Rev. Letters 39, 1336 (1977). P. Mulser and C. van Kessel, Phys. Rev. Letters 38, 902 (1977). O. Willi and P.T. Rumsby, Opt. Comm. 37, 45 (1981). W.B. Mori, C. Joshi, J.M. Forslund, and J.M. Kindel, Phys. Rev. Letters 60, 1298 (1988).

[19] Such a decomposition is not unique in rigorous mathematical terms, however, the arbitrariness is small as long as the amplitude $\hat{\vec{E}}(x)$ changes slowly over one wavelength and v_o is much less than the phase velocity of the wave. In the opposite case the concept of ponderomotive force becomes meaningless.

[20] L.D. Landau and E.M. Lifshitz, *The Classical Theory of Fields*, Pergamon, Oxford, 1980, p. 118.

[21] M.L. Sawley, J. Plasma Phys. 32, 487 (1984).

[22] M. Kruskal, J. Math. Phys. 3, 806 (1962).

[23] W. Schneider, Elektronenbeschleunigung durch inhomogene Langmuirwellen hoher Amplitude, Thesis, Tech. Hochschule Darmstadt, 1984 (unpublished). Part of it is published in P. Mulser and W. Schneider, Excitation of Nonlinear Electron Plasma Waves and Particle Acceleration by Laser, in *Twenty Years of Plasma Physics*, B. McNamara ed., World Scientific, Philadelphia, 1985, p. 280.

[24] W.K.H. Panowsky and Melba Phillips, *Classical Electricity and Magnetism*, Addison-Wesley, Reading Mass., 1962 Sec.6-6.

[25] L.D. Landau and E.M. Lifshitz, *Electrodynamics of Continuous Media*, Pergamon, Oxford 1981, Secs.15, 16.

[26] R. Becker and F. Sauter, *Electromagnetic Fields and Interactions*, Dover, New York, 1982, Vol.I, Sec.35.

[27] J.A. Stratton, *Electromagnetic Theory*, McGraw-Hill, New York, 1941, Sec.2.22.

[28] P. Penfield and H.A. Haus, *Electrodynamics of Moving Media*, M.I.T. Press, Cambridge, Mass., 1967, chaps. 7 and 8.

[29] V.P. Silin, Sov. Phys. JETP 21, 1127 (1965).

[30] F.F. Chen, *Introduction to Plasma Physics*, Plenum Press, New York, 1976, p.264.

[31] There exists another qualitative explanation in the literature which at first glance works in the OTSI case; however, it is inconsistent and when applied to the PDI it fails. Therefore we do not follow this attempt further.

[32] K. Nishikawa, J. Phys. Soc. Japan 24, 916 and 1152 (1968).

[33] P. Mulser, A. Giulietti, and M. Vaselli, Phys. Fluids 27, 2035 (1984).

[34] P. Mulser, Ponderomotive Force Effects in Laser-Plasma Interaction, in *Inertial Confinement Fusion*, EUR 11930 EN, eds. A. Caruso and E. Sindoni, Bologna, 1989, p.54.

Part IV

Guiding Structures: Surfaces and Fibres

Light Scattering from Solid Surfaces

F. Nizzoli

Department of Mathematics and Physics, University of Camerino, I-62032 Camerino, Italy

Abstract. We review the field of Brillouin and Raman scattering in opaque media, where the surfaces, or the interfaces, play an essential role, both in activating new scattering channels and in displaying the density of states of surface phonons. We illustrate the main features of the experimental techniques and the bases of the theoretical approach. Several examples are presented, by comparing experimental and calculated spectra. The cases considered include light scattering from surfaces, supported and unsupported films.

1. Introduction

The interaction between light and vibrational excitations in condensed matter has been investigated early in this century by Brillouin [1], Raman [2] and Landsberg abd Mandelstam [3]. The subject received great attention with the advent of the laser. Since then light scattering has been used to study the elastic, vibrational and optical properties of transparent solids. A step forward occurred with the refinement of the experimental techniques, allowing measurements on opaque samples and the detection of surface excitations.

These latter developments opened new fields: the small penetration depth of light in opaque media makes light scattering sensitive to surface excitations, like surface acoustic waves, surface optical phonon polaritons and spin waves. These spectroscopies have been widely used since the 70's to study the physical properties of surfaces, thin films, interfaces and multilayer materials.

We will deal exclusively with the detection of thermal excited phonons. The external generation of surface modes is not considered here. In light scattering from thermal noise the modes are excited by statistical fluctuations and there is no net transport of energy via these modes from one part of the sample to another. Hence these modes occur as standing waves. A second major difference, with respect to externally generated phonons, is that it is not the propagation distance that one can detect experimentally, but rather the phonon lifetime. Since these modes are excited by the fluctuations, they correspond to the true normal modes of the sample. In this way the detail of the phonon density of states contributes to the information obtained from a Brillouin or Raman scattering experiment.

In this paper we focus our attention on Brillouin scattering from acoustic phonons and Raman scattering from surface phonon polaritons, with a special emphasis on the former case. Section 2 is devoted to the illustration of the experimental techniques used in surface Brillouin scattering. In section 3 the surface acoustic modes of semi-infinite crystals and the guided modes of films are reviewed. The main differences between Brillouin scattering from bulk modes in transparent and opaque media are summarized in section 4. The basic theory of Brillouin scattering from surface modes in opaque crystals is presented in section 5. The generalization of the theory to supported films is also outlined. A few representative experimental results are reviewed and explained in terms of the theory in section 6. The surfaces of semiconductors and metals are considered. Several case studies for a variety of films, including layered and

Springer Series in Wave Phenomena, Vol. 9 **Nonlinear Optics in Solids**
Editor: O. Keller

free-standing films are presented. Section 7 covers Brillouin scattering in the presence of surface plasmons when the strong electric field in the proximity of the boundary causes an enhancement of the scattered light intensity. Finally surface Raman scattering is considered in section 8.

2. Experimental Techniques for Surface Brillouin Scattering

Brillouin scattering is generally carried out using a Fabry-Perot interferometer. In fact it offers the best combination of high resolution and good throughput. A high resolution spectrometer is required because the scattered light lies close in frequency to the elastically scattered contribution (Brillouin doublets have frequencies up to about 150 GHz). We will discuss in this section how to overcome difficulties associated with this instrument, such as the limited contrast and the overlapping of neighbouring interference orders. The problems related to the latter point are especially relevant in light scattering from films, where the number of measurable excitations can be very large due to the presence of guided modes.

2.1 The Fabry-Perot Interferometer

The Fabry-Perot interferometer is used as a scanning spectrometer. In most applications, a plane parallel Fabry-Perot (PFP) is used.

The PFP consists of two very flat mirrors mounted accurately parallel to each other with a spacing L_1 which may be varied. The instrument will transmit light of wavelength λ_1 if the spacing L_1 is such that

$$L_1 = \frac{m\lambda_1}{2},$$

where m is an integer. The instrument acts as a tunable filter whose peak transmission is close to unity over a narrow spectral interval, falling to a very low value outside this interval. Two incident signals of wavelength λ_1 and $\lambda_1+\Delta\lambda$ will be simultaneously transmitted (in adjacent interference orders) if

$$m\lambda_1 = (m-1)(\lambda_1+\Delta\lambda).$$

In other words, neighboring orders of interference are separated in frequency by $1/2L$ cm^{-1}. This interorder spacing is called the free spectral range (FSR). The width of the trasmission peak determines the resolution of the instrument. The ratio of the FSR to width is known as the finesse F.

The finesse is primarily a function of the mirror reflectivity, although instrumental aperture and mirror flatness are also important parameters. For a discussion of the Fabry-Perot, see, for example, the review article by Jacquinot [4]. In practice, the finesse is limited to values less than about 100 and this places an upper limit on the possible contrast, where the contrast C is the ratio of maximum to minimum transmission given by

$$C = 1 + \frac{4F^2}{\pi^2} \simeq \frac{4F^2}{\pi^2} \leq 10^4.$$

It is apparent that this contrast will be insufficient for measuring in situations where the elastically scattered component of the scattered light exceeds the intensity of the Brillouin component by more than a factor of 10^4 to 10^5. For backscattering measurements on opaque materials, this is generally the case and so some means of increasing the spectral contrast is essential.

2.2 High-contrast Multipass Interferometer

In a Brillouin spectrum it is the elastically scattered peak which is intense. If this peak were not present, a low contrast spectrometer would be adequate for resolving the spectrum. Various approaches have been proposed to improve the contrast [5]. The best solution lies in the use of multiple or multiply-passed interferometers. This is achieved by placing two or more interferometers in series, but it is not straightforward to synchronize the scans of the separate units. An equivalent and more elegant technique is to pass the light two or more times through the same interferometer - in this case, provided that the mirrors remain parallel, all passes are identical and so the problem of synchronization is obviated. The design aspects of a multipass interferometer have been discussed by many authors [6-8]. It has been demonstrated that a five pass interferometer can achieve a contrast of greater than 10^9, five or six orders of magnitude greater than that of a single interferometer. At the same time, the peak transmission (~50%) and finesse (50-100) stay comparable.

2.3 The Problem of Adjacent Interference Orders: Tandem Interferometer

In opaque materials the problem arises of measuring spectra containing many features, extending out to some maximum frequency Ω_{max}. The peaks can be unambigously assigned only if the free spectral range is chosen such that $FSR > 2\Omega_{max}$. This upper bound placed on the resolution combined with the fixed finesse means that low energy peaks in the spectrum may not be resolved from the elastically scattered light. If the FSR is reduced so that the low energy peaks can be resolved, the higher energy peaks overlap with the equivalent ones measured in the next interference order so that interpretation of the spectrum becomes difficult.

Since the finesse cannot be significantly increased (typically $F \leq 100$) without suffering a serious drop in throughput, the only way to solve the problem is to artificially increase the FSR by combining two or more interferometers of unequal mirror spacing. The first interferometer of spacing L_1 transmits wavelengths

$$\lambda_1 = \frac{2L_1}{m_1} \qquad \text{for integral } m_1,$$

while the second interferometer of spacing L_2 transmits wavelenghts

$$\lambda_2 = \frac{2L_2}{m_2} \qquad \text{for integral } m_2.$$

Only if $\lambda_1 = \lambda_2$ will light be transmitted through the combination. In this way it is possible to increase the FSR by a factor 10 to 20 over that of the single interferometer, although as shown in fig.1 small ghosts remain of the suppressed orders.

The practical limitation in the use of tandem interferometers has been due to the problem of scan synchronization [5,9]. Thus, to scan the transmitted wavelength, it is necessary to increment the mirror spacings L_1 and L_2 by δL_1 and δL_2 such that

$$\frac{\delta L_1}{\delta L_2} = \frac{L_1}{L_2}. \tag{1}$$

A simple and flexible technique is described below which achieves synchronization of two or more interferometers based on a single translation stage common to all interferometers. Such a design is due to Sandercock [5]. The principle of the construction of the Sandercock-type interferometer is illustrated in fig.2 and fig.3. A scanning stage consisting of a deformable parallelogram rides on top of a roller translation stage. The former, actuated by

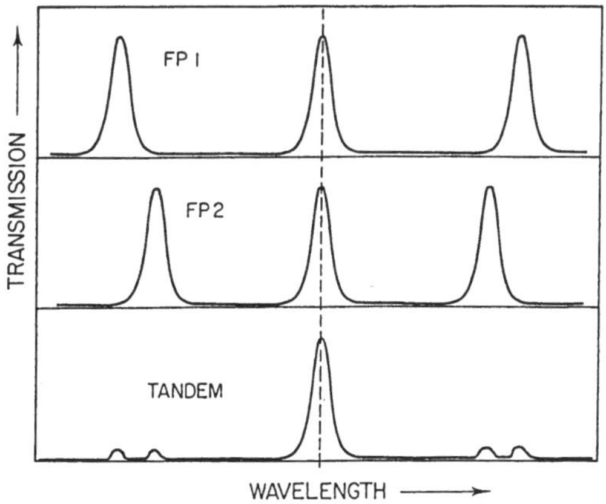

Fig.1 Elimination of neighboring interference orders in a tandem arrangement of two unequal Fabry-Perot interferometers. Only the ghosts of neighboring orders remain. After Sandercock [5]

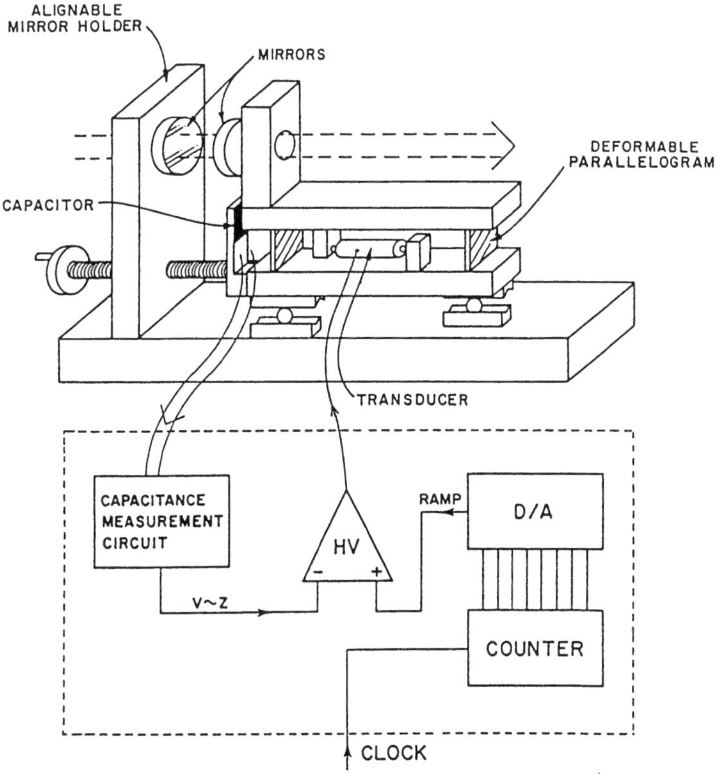

Fig.2 The construction principle of a practical design of interferometer for Brillouin scattering. After Sandercock [5]

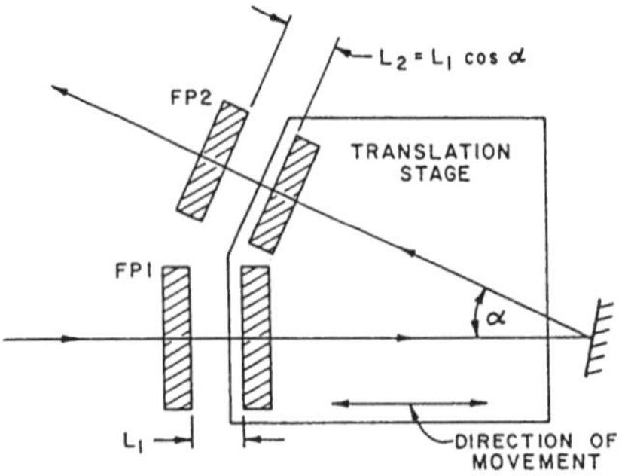

Fig.3 A translation stage designed to synchronize automatically the scans of two tandem interferometers, viewed from above. After Sandercock [5]

a piezoelectirc transducer, provides completely tilt-free movement of the interferometer mirrors over scan lengths up to 10 µm or more. The latter enables the coarse mirror spacing to be set to the desired value in the range 0–50 mm. The roller translation stage is sufficiently collinear that a movement of several mm leaves the mirror aligned parallel to better than 1 fringe across the total mirror diameter. A novelty of the construction is the use of a small parallel plate capacitor to measure the scan displacement. The associated electronics produces a voltage accurately proportional to the capacitor spacing, this voltage being used in a feedback loop in order to linearize the scan displacement with respect to the applied scan voltage. Furthermore, it is seen from fig.2 that the mirror spacing is related to the capacitor spacing only through the length of the translation stage screw. Thus, provided this screw is made of low-expansion material, the mirror spacing will be held by the feedback loop invariant to temperature changes, even though expansion occurs in other parts of the system. The instrument is built entirely of aluminium and cast iron and yet is highly stable. The advantages of this construction system are the following: completely tilt-free scan, highly linear scan (less than 5 Å nonlinearity over 5 µm scan), ability to change mirror spacing without losing alignment, stable against temperature change despite simple construction of aluminium and cast iron.

By a slight modification to the scanning stage of the instrument depicted in fig.2, it is possible to make a synchronously scanning tandem interferometer [5]. The resulting scanning stage viewed from above is shown in fig.3. The scanning mirrors of the interferometers are mounted on the same scanning stage, one with the mirror axis parallel to the scan direction, the other offset by an angle α. The scan of the second interferometer therefore has a light shear although this has no effect on its use as an interferometer. It is clear that the spacings of the two interferometers satisfy

$$L_2 = L_1 \cos\alpha$$

and that the synchronization condition (1) is satisfied. Other technical problems related to the Fabry-Perot interferometer described here have been discussed in the literature, such as: vibration isolation [10], alignment [11] and stabilization [12].

3. Surface Acoustic Modes

Brillouin scattering probes long wavelength acoustic phonons. In this limit, the theory of elasticity is fully adequate to describe the dispersion relations and the vibration amplitudes of the modes. Surface waves in solids are reviewed in detail in many books and articles, e.g. by Auld [13], Farnell [14] and Farnell and

Adler [15]. In the literature emphasis is placed on genuine surface modes, localized at a surface or in a thin supported film. More recently the measurement of Brillouin spectra, directly related to the phonon density of states, has aroused new interest upon the so-called "mixed modes" (combination of evanescent waves and bulk modes) which form a continuous spectrum. The pseudo Rayleigh waves well known in acoustics [14] actually belong to this part of the spectrum. In this section we present a derivation of the surface acoustic modes, belonging to both the discrete and the continuous spectra.

3.1 Semi-Infinite Medium

Let us consider a semi-infinite medium in the half space z<0 with the surface in the x-y plane (fig.4). We assume the surface phonon wavevector **Q** directed along the x-axis. In this reference frame we have phonons of longitudinal polarization L (along the x-axis), of shear-vertical polarization SV (along the z-axis) and of shear-horizontal polarization SH (along the y-axis). In the following we will consider very often, for simplicity, wave propagation along a high symmetry direction of the crystal, such that the sagittal modes (mixture of L and SV modes) are decoupled from the SH modes. Actually this happens in all the applications presented in the sections below. Therefore our treatment is confined in the x-z plane, since the strongest and most interesting features in Brillouin spectra are due to sagittal excitations.

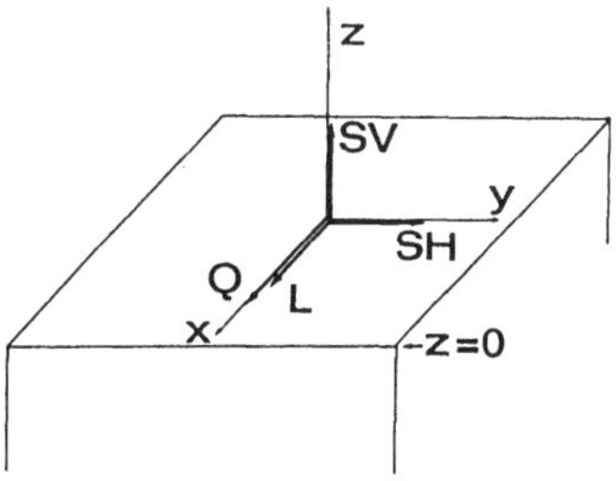

Fig.4 A sketch of the surface of a semi-infinite crystal and of the reference frame. Q is the surface phonon wavevector. Phonon polarizations: L (longitudinal), SV (shear-vertical), SH (shear-horizontal)

The polarization vector **w** of the semi-infinite crystal can be found with the partial wave method [14]. In other words **w** is written as a linear combination of the polarization vectors **e** of the infinite medium. This assumption is perfectly valid in the long wavelength limit, where the bulk phonons "feel" the surface only as a boundary condition, being not affected by microscopic changes of the interatomic forces close to the surface. The polarization vector **e** of the infinite medium is solution of the wave equation of elasticity [13,14] written for the Fourier component $(\mathbf{Q},q_z)$, i.e.

$$[\rho\,\Omega^2 \delta_{\alpha\beta} - c_{\gamma\alpha\beta\delta}\, q_\gamma q_\delta]\, e_\beta(\mathbf{Q},q_z,\Omega)=0. \qquad (2)$$

Because of the translational symmetry in the surface plane, the component of the phonon wavevector **Q** parallel to the surface is fixed by the scattering geometry and is real, whereas the normal component q_z can assume either real or complex values. Ω is the phonon frequency and ρ is the mass density of the medium. The $c_{\gamma\alpha\beta\delta}$ are the components of the elastic tensor referred to the surface frame. In bulk problems the above equation is solved for a fixed wavevector $(\mathbf{Q},q_z)$ and gives three eigenvalues Ω_n corresponding to as many acoustic branches. Instead,

In the surface case, we fix $\mathbf{Q}$ and Ω in order to find the (6 in general) values of q_z which satisfy the secular problem derived from eq.(2). In the following the solutions q_z are labelled by λ. For bulk modes the $q_z{}^{\lambda}$ are real and represent transverse and longitudinal travelling waves. In surface problems however we have to consider transverse and longitudinal cut-off frequencies (Ω_T and Ω_L respectively) below which eq.(2) allows solutions with complex $q_z{}^{\lambda}$. This corresponds to evanescent waves localized at the surface and decaying into the bulk, i.e. with $\mathrm{Im}(q_z{}^{\lambda})<0$. Exponential waves with $\mathrm{Im}(q_z{}^{\lambda})>0$ must be discarded because they grow exponentially inside the medium and do not satisfy the required boundary conditions. By considering for simplicity, as already stated, modes of sagittal polarization, we have the following cases of superposition of partial waves:

a) $\Omega<\Omega_T<\Omega_L$

Allowed values of $q_z{}^{\lambda}$:
one from the quasi-transverse branch with $\mathrm{Im}(q_z{}^{T})<0$,
one from the quasi-longitudinal branch with $\mathrm{Im}(q_z{}^{L})<0$.
A solution in this frequency range represents obviously a surface mode (Rayleigh wave) because its amplitude goes to zero for $z\to-\infty$.

b) $\Omega_T<\Omega<\Omega_L$

Allowed values of $q_z{}^{\lambda}$:
two from the quasi-transverse branch with $\mathrm{Im}(q_z{}^{T})=0$ (bulk waves with $q_z{}^{T}, -q_z{}^{T}$),
one from the quasi-longitudinal branch with $\mathrm{Im}(q_z{}^{L})<0$.
A solution in this frequency range is a superposition of bulk and localized modes. For this reason these modes are called "mixed modes".

c) $\Omega_T<\Omega_L<\Omega$

Allowed values of $q_z{}^{\lambda}$:
two from the quasi-transverse branch with $\mathrm{Im}(q_z{}^{T})=0$ (bulk waves with $q_z{}^{T}, -q_z{}^{T}$),
two from the quasi-longitudinal branch with $\mathrm{Im}(q_z{}^{L})=0$ (bulk waves with $q_z{}^{L}, -q_z{}^{L}$).
This solution is bulk-like and the only effect of the surface is to modify the density of states through the boundary conditions.

The effect of the surface is to couple the modes so that the resulting displacement field satisfies the surface boundary condition. The superposition of partial waves can be written as

$$\mathbf{w}(\mathbf{Q},\Omega_n,z)=\sum_{\lambda} a(\mathbf{Q},q_z{}^{\lambda},\Omega_n)\,\exp(iq_z{}^{\lambda}z)\,e(\mathbf{Q},q_z{}^{\lambda},\Omega_n) \tag{3}$$

where the unknown coefficients $a(\mathbf{Q},q_z{}^{\lambda},\Omega_n)$ are determined by imposing the stress-free surface boundary conditions [14]:

$$c_{z\beta\mu\nu}\,\eta_{\mu\nu}\Big|_{z=0}=0 \qquad (\beta=x,y,z) \tag{4}$$

where the strain components $\eta_{\mu\nu}$ are defined, as usual, as symmetric combinations of the displacement gradients:

$$\eta_{\mu\nu}=\frac{1}{2}\left[\frac{\partial u_\mu}{\partial x_\nu}+\frac{\partial u_\nu}{\partial x_\mu}\right]. \tag{5}$$

Eq.(4) may be also written as

$$c_{z\beta\mu\nu}\,q_\mu w_\nu\Big|_{z=0}=0. \qquad (\beta=x,y,z) \tag{6}$$

In the simple case of only sagittal modes, the condition (6) forms a set of two linear equations in the unknowns $a(\mathbf{Q},q_z{}^{\lambda},\Omega_n)$. With reference to the frequency ranges a), b), c) introduced above, we have three regimes of solution.

a) $\Omega<\Omega_T<\Omega_L$
(6) is an homogeneous set of two equations with two unknowns because there are only two partial waves in the expansion (3). The solution corresponds to the frequency Ω_R of the Rayleigh wave. The spectrum is discrete. The orthonormality condition for **w** in the discrete spectrum is given in terms of a Kronecker delta as

$$\sum_{\alpha}\int_{-\infty}^{0} w_{\alpha}(\mathbf{Q},\Omega_n,z)^* \, w_{\alpha}(\mathbf{Q},\Omega_{n'},z)\, dz = \delta_{n,n'} \qquad (7)$$

where n and n' are the mode indices.

b) $\Omega_T<\Omega<\Omega_L$
(6) becomes an inhomogeneous set of equations, the partial waves and the corresponding unknowns being three. We have a solution for every frequency in this range and the spectrum (of mixed modes) is continuous.

c) $\Omega_T<\Omega_L<\Omega$
Similar to case b), but the unknowns are four and hence there are two linearly independent solutions.

In the cases b) and c), the continuous vibrational spectrum is characterized by a certain density of states which must be properly normalized [16]. Introducing a discrete mode index j (referring to the independent solutions of the equation of motion for a given frequency Ω) the orthonormality condition reads

$$\sum_{\alpha}\int_{-\infty}^{0} w_{\alpha}(\mathbf{Q},\Omega,j,z)^* \, w_{\alpha}(\mathbf{Q},\Omega',j',z)\, dz = \delta_{j,j'}\,\delta(\Omega,\Omega'). \qquad (8)$$

Here $\delta(\Omega,\Omega')$ is the Dirac delta function. Explicit expressions for the normalization of the modes both in the discrete and in the continuous spectra can be found in Bortolani et al. [17]. The density of states of the continuous spectrum contains more or less pronounced structures that may appear in Brillouin spectra as separated peaks. Such peaks are related to the existence of resonances or leaky modes [14], so called because energy flows into the substrate, due to the presence of bulk modes in the expansion (3). Although localized and leaky modes are different in principle, they appear in Brillouin spectra more or less the same, the only difference being the width of their peaks, usually larger for leaky modes, due to the shorter intrinsic lifetime. Until now the continuous spectra of surface modes has not received much attention, compared to the well studied Rayleigh mode [13,14]. It is therefore useful to review the main features of the continuous spectrum. In the simple case of an isotropic material the density of states of surface phonons can be calculated analytically [18-21]. The medium is characterized by a transverse velocity v_T and a longitudinal velocity v_L. The density of surface phonons of L and SV polarizations are plotted in fig.5 for three different values of the Poisson ratio σ. The shear-vertical displacements are more or less characterized by the same structure: a plateaux between v_T and v_L. The longitudinal displacements present a sharp peak at $v \approx v_L$ for $\sigma \leq 1/3$, while for $\sigma \geq 1/3$ there is a broad structure at a velocity $v \approx 2v_T$. Glass and Maradudin [22] showed that this structure in the longitudinal density of states corresponds to a "leaky" longitudinal mode. This mode is usually called "longitudinal resonance". The continuous spectrum of surface phonons has also been calculated for anisotropic materials [21,23].

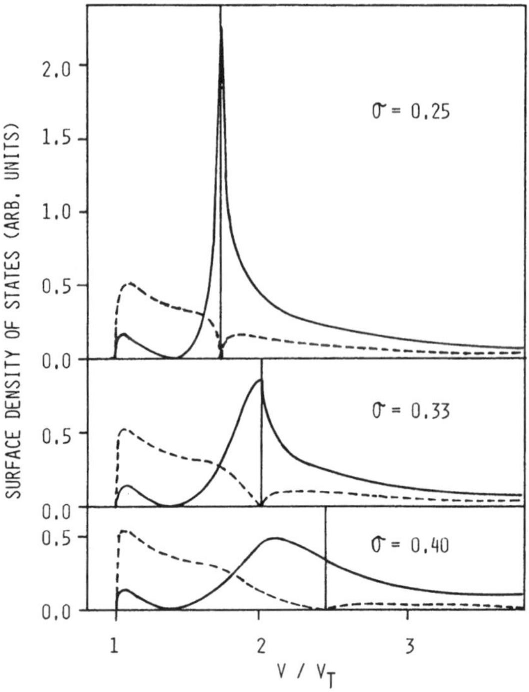

Fig.5 The surface density of states of the continuous spectrum of acoustic phonons, versus the adimensional velocity v/v_T at fixed Q. Full lines: longitudinal components. Broken lines: shear-vertical components. The Poisson ratio σ is equal to 0.25 (a), 0.336 (b), 0.40 (c). The vertical lines mark the longitudinal velocity v_L. After Camley and Nizzoli [21]

3.2 Unsupported and Supported Films

The vibrational spectra of unsupported films are well known [15]. They can be obtained by imposing the stress-free surface boundary conditions (6) at both surfaces of the film. The spectrum is discrete. In the case of decoupling of the sagittal and SH modes, the modes of a free standing film are called Lamb modes (sagittal) and Love modes (SH). The modes of unsupported films have been observed with Brillouin scattering, although the supported films have been measured much more frequently.

The partial wave method is used also for supported films, by generalizing the theory outlined in the previous section for a single medium. This means that the phonon displacement field in each medium can be written as a linear combination of acoustic waves of the corresponding infinite medium. Within the partial wave method the phonon polarization field, calculated for a given parallel wavevector **Q** and for a given frequency Ω_n, is given by

$$\mathbf{w}(\mathbf{Q},\Omega_n,z) = \sum_\lambda a^\lambda(\mathbf{Q},q_z{}^\lambda,\Omega_n)\; e^{(f)}(\mathbf{Q},q_z{}^\lambda,\Omega_n)\; e^{iq_z{}^\lambda(\mathbf{Q},\Omega_n)z} \qquad (0\le z\le h) \qquad (9a)$$

$$= \sum_\mu a^\mu(\mathbf{Q},q_z{}^\mu,\Omega_n)\; e^{(s)}(\mathbf{Q},q_z{}^\mu,\Omega_n)\; e^{iq_z{}^\mu(\mathbf{Q},\Omega_n)z} \qquad (-\infty<z\le 0) \qquad (9b)$$

The superscripts (f) and (s) mean respectively film and substrate. The sum over λ

refers to the film and the sum over μ refers to the substrate. The coefficients a^λ and a^μ can be found by imposing the (six) acoustic boundary conditions, e.g. continuity of displacements and stresses at the interface and vanishing of stresses at the film free surface. The elastic constants of the two media affect the entire calculation because they determine the boundary conditions and the bulk quantities $\mathbf{e}^{(f)}$, $\mathbf{e}^{(s)}$, $q_z{}^\lambda$ and $q_z{}^\mu$. The expansion (9) must be substituted into the acoustic boundary conditions, written for the $(\mathbf{Q}, \Omega_n)$ Fourier component of the phonon polarization field. Because these equations are linear in $\mathbf{w}$, the mathematical problem eventually reduces to a set of linear equations in the unknowns a^λ and a^μ. The index λ runs over four partial waves in the film (two transverse and two longitudinal) while μ can account for two to four partial waves in the substrate, depending on the value of the frequency Ω_n with respect to the cut-off transverse frequency $\Omega_T{}^{(s)}$ of the substrate. In analogy to the case of the semi-infinite medium we have a discrete spectrum of surface modes for $\Omega < \Omega_T{}^{(s)}$, while the continuous spectrum is characterized by $\Omega > \Omega_T{}^{(s)}$. The discrete spectrum contains modes guided in the film with a displacement field exponentially decaying into the substrate. In order to actually have propagation of guided modes in the film the transverse cut-off frequency of the film $\Omega_T{}^{(f)}$ must be smaller than the corresponding quantity of the substrate $\Omega_T{}^{(s)}$ [15]. The SH guided modes of a supported film are called generalized Love modes, while the sagittal modes are called generalized Lamb (or Sewaza) modes. By increasing the value of Qh, the number of guided modes also increases, so that we expect to find many peaks in Brillouin spectra of thick supported films: in section 6 a few examples of this behaviour will be shown.

4. Brillouin Scattering from Bulk Modes: the Elasto-Optic Coupling

Before considering Brillouin scattering from surface excitations, it is convenient to briefly review the concepts in light scattering from bulk acoustic phonons. In quantum mechanics light scattering is described as creation and annihilation of phonons of wavevector $\mathbf{q}$ and frequency Ω, with energy and momentum conservation. In the long wavelength limit where umklapp processes are not important we have

phonon creation (Stokes events) $\begin{cases} \omega = \omega_0 - \Omega \\ \mathbf{k} = \mathbf{k}_0 - \mathbf{q} \end{cases}$

phonon annihilation (anti Stokes events) $\begin{cases} \omega = \omega_0 + \Omega \\ \mathbf{k} = \mathbf{k}_0 + \mathbf{q}. \end{cases}$

$\mathbf{k}_0$ and ω_0 are the wavevector and frequency of the incident light, whereas $\mathbf{k}$ and ω are the corresponding quantities of the scattered light. From energy and wavevector conservation the scattering geometry in transparent solids completely specifies the acoustic wavevector $\mathbf{q}$, whose modulus is given by

$$q \simeq 2 k_0 \sin(\theta/2) \tag{10}$$

where θ is the scattering angle. Therefore in general the light spectrum from bulk sound waves in a solid is composed of three Brillouin doublets, one for each acoustic branch. Symmetry reasons may cause doublets to disappear or to be degenerate. Note that eq. (10) is not valid for optically anisotropic media. The physical reason for light scattering is to be found in the inhomogeneities of the medium due to fluctuations in the dielectric constant, which in turn are induced by the thermally excited phonons. The unperturbed dielectric constant of the crystal is indicated by ε_0. Its fluctuation $\delta\varepsilon_{\alpha\beta}$ is a linear combination of the strains $\eta_{\gamma\delta}$ through the elasto-optic (Pockels) coefficients $k_{\alpha\beta\gamma\delta}$

$$\delta\varepsilon_{\alpha\beta} = \sum_{\gamma\delta} k_{\alpha\beta\gamma\delta}\, \eta_{\gamma\delta}. \qquad (11)$$

As derived by Landau and Lifshitz [24], the electric field E of the scattered light, at large distance R from the scattering volume V and to first order Born approximation, is given by

$$E=-\frac{e^{ikR}}{4\pi\varepsilon R}\, k\times \left[\mathbf{k}\times \int_V \boldsymbol{\delta\varepsilon}\; E_0 e^{-i(\mathbf{k}-\mathbf{k}_0)\cdot\mathbf{r}}\, d^3r \right] \qquad (12)$$

where the second rank tensor $\boldsymbol{\delta\varepsilon}$ is defined by eq.(11). Integrals of the type (12) immediately lead to momentum conservation. In fact for a given excitation of wavevector $\mathbf{q}$, $\delta\varepsilon_{\alpha\beta}$ turns out to be proportional to $\exp(i\mathbf{q}\cdot\mathbf{r})$. If V is large compared to q^{-1}, the integral in eq.(12) becomes proportional to $\delta(\mathbf{k}-\mathbf{k}_0\pm\mathbf{q})$. The conservation of momentum is therefore recovered. The integrals appearing in eq.(12) can be developed and explicit expressions for the scattered electric field and the cross-section can be found. For isotropic media see, for example, Sandercock [5].

4.1 Bulk Scattering in Opaque Media

The basic theory outlined in the previous section must be extended to include scattering from opaque materials. There are several ways in which the opacity makes itself felt. Firstly, the high optical absorption limits the scattering to a volume close to the sample surface. This has an influence on the wavevector conservation rule. Secondly, the effect of the surface on the excitations cannot be ignored. This manifests itself in the reflection of bulk excitations at the surface and indeed in the appearance of new excitations associated with the surface. Thirdly, the ripples produced on the surface by these excitations scatter light and, for highly opaque materials, may be the dominant scattering source. The last two effects will be considered in the next section.

We first consider the confinement of the elasto-optic coupling to a limited volume close to the surface [25,26]. When this happens, the argument presented in the previous section to show that (12) is proportional to $\delta(\mathbf{k}-\mathbf{k}_0\pm\mathbf{q})$ is no longer valid. Instead $(\mathbf{k}-\mathbf{k}_0)=\mathbf{k}'-i\mathbf{k}''$ is now complex so that we are left with integrals of the type

$$\int_V \eta_{\alpha\beta}\; e^{-i(\mathbf{k}'-i\mathbf{k}'')\cdot\mathbf{r}}\, d^3r. \qquad (13)$$

By considering a phonon of wavevector $\mathbf{q}$, $\eta_{\alpha\beta}$ is proportional to $\exp(i\mathbf{q}\cdot\mathbf{r})$. Carrying out the integration over the surface coordinates x,y we have that the scattering amplitude F is proportional to

$$\int_{-\infty}^{0} e^{[i(q_z-k_z')z-k_z''z]}\, dz \propto \frac{1}{[i(q_z-k_z')-k_z'']} \qquad (14)$$

where we have assumed a semi-infinite medium extending from $z=-\infty$ to $z=0$. Let us consider, for simplicity's sake, a backscattering geometry with $\mathbf{k}'-i\mathbf{k}''$ perpendicular to the surface. Under this simplifying hypothesis $\mathbf{k}'-i\mathbf{k}''= 2\mathbf{k}_0(\tilde{n}+i\tilde{\kappa})$, where $\tilde{n}+i\tilde{\kappa}=\varepsilon_0^{-1/2}$ is the complex refractive index of the medium. We obtain

$$|F|^2 \propto \frac{2\tilde{\kappa}k_0}{[(q-2\tilde{n}k_0)^2+(2\tilde{\kappa}k_0)^2]}. \qquad (15)$$

This equation shows that the presence of optical absorption allows interaction with phonons not only at the frequency $\Omega=2\tilde{n}k_0v$ but rather within a finite frequency range. From eq.(15) the broadening has a full width at half maximum $\Delta\omega=4\tilde{k}k_0v$. An effect of this nature was observed experimentally by Pine [27] in CdS and by Sandercock [28] in Si and Ge.

Dresselhaus and Pine [29] observed in addition that the opacity broadened bulk lines were asymmetric. This effect was explained by Dervisch and Loudon [30]: in opaque materials the scattering occurs in a region close to the surface so that the effect of reflection and interference of phonons at the boundary must be considered.

5. Brillouin Scattering from Surface Modes

The first attempts to measure surface waves in opaque solids by Brillouin scattering were confined to the case of plates much thicker than the phonon wavelength, so that they can be considered semi-infinite crystals. In this section we address ourselves to opaque materials, so that the scattered light could only be detected in backscattering. Other scattering geometries are applicable to transparent media [9,31]. As already stated in section 4.1, two mechanisms contribute to the scattering by surface acoustic excitations: the modulation of the dielectric function and the surface dynamical corrugation induced by the SV component of the phonons (ripple effect). Another scattering mechanism has been proposed [32]. By considering nonlocal effects, the irrotational part of the incoming field adds to the cross-section a new elasto-optic term. This term should enable one to pick up acoustic phonons far out in the Brillouin zone. However it would require non visible light, with further problems from the point of view of the instrumentation. In the following sections we outline the theory of Brillouin scattering in opaque crystals. The Green functions have been used by Loudon [18,19,33], Subbaswamy and Maradudin [20] and Velasco and Garcia-Moliner [23,34]. A treatment in terms of a direct matching of the electro-magnetic field at the surface has been proposed by Rowell and Stegeman [35] and by Marvin et al. [16,36]. We will follow the latter approach.

5.1 Theory of Brillouin Scattering in Semi-Infinite Opaque Crystals

The results for the Brillouin cross-section depend on the polarization of the light. To fix the ideas let us consider p(TM) incident and scattered light. In this case the derivation of the scattered electromagnetic field is more easily carried out in terms of the magnetic field **B**. The Maxwell equations for **B** can be obtained by expressing $(\varepsilon_{\alpha\beta})^{-1}$ in terms of ε_0 and $\delta\varepsilon_{\alpha\beta}$. It can be easily shown that [16]

$$\left[\nabla^2-\frac{1}{c^2}\frac{\partial^2}{\partial t^2}\right]\mathbf{B}=0 \qquad (z>0) \qquad (16a)$$

$$\varepsilon_0\left[\nabla^2-\frac{\varepsilon_0}{c^2}\frac{\partial^2}{\partial t^2}\right]\mathbf{B}=-\nabla\times(\delta\boldsymbol{\varepsilon}\,\nabla\times\mathbf{B})\,. \qquad (z<0) \qquad (16b)$$

The solution of eqs. (16a) and (16b) to zeroth order in the displacement field simply gives the well known Fresnel coefficients for refraction of light at the surface of a dielectric medium. We are interested here in a solution for **B** to first order in the displacement field **u**. This is justified by the very small vibration amplitude with respect to the phonon wavelength. The proper solution will be the sum of a particular solution of the inhomogeneous Maxwell equation (16b) with the general solution of the homogeneous equations (obtained with $\delta\varepsilon_{\alpha\beta}=0$) to satisfy the boundary conditions:

a) continuity of **B** at the surface

b) continuity at the surface of the tangential component of the electric field E, i.e. of $\boldsymbol{\varepsilon}^{-1}\nabla\times\mathbf{B}$.

The electromagnetic boundary conditions are to be written on the corrugated surface, whose profile is just u_z. By assuming a small corrugation, an expansion in u_z can be made and only first order terms can be retained. The general form of the differential cross-section is given by [16,36]

$$\left[\frac{d^2\sigma}{d\Omega_a d\omega}\right] = \frac{\omega^2}{(2\pi)^2c^2}\frac{\cos^2\theta_s}{\cos\theta_i}\frac{k_B T}{2\rho|\omega-\omega_0|^2}\sum_n\left|D(\underline{K}-\underline{K}_0,\Omega_n)\right|^2\delta[\Omega_n-|\omega-\omega_0|]. \tag{17}$$

Ω_a is the solid angle. θ_i and θ_s are respectively the angles of incoming and scattered light with respect to the surface normal. k_B is the Boltzmann constant and T is the temperature. D is a scattering amplitude which depends on the polarization of the incoming and scattered light and on the scattering geometry. Long calculations are needed to derive D. Explicit expressions may be found in Marvin et al. [16,36] for the various combinations of polarization. D contains both surface contributions (coupling of light to surface modes through the ripple and the elasto-optic mechanisms) and bulk contributions (bulk resonancies via the elasto-optic mechanism). The summation index n in eq. (17) is a discrete one in the case of localized surface modes. For the continuous spectrum the expression (17) is still valid with the substitution

$$\sum_n \delta[\Omega_n-|\omega-\omega_0|] \rightarrow \left[\frac{dn}{d\Omega_n}\right]_{\Omega_n=|\omega-\omega_0|}\sum_j \tag{18}$$

where on the right hand side the integer index j runs over the linearly independent solutions of the continuous spectrum, with the same meaning as in eq. (8).

It is useful to write the cross section for p→p in-plane scattering in the limit of large ε_0. As we will see in the next section many surface Brillouin scattering experiments in opaque materials can be interpreted on the basis of this formula:

$$\left[\frac{d^2\sigma}{d\Omega_a d\omega}\right]_{p\to p} \simeq \frac{\omega^4 k_B T}{2\pi^2c^4\rho(\omega-\omega_0)^2}\cos^2\theta_s\frac{\cos\theta_i}{[(\cos\theta_s+\varepsilon_0^{-1/2})(\cos\theta_i+\varepsilon_0^{-1/2})]^2}$$

$$\times\sum_n\Big|\varepsilon_0^{-3/2}(k_{11}-k_{13}c_{13}/c_{33})(\sin\theta_i-\sin\theta_s)\,w_x(0,\underline{Q},\Omega_n)/2$$

$$+(1-\sin\theta_s\sin\theta_i)\,w_z(\mathbf{Q},\Omega_n,0)\Big|^2\delta[\Omega_n-|\omega-\omega_0|]. \tag{19}$$

Note that the ripple contribution is proportional to the shear-vertical displacement component w_z while, at least in this approximation, the elasto-optic contribution is simply proportional to the longitudinal displacement component w_x. The latter contribution is obviously proportional to the elasto-optic coefficients k_{ij}. The displacement ellipse has a vertical-to-longitudinal axis ratio $|w_z/w_x|$ usually in the range 1÷2, giving the RW a dominant shear-vertical character. In opaque materials $\varepsilon_0^{-3/2}(k_{11}-k_{13}c_{13}/c_{33})$ is usually smaller than 1, so that the ripple term in eq. (19), proportional to w_z, is the relevant one; however in many cases the elasto-optic contribution is not negligible and can give rise to interference effects.

The eq. (19) is valid for the (001) surface of cubic materials and Q along [100]. It is also formally valid, after a rotation of the reference system, for the [110] direction on the (001) surface and for the [110] surface, Q along [001] and [1$\bar{1}$0]. In these cases the tensor quantity $k_{11}-k_{13}c_{13}/c_{33}$ must be properly rotated [36].

5.2 Theory of Brillouin Scattering in Supported Films

The theory of Brillouin scattering in supported films has been formulated by Bortolani et al. [17]. It applies to crystals of cubic symmetry. They consider both ripple and elasto-optic mechanisms and deal with a film of any thickness. Scattering for Lamb and Love modes of unsupported films can be also obtained as limiting cases of the general results, by letting (for the substrate) $c_{ij}\to 0$, $\rho\to 0$, $\varepsilon_0\to 1$.

The theory of Bortolani et al. [17] requires numerical calculations for the phonon modes, otherwise analytical expressions for the scattered electromagnetic field are provided. The final expressions are very lengthy. The Brillouin intensity for p→p scattering has the following form:

$$\left[\frac{d^2\sigma}{d\Omega_a}d\omega\right]_{p\to p} \propto \Big| \sum_{\lambda} a^{\lambda}[A_{\lambda,x}e_x^{(f)}+A_{\lambda,z}e_z^{(f)}]+A^{(f)}w_z(\mathbf{Q},\Omega,h)+$$

$$+\sum_{\mu} a^{\mu}[A_{\mu,x}e_x^{(s)}+A_{\mu,z}e_z^{(s)}]+A^{(s)}w_z(\mathbf{Q},\Omega,0)\Big|^2 \qquad (20)$$

The first term in the square modulus refers to the elasto-optic coupling in the film and the third one to the elasto-optic coupling in the substrate. The quantities a^{λ}, $\mathbf{e}^{(f)}$, a^{μ}, $\mathbf{e}^{(s)}$ are related to the phonon displacement field in the two media, according to eqs. (9a,b), and require numerical calculations in anisotropic crystals. The weighing factors $A_{\alpha,j}$ are linear combinations of the elasto-optic constants and are determined by the electromagnetic boundary conditions. Note that the elasto-optic terms depend on both the transverse and the longitudinal components of the partial waves of the film $e_z^{(f)}$, $e_x^{(f)}$ and of the substrate $e_z^{(s)}$, $e_x^{(s)}$. The term $A^{(f)}w_z(z=h)$ represents ripple scattering from the free surface of the film and is proportional to the normal component of the displacement field (9a) calculated at the free surface (z=h). $A^{(f)}$ is a function of the scattering angles, of the film thickness h and of the dielectric functions of the media. The term $A^{(s)}w_z(z=0)$ represents ripple scattering from the film-substrate interface and is therefore proportional to the vertical displacements at the interface (z=0). The form of eq. (20) leads to an immediate important conclusion: the four contributions to the cross-section coherently interfere and, whenever their amplitudes are comparable, modulate the cross-section in a complicated way with strong interference effects. The next sections are devoted to the illustration of experimental results for surfaces and films, and a few applications of the theory will be presented.

6. Experimental Results and Comparison with Theory

The next three subsections are devoted to the comparison between measured and calculated surface Brillouin spectra. Semi-infinite crystals, supported and free-standing films are treated separately. Surface Brillouin scattering turns out to be very useful for the investigation of the physical properties of opaque and transparent solids in two different ways, as described below.

When the elastic and photo-elastic properties of a material are known, Brillouin scattering is a unique tool for studying the density of states of surface acoustic phonons in thermal equilibrium. It allows especially to investigate the "leaky" modes and the continuous spectrum. The "longitudinal resonance" described in section 3.1 was analyzed in details with this technique. From this point of view surface Brillouin scattering is complementary to inelastic scattering of He atoms, which can give the same kind of information, far out in the Brillouin zone [37]. A class of guided modes of longitudinal polarization existing in supported films has been discovered with surface Brillouin scattering [38].

When the properties of a material are unknown, surface Brillouin scattering can be used as a characterization technique. By fitting the theory to the experimental spectra, it is possible to determine unknown macroscopic coefficients as the elastic constants, the elasto-optic constants and the optical constants. Their dependence on the distance from the surface can be studied. A few examples will be presented below.

6.1 Metal and Semiconductor Surfaces

Semiconductors are partially transparent to the light, i.e. the penetration depth is comparable with the phonon wavelength. Therefore Brillouin scattering by bulk and surface phonons can occur in the same spectrum. The (110) surface of GaAs offers a typical example. The spectrum in fig.6 refers to such a case and is particularly rich in information. Its prominent feature is the sharp peak (at about 10 GHz) due to scattering from the surface Rayleigh wave. In GaAs, with a wavelength of the laser beam λ=5145 Å, both the elasto-optic and the ripple mechanisms contribute to the scattering strength of the Rayleigh wave. All the other features of fig.6 belong to the continuous spectrum, i.e. have an intrinsic linewidth.

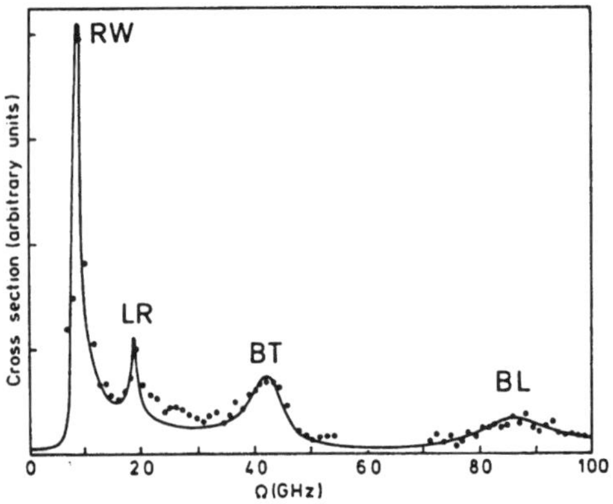

Fig.6 Calculated Brillouin cross-section (full line) compared with the experimental data (dots) for the (110) surface of GaAs, with **Q** along [1$\bar{1}$0]. RW: Rayleigh wave, LR: longitudinal resonance, BT: bulk transverse mode, BL: bulk longitudinal mode. After Marvin et al. [36]; the original experimental data are from Loudon and Sandercock [33]

The two high-frequency broad structures of fig.6 (at about 42 and 87 GHz) are related to quasi-transverse and quasi-longitudinal bulk phonons respectively. The large intrinsic linewidth of these peaks is due to the skin depth in GaAs, which is smaller than the wavelength of the light. These are the opacity broaden bulk lines, already mentioned in section 4.1. The opacity broadened bulk lines are obviously detected in Brillouin scattering through the elasto-optic mechanism only.

Finally we have to explain the continuum of modes between the Rayleigh wave and the transverse bulk wave in fig.6, especially the peak at about $\Omega_L \simeq 19$ GHz. This structure comes from those bulk and evanescent waves whose total wavevector $\mathbf{q}=\mathbf{Q}+q_z\hat{z}$ has the same surface wavevector Q (mixed modes). Note that, although this continuous band mainly originates from bulk waves, it exhibits a real surface character: in other words this part of the spectrum has dispersion relations of the kind ω=Qv, where v is the propagation velocity, in complete analogy to the Rayleigh wave and unlike the opacity broadened bulk lines, whose frequency is proportional to the total wavevector q.

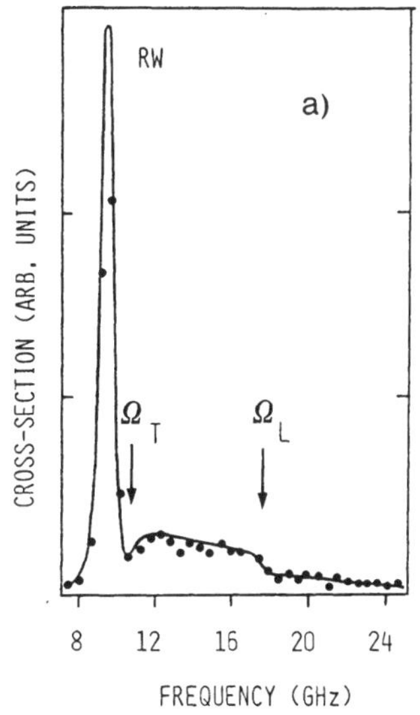

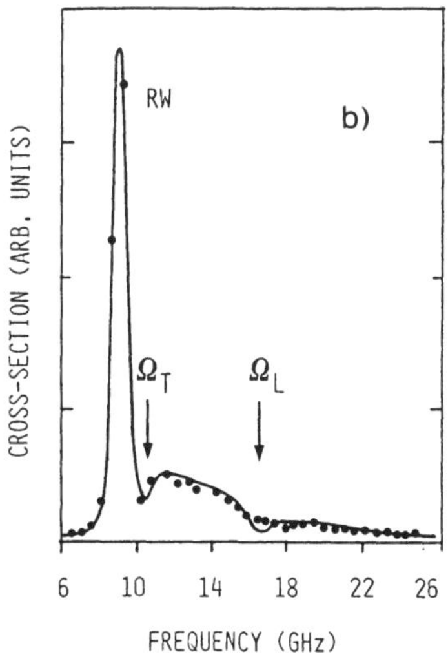

Fig.7 Calculated Brillouin cross-section (full line) compared with the experimental data (dots) for the (001) surface of GaAs, with **Q** along [110] (a) and [100] (b). After Bortolani et al. [40]

The lineshape of the mixed-modes spectrum is today well understood in terms of the density of phonons states projected onto the plane of the surface and already illustrated in section 3.1. The peak of fig.6 at about 19 GHz has been explained by Bortolani et al. [39,40] as due to the leaky longitudinal surface mode and the elasto-optic coupling. In fact the calculated continuous spectrum of the longitudinal modes shows a prominent peak at $\Omega \simeq \Omega_L$ when $v_L/v_T \leq 2$, as shown in fig.5. This is actually the case for GaAs. In addition the elastic and optical parameters of GaAs are favourable to magnify in eq.(19) the elasto-optic contribution, proportional to the longitudinal displacement w_x. It should be remarked that a peak at $\Omega \simeq \Omega_L$ in semiconductors is not always present: for example, spectra taken on GaAs in scattering geometries different from that considered in fig.6 do not exhibit any peak. We show in fig.7 the measured and calculated Brillouin spectra for the (001) surface of GaAs and two surface directions. It can be seen, by comparing the spectra with the calculations of fig.5, that the mixed-mode region (10-18 Ghz) shows a SV-type density of states. Therefore we conclude that in these scattering geometries the spectra are dominated by the ripple effect.

Metals are usually more opaque than semiconductors for visible light. Therefore the surface Brillouin spectra of metals look very similar to those of fig.7a,b. This is the case for Al, Ni, Cr, Mo [41].

Gold represents an exception and is considered here in some detail. The elasto-optic mechanism turned out to be essential to explain the data of gold, due to the relatively large penetration depth of the light in this metal for λ=5145 Å. The Poisson ratio σ of gold is unusually large, about 0.43. Therefore, from the analysis of fig.5, one expects a longitudinal surface mode of short lifetime, with a broad lineshape and a propagation velocity much lower than v_L. Bassoli et al. [42] studied supported films and bulk gold in polycrystalline form. By fitting the film spectra they derived the elasto-optic constants of gold and then obtained a good agreement between the calculated and the measured Brillouin spectra of the gold surface, as can be seen in fig.8. They also showed that the approximated expression (19) is capable of explaining only qualitatively the surface spectrum of gold. Because ε_0 is small, the exact

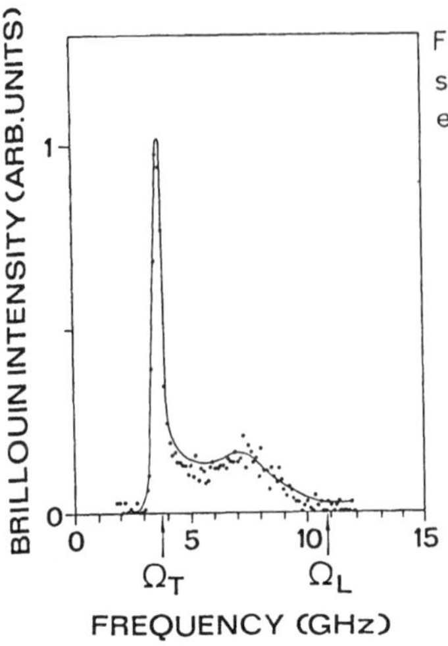

Fig.8 Measured and calculated surface Brillouin spectra of polycrystalline gold. After Bassoli et al. [42]

expression for the cross-section, as given for example by Marvin et al. [16,36], is needed to obtain a quantitative agreement. In gold large interference effects between the ripple and the elasto-optic scattering amplitudes were also found.

6.2 Supported Films

The first paper showing the possibility of detecting with Brillouin scattering guided acoustic modes in supported metal films has been published by Bortolani et al. [43]. A spectrum for a 400 Å film of aluminium on silicon was satisfactorily explained in terms of ripple scattering at the film surface. The spectrum gave indication of the Rayleigh wave, of the first order Sezawa wave and of the continuum of mixed-modes due to leaky Sezawa waves. Ripple calculations for Al films of various thicknesses allowed Sandercock et al. [44] to reproduce the experimental relative scattering intensities of the guided modes very well using the literature values of the elastic constants of polycrystalline aluminum. Also the dispersion relations of the modes were satisfactorily explained.

In films of molybdenum with thicknesses 4000 and 8000 Å, deposited onto a glass substrate by sputtering, Bell et al. [45] measured for the first time by Brillouin scattering the Stoneley wave of a solid interface. Mo films in the thickness range 300-8000 Å turned out to be homogeneous and with the same elastic properties. The Stoneley wave has been also measured by Jorna et al. [46] at the nickel-glass interface: in this case, however, thick Ni films ($h>5000$ Å) were found to be not homogeneous.

The elastic properties of a stratified medium composed of two alternate layers of thicknesses d_1 and d_2 have been studied about 50 years ago by Bruggeman [47]. Since then and for a long time the problem has attracted interest mainly in seismology. More recently Rytov [48], in the context of physical acoustics, calculated explicitly the elastic constants of a stratified medium, made of two isotropic materials, in the long wavelength limit when the excitation wavelength is larger than the modulation wavelength of the superlattice $\Lambda=d_1+d_2$, viz. $q\Lambda<1$. In practical Brillouin experiments this condition is satisfied for a modulation wavelength of a small number of hundreds of Angstroms. The Rytov approximation leads to the concept of "effective elastic constants" that is, in this limit, the superlattice behaves as a homogeneous medium with a well defined elastic constant tensor. The symmetry of the effective medium depends on the two constituents [49]: for example two isotropic media give rise to a hexagonal effective crystal. An important feature of the method is that the effective elastic constants do not depend on the modulation wavelength Λ, but rather on the relative thicknesses d_n/Λ. For example, the effective c_{44} of a stratified medium made of two orthorombic crystals with a principal axis along the superlattice normal is given by

$$c_{44} = \left[\frac{d_1}{\Lambda c_{44}(1)} + \frac{d_2}{\Lambda c_{44}(2)}\right]^{-1} . \tag{21}$$

The experimental measurements of the Rayleigh wave in superlattices have been initiated by Kueny et al. [50] in metal systems (Nb/Cu) and by Sapriel et al. [51] in semiconductor systems ($GaAs/Ga_{1-x}Al_xAs$). The result obtained by Kueny et al. [50] was surprising: the Rayleigh wave velocity as a function of the modulation wavelength Λ showed a marked anomaly (a dip for $20 \leq \Lambda \leq 40$), instead of being independent of Λ, as predicted by eq. (21). On the other hand there is no reason why eq. (21) is not valid in this range of Λ. The explanation may be found in a change of the elastic properties of at least one material, when produced in form of a thin film in the superlattice with a certain critical thickness [49].

Let us consider now to the problem of determining the unknown elastic constants of a supported film, in this case a metal superlattice. Bell et al. [52,53] measured several (up to nine) Sezawa modes in Mo/Ta and Nb/Cu superlattices. As an example, in fig.9 we show the experimental dispersion relations of the guided modes in a 4480 Å-thick Mo/Ta superlattice on sapphire. The calculated best-fit curves are also shown. This figure offers a nice example of dispersion relations of Sezawa modes in a supported film because it illustrates the behaviour of the guided modes both in the discrete spectrum and in the continuum ($v > v_T^{(s)}$), where the modes are actually "leaky". Bell et al. [52] tried to fit the four independent elastic constants of the hexagonal Mo/Ta superlattice. The best-fit results obtained for c_{33} and c_{44} versus the modulation wavelength show that c_{44} exhibits a softening, while c_{33} is nearly constant.

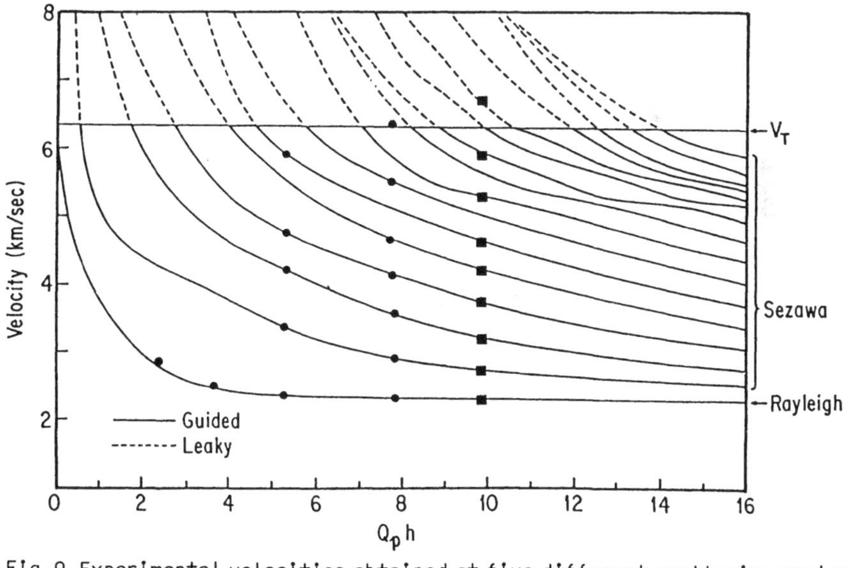

Fig.9 Experimental velocities obtained at five different scattering angles for a 4480 Å-thick Mo/Ta superlattice on sapphire with modulation wavelength Λ=160 Å. The solid curves are the best fit to the data obtained at Qh=9.83 (squares). For velocities greater than the shear velocity of the substrate, the guided modes become leaky and couple to the bulk substrate waves. After Bell et al. [52]

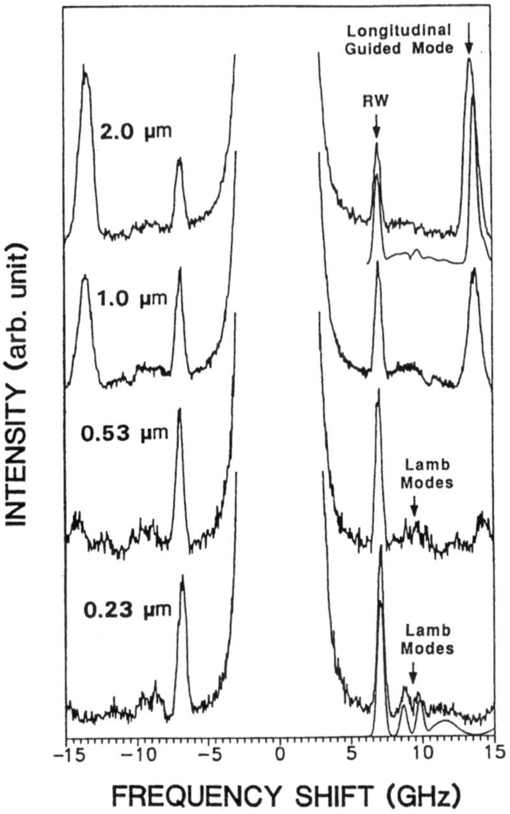

Fig.10 Brillouin spectra of (001) ZnSe films on a (001) GaAs substrate. **Q** is along the [100] direction. Angle of incidence $\theta_I=60°$. The film thicknesses h are as indicated. In two cases (h= 2 and 0.23 μm) a comparison is made with the calculated intensities (thin lines). RW: Rayleigh wave. After Hillebrands et al. [38]

An interesting application of surface Brillouin scattering to transparent films on opaque substrates is now presented. In transparent films Hillebrands et al. [38] reported theoretical as well as experimental evidence for a new class of long wavelength acoustical excitations, called "longitudinal guided modes" (LGM). In ZnSe films grown on (001) GaAs they observed, for film thicknesses in the range 0.5–2 μm, a pronounced peak of magnitude comparable to the Rayleigh wave line. Its linewidth is only slightly larger than that of the RW peak. The spectra are shown in fig.10. With decreasing film thickness h, the intensity of the LGM decreases, accompanied by a linewidth broadening and a slight frequency increase. The frequency of the LGM is just above that of the longitudinal bulk wave. Using data taken from the literature and the elastic constants of ZnSe determined from the dispersion relations of the guided modes themselves [54], Hillebrands et al. [38] reproduced very accurately the scattering intensity of the measured spectra. It has been found that the LGM is strongly localized in the film and couples with light through the elasto-optic mechanism solely. The authors conclude that the classification of acoustical sagittal surface and film excitations now exhibits complete symmetry with respect to shear and longitudinal waves and include surface modes (Rayleigh wave, longitudinal resonance) and film modes (Sezawa modes, longitudinal guided modes). The LGM offers, as a major application, the determination of the longitudinal sound velocity in transparent films, a parameter which is otherwise difficult to evaluate.

Nizzoli et al. [55] measured guided modes in both thin and thick polymeric films deposited on a molybdenum substrate with the Langmuir-Blodgett technique. The films were strongly anisotropic. The complete set of five elastic constants was determined from the spectra.

6.3 Free-standing Films

Sandercock [56] observed with Brillouin scattering discrete values of the phonon wavevector perpendicular to the surface in the spectra of polymer and crystal unsupported films. Film thicknesses were in the range 1-2 μm. The great advantage of these observations is that the measured frequencies depend only on the elastic properties of the film even if the thicknesses are much less than the wavelength of the excitations. The measurements of Sandercock [56] were performed in normal incidence, so that the dispersion of the modes of the film was not measured.

Gold has been produced in the form of thin unsupported films by Grimsditch et al. [57] and Brillouin spectra were recorded. Typically, free-standing films could be obtained over surface areas of about 1 mm^2. The quadratic dispersion of the first order, low frequency, Lamb bending mode was measured. Grimsditch et al. [57] were able to derive information on a number of elastic constants of the film.

7. Enhanced Brillouin Scattering

It is well known [58] that surface plasmon polaritons (or surface electromagnetic waves: SEW) can be excited at the air-metal interface of an air-metal-prism system by using the attenuated total reflection (ATR) method in the so-called Kretschmann configuration [59]. The electric field associated with the SEW is strongly enhanced with respect to the incident field. Therefore also the cross-section is enhanced. This property has been widely applied to several kinds of light spectroscopies, as surface enhanced Raman scattering and non-linear phenomena.

Fukui et al. [60] and Marvin et al. [61] have derived a theory of enhanced Brillouin scattering from metal films involving surface plasmons. These authors calculated the scattered field in forward and backscattering geometries (respectively on the air side and on the prism side in the ATR Kretschmann configuration) varying the incidence and scattering angles around the condition of surface plasmon coupling. Coupling of volume electromagnetic waves (VEW) and SEW, and viceversa, occurs when the usual ATR condition $(\omega/c)n_p \sin\theta_{ATR}=K_s$ is verified, where n_p is the refractive index of the prism, K_s is the SEW wavevector, ω is the frequency of the light and θ_{ATR} the ATR incidence angle. Also a grating at the active metal-air interface can convert VEW into SEW, and viceversa. In Brillouin scattering this grating is provided by the dynamical corrugation created by a surface acoustic wave (SAW): the ripple effect. We have the following scattering precesses.

a) Forward scattering. The VEW is converted into a SEW by the prism coupling. The SEW is radiated out in air by the SAW.
b) Backward scattering. After conversion of the VEW into a SEW by the prism coupling, the SEW of surface wavevector **Q** is backscattered in a state of wavevector **-Q** by a SAW. Finally the prism coupling transforms the latter SEW into a VEW on the prism side.

In forward scattering geometry Moretti et al. [62] found an enhancement factor of about 25. Recently Robertson et al. [63] measured Brillouin enhanced signals from the Rayleigh wave in backscattering, obtaining an enhancement of

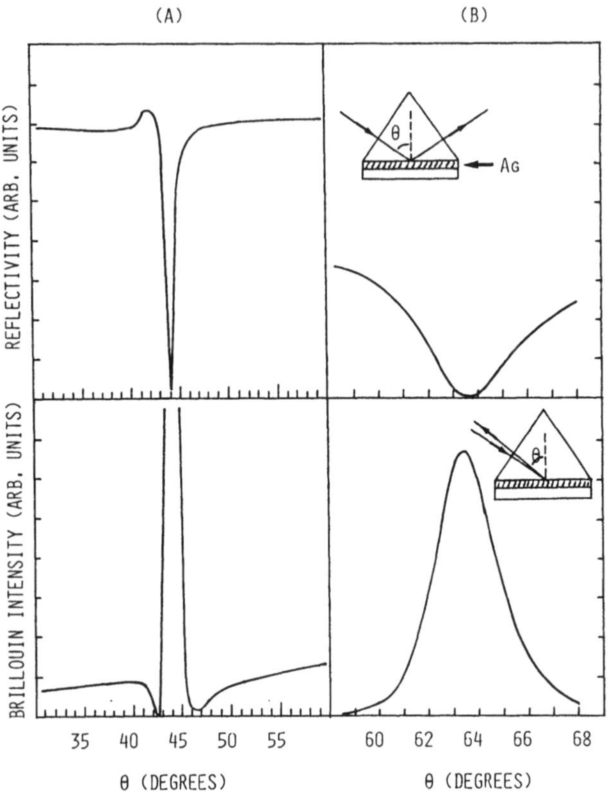

Fig.11 Calculated reflectivity (upper curves) and Brillouin scattering intensity from a surface wave (lower curves), versus the incidence angle of the light θ, for a silver film deposited onto a glass prism. The scattering geometries are shown in the insets. The coupling to the surface plasmon polariton produces minima in the reflectivity and maxima in the Brillouin intensity. Bare silver film of thickness 500 Å (a). The silver film is coated with a layer of thickness 482 Å and dielectric constant ε=2.24 (b)

about 750 due to the double prism coupling. Both transmission and reflection measurements were quantitatively analyzed with the theory of Marvin et al. [61], which includes the ripple scattering from the dynamical corrugations induced by the Rayleigh wave at both the film interfaces (film-vacuum and film-prism). The agreement between the measured and calculated Brillouin intensities versus the scattering angle θ is very good. In particular strong maxima are found when θ equals the attenuated total reflection angle θ_{ATR}. This behaviour is illustrated in fig.11a for a typical case (silver film). The small lateral minima in the Brillouin scattering intensity are due to interference effects at the two interfaces of the metal film. Very recently enhanced Brillouin scattering has been investigated in a system prism/metal/dielectric/air from both the experimental [64] and the theoretical [65] points of view. The results of a model calculation are shown in fig.11b. The addition of the dielectric layer shifts the angles of minimum reflectivity and of maximum Brillouin scattering of nearly the same amount. Strong broadening effects also occur. This technique has been used to study the dispersion relations of the guided acoustic modes of a dielectric Langmuir-Blodgett film deposited on silver [64].

8. Surface Raman Scattering

In surface Brillouin scattering we considered the coupling between light and acoustic waves. In that case the interaction does not modify the dispersion relations because there is no mode crossing between the surface elecromagnetic wave and the acoustic phonons, due to the great difference in the propagation velocity (light versus sound). On the contrary the surface optical lattice vibrations are coupled in the long wavelength regime to the surface electromagnetic waves giving rise to surface polariton branches [66]. The form of these branches depends on the geometry of the sample. In any case the dielectric permittivities of the media must obey certain conditions to allow surface polaritons to propagate. We may distinguish three cases:

a) single interface, ideally a semiinfinite medium. Only a single branch of surface polaritons is expected, extending in the frequency range between the transverse (Ω_T)and the longitudinal (Ω_L)phonon frequencies.Practically the measurements are made in trasmission through a thick film and in small angle scattering geometry ($\theta<4°$). In a 20 μm slab of GaP Valdez and Ushioda [67] found in the Raman spectra a small shoulder due to the surface phonon polariton, in addition to the strong LO bulk phonon peak. A sophisticated apparatus is needed to detect the feeble surface signal, approximately 2≈3 photon counts per second. Thus the main effort is directed towards reduction of the elastically scattered light, minimizing the photomultiplier dark noise and maximizing the Raman signal. A long collection time is needed, which requires a spectrometer interfaced to a minicomputer as the experiment controller and data accumulator. Such a system has been developed by Ushioda et al. [68]. The measured surface polariton dispersion was in good agreement with the theoretical calculation based on the known dielectric data of GaP. A single smooth polariton branch was obtained.

b) double interface geometry, typically a film of thickness h deposited on a semiinfinite substrate. In the limit of a thick film two branches are found, with a dispersion depending on the properties of the two non-interacting interfaces. In a thin supported film the two polaritons interfere and the mode frequencies move further apart giving rise to an upper mode (UM) and to a lower mode (LM) whose separation depends on the film thickness and the dielectric properties of the media. The UM and the LM exist in the gap between Ω_T and Ω_L of the supported film. The dispersion relations of the UM and LM are obtained subtracting the backward spectrum (containing the bulk features only) from the forward spectrum (containing surface and bulk features). The subtracting procedure requires spectra measured simultaneously and in the same experimental conditions. A suitable scattering geometry based on the use of moving mirrors has been designed for this purpose [69]. Prieur and Ushioda [69] measured the dispersion relations of the UM and LM in thin films of GaAs on sapphire.

c) The modes previously considered are localised at the boundaries. We may also consider guided-wave polaritons decaying exponentially outside a supported film and exhibiting an oscillatory behaviour across the film [70]. Many branches of guided-wave exist outside the gap defined by Ω_T and Ω_L. The number of modes increases with Qh and in the limit of $Qh\to\infty$ two continua are generated [66]. Note the analogy with the Sezawa modes of a slab discussed in sections 3.2 and 6.2. Guided-wave polaritons are either TM or TE polarized.

Acknowledgements

The experimental results shown in this paper are due to J.R. Sandercock and to G.I.Stegeman and coworkers: the very fruitful cooperation in joint research projects is gratefully acknowledged. This work has been partially supported by Ministero Universita' e Ricerca (grant 40%) and by Consiglio Nazionale delle Ricerche.

References

1. L. Brillouin: Ann. Phys. (Paris) **17**, 88 (1922)
2. C.V. Raman: Ind. J. Phys. **2**, 387 (1928)
3. G. Landsberg, L. Mandelstam: Naturwiss. **16**, 557 (1928)
4. P. Jacquinot: Rep. Prog. Phys. **23**, 268 (1960)
5. J.R. Sandercock: In Light Scattering in Solids III, ed. by M. Cardona, G. Güntherodt (Springer, Berlin, Heidelberg 1982) p.173
6. J.R. Sandercock: In 2nd Int. Conf. on Light Scattering in Solids, ed. by M. Balkanski (Flammarion, Paris 1969) p.9
7. S.M. Lindsay, I.W. Shephard: J. Phys. E **10**, 150 (1977)
8. C. Roychoudhuri, M.Hercher: Appl. Opt. 16, 2514 (1977)
9. J.G. Dil, N.C.J.A. van Hijningen, F. van Dorst, R.M. Aarst: Appl. Opt. **20**, 1374 (1981)
10. J.R. Sandercock: In Vibration Control in Optics and Metrology, SPIE **732**, 157 (1987)
11. R. Mock, B.Hillebrands, J.R.Sandercock: J. Phys. E **20**, 656 (1987)
12. J.R. Sandercock: J. Phys. E **9**, 566 (1976)
13. B.A. Auld: Acoustic Fields and Waves in Solids (J.Wiley, New York 1972) Vol.1 and Vol.2
14. G.W. Farnell: In Physical Acoustics, Vol.6, ed. by W.P. Mason, R.N. Thurston (Academic Press, New York 1970) pag.109
15. G.W. Farnell, E.L. Adler: In Physical Acoustics, Principles and Methods, Vol.9, ed. by W.P. Mason, R.N. Thurston (Academic Press, New York 1972) pag.35
16. A.M. Marvin, V. Bortolani, F. Nizzoli: J. Phys. C **13**, 299 (1980)
17. V. Bortolani, A.M. Marvin, F. Nizzoli, G. Santoro: J. Phys. C **16**, 1757 (1983)
18. R. Loudon: Phys. Rev. Lett. **40**, 581 (1978)
19. R. Loudon: J. Phys. C **11**, 403 (1978)
20. K.R. Subbaswamy, A.A.Maradudin: Phys. Rev. B **18**, 4181 (1978)
21. R.E. Camley, F. Nizzoli: J. Phys. C **18**, 4795 (1985)
22. N.E. Glass, A.A. Maradudin: J. Appl. Phys. **54**, 79 (1983)
23. V.R. Velasco, F. Garcia-Moliner: J. Phys. C **13**, 2237 (1980)
24. L.D. Landau, E.M. Lifshitz: Electrodinamics of Continuous Media (Pergamon Press, New York, 1958) chap.14
25. D.L. Mills, A.A. Maradudin, E.Burnstein: Ann. Phys. N.Y., **56**, 504 (1970)
26. B.I. Bennett, A.A.Maradudin, L.R. Swanson: Ann. Phys. N.Y. **71**, 357 (1972)
27. A.S. Pine: Phys. Rev. B **5**, 3003 (1972)
28. J.R. Sandercock: Phys. Rev. Lett. **28**, 237 (1972)
29. G. Dresselhaus, A.S. Pine: Solis State Commun. **16**, 1001 (1975)
30. A. Dervisch, R. Loudon: J. Phys. C **9**, L669 (1976)
31. N.L. Rowell, G.I. Stegeman: Phys. Rev. Lett. **41**, 970 (1978)
32. O. Keller: Phys. Rev. B **29**, 4659 (1984)
33. R. Loudon, J.R. Sandercock: J. Phys. C **13**, 2609 (1980)
34. V.R. Velasco and F. Garcia-Moliner: Solid State Commun. **33**, 1 (1980)
35. N.L. Rowell, G.I. Stegeman: Phys. Rev. B **18**, 2598 (1978)
36. A.M. Marvin, V. Bortolani, F. Nizzoli, G. Santoro: J. Phys. C **13**, 1607 (1980)
37. V. Bortolani, A. Franchini, F. Nizzoli, G. Santoro: Phys. Rev. Lett. **52**, 429 (1984)
38. B. Hillebrands, S.Lee, G.I.Stegeman, H.Cheng, J.E.Potts, F.Nizzoli: Phys. Rev. Lett. 60, 832 (1988)
39. V. Bortolani,, F. Nizzoli, G. Santoro: Phys. Rev. Lett. **41**, 39 (1978)
40. V. Bortolani, F. Nizzoli, G. Santoro: In Ab Initio Calculation of Phonon Spectra, ed. by J.T. Devreese (Plenum Publ. Co., New York 1983) p.201
41. J.R. Sandercock: Solid St. Commun. **26**, 547 (1978)
42. L. Bassoli, F.Nizzoli, J.R.Sandercock: Phys. Rev. B **34**, 1296 (1986)
43. V. Bortolani, F.Nizzoli, G.Santoro, A.Marvin, J.R.Sandercock: Phys. Rev. Lett. 43, 224 (1979)

44. J.R. Sandercock, F. Nizzoli, V. Bortolani, G. Santoro, A.M. Marvin: In Proc. 3rd Europ. Conf. on Surface Science (Societe du Vide, Paris 1980) p.754
45. J.A. Bell, R.J.Zanoni, C.T.Seaton, G.I.Stegeman, J. Mokous, C.M.Falco: Appl. Phys. Lett. **52**, 610 (1988)
46. R. Jorna, D. Visser, V. Bortolani, F. Nizzoli: J. Appl. Phys. **65**, 718 (1989)
47. D.A.G. Bruggeman: Ann. Physik (Leipzig) **29**, 160 (1937)
48. S.M. Rytov: Akust. Zh. **2**, 71 (1956) [Sov. Phys.-Acoust. 2, 68 (1956)]
49. M. Grimsditch: In Light Scattering in Solids V, ed. by M. Cardona, G. Güntherodt (Springer, Berlin, Heidelberg 1989) p.285
50. A. Kueny, M. Grimsditch, K. Miyano, I. Banerjee, C. Falco, I.K. Schuller: Phys. Rev. Lett. **48**, 166 (1982)
51. J. Sapriel, J.C.Michel, J.C.Toledano, R.Vacher, J.Kervarec, A. Regreny: Phys. Rev. B **28**, 2007 (1983)
52. J.A. Bell, W.R.Bennett, R.Zanoni, G.I.Stegeman, C.M.Falco, F.Nizzoli: Phys. Rev. B **35**, 4127 (1987)
53. J.A. Bell, W.R.Bennett, R.Zanoni, G.I.Stegeman, C.M.Falco, C.T.Seaton: Solid State Commun. **64**, 1339 (1987)
54. S. Lee, B.Hillebrands, G.I.Stegeman, H.Cheng, J.E.Potts, F.Nizzoli: J. Appl. Phys. **63**, 1914 (1988)
55. F. Nizzoli, B. Hillebrands, S. Lee, G.I. Stegeman, G. Duda, G. Wegner, W. Knoll: Phys. Rev. B **40**, 3323 (1989)
56. J.R. Sandercock: Phys. Rev. Lett. **29**, 1735 (1972)
57. M. Grimsditch, R.Bhadra, I.K.Schuller: Phys. Rev. Lett. **58**, 1216 (1987)
58. F. Abeles: In Electromagnetic Surface Excitations, ed. by R.F. Wallis, G.I. Stegeman, Springer Ser. Wave Phen. Vol.3 (Springer, Berlin, Heidelberg 1986) p.8
59. E. Kretschmann: Z. Phys. **227**, 412 (1969); **241**, 313 (1971)
60. M. Fukui, O. Toda, V.C-Y. So, G.I. Stegeman: Solid State Commun. **36**, 995 (1980); J. Phys. C **14**, 5591 (1981)
61. A.M. Marvin, V. Bortolani, F. Nizzoli, G. Santoro, V. Celli: J. Phys. C **15**, 3273 (1982)
62. A.L. Moretti, W.M. Robertson, B. Fisher, R. Bray: Phys. Rev. B **31**, 3361 (1985)
63. W.M. Robertson, A.L. Moretti, R. Bray: Phys. Rev. B **35**, 8919 (1987)
64. S. Lee, B. Hillebrands, J. Dutcher, G.I. Stegeman, W. Knoll, F. Nizzoli: to be published
65. N. Marucci, F. Nizzoli: to be published
66. S. Ushioda, R. Loudon: In Surface Polaritons, ed. by V.M. Agranovich, D.L. Mills (North-Holland, Amsterdam, 1982) pag.535
67. J.B. Valdez, S. Ushioda: Phys. Rev. Lett. **38**, 1098 (1977)
68. S. Ushioda, J.B. Valdez, W.H. Ward, A.R. Evans: Rev. Sci. Instrum. **45**, 479 (1974)
69. J.Y. Prieur, S. Ushioda: Phys. Rev. Lett. **34**, 1012 (1975)
70. K.L. Kliewer, R. Fuchs: Phys. Rev. **144**, 495 (1966).

Second Harmonic Generation in Optical Fibres

D.A. Weinberger

Department of Electrical Engineering and Computer Science,
University of Michigan, Ann Arbor, MI48109, USA

Abstract. The recent observations of efficient second harmonic generation in optical fibers have stimulated great interest, since in theory all such second-order nonlinear optical processes are disallowed. This article reviews progress to date in the understanding of this intriguing phenomenon.

1. Introduction

Optical fibers are attractive media for the observation and study of many nonlinear phenomena. By virtue of their small core size and waveguide geometry which affords long interaction lengths, fibers can exhibit nonlinearities at modest input powers. The third order nonlinear susceptibility $\chi^{(3)}$ gives rise to such effects as stimulated Raman and Brillouin scattering, four-photon parametric processes, and intensity-dependent refractive index phenomena, all of which have been studied extensively and utilized to achieve important functions for optical signal processing and communications applications [1-3]. For example, fiber Raman oscillators and amplifiers hold promise for sources and repeaters; self-phase modulation in fibers allows compression of optical pulses into the femtosecond regime; and the combination of self-phase modulation, group velocity dispersion, and Raman gain was used to produce solitons which propagated over more than 4000 km of fiber with no discernible attenuation [4].

The second-order nonlinear susceptibility $\chi^{(2)}$, however, vanishes in all media with inversion symmetry. Recent observations of efficient second harmonic generation (SHG) in short fiber lengths were therefore quite surprising, in light of the centrosymmetry of fused silica. The operative mechanism is still not fully understood.

During the period 1980-84, several groups reported the first observations of weak SHG in fibers with Ge-doped cores [5-7]. In these experiments, mode locked, Q-switched Nd:YAG pulses at 1.06 μm with peak powers ~1-100 kW were coupled into long fiber lengths ~5-400 m and produced SH output at 532 nm with typical conversion efficiencies $\leq 10^{-3}$. In similar experiments in 1986 *Østerberg and Margulis* [8] achieved ~5% conversion efficiency with 5-40 kW peak input power, in fibers co-

Springer Series in Wave Phenomena, Vol. 9 **Nonlinear Optics in Solids**
Editor: O. Keller

doped with Ge and P with lengths as short as 50 cm. The output at 532 nm was sufficiently intense to pump a commercial dye laser [9]. They further observed that the fiber core needed to be illuminated with the Nd:YAG input at 1.06 μm over a period of many hours to produce a measurable SH output. During this preparation period the SH signal grew exponentially before finally saturating after 12 h at an average power level of ~1 mW. Apparently the fibers were somehow being permanently conditioned by the incident light to produce a non-zero $\chi^{(2)}$ in the fiber core, presumably through the creation of internal defects.

These experiments generated a great deal of excitement and raised hopes that short pieces of fiber could replace expensive, cumbersome, easily-damaged nonlinear crystals in many applications. Unfortunately, even though a record 13% SH efficiency has recently been attained [10], the prospects in the near future for efficient fiber doublers at arbitrary wavelengths are remote. Nonetheless, progress has been made in elucidating certain aspects of SHG in fibers: in particular it is now well-accepted that the SHG arises due to a permanent $\chi^{(2)}$ grating written into a short section of fiber. In the following sections we discuss the SHG phase matching problem in fibers, quadrupolar contributions to SHG, the mixing model for self-organization of the $\chi^{(2)}$ grating, dynamics of the grating growth, evidence for defect structures, and electric-field-induced SHG and related experiments.

2. Phase Matching Considerations

In a medium with a bulk $\chi^{(2)}$, the nonlinear polarization at frequency 2ω produced by a pump field at ω is of the form

$$\mathcal{P}(2\omega) = \epsilon_0 \chi^{(2)} |E(\omega)|^2 . \quad (1)$$

The coupled mode equations yield a simple result for the growth of the SH field with distance z:

$$dE(2\omega)/dz \sim \chi^{(2)} |E(\omega)|^2 \exp(i\Delta\beta z) . \quad (2)$$

Here $\Delta\beta = \beta(2\omega) - 2\beta(\omega)$ is the phase mismatch due to the different phase velocities of the fundamental and second harmonic; in fibers $\Delta\beta$ can incorporate both material and waveguide dispersion. After a distance

$$\mathcal{L} = 2\pi/\Delta\beta , \quad (3)$$

known as the coherence length, the phase mismatch leads to destructive interference of the SH generated at different axial positions in the fiber. Thus $\mathcal{L}$ is a measure of the maximum useful distance for producing SH power. The effective index N is defined by $\beta = Nk$, where k is the free-space propagation constant. The phase matching problem can then be regarded as matching the effective indices $N(\omega) = N(2\omega)$ so that $\Delta\beta \rightarrow 0$, and the effective $\mathcal{L}$ becomes infinite. Under these conditions the SH power then depends quadratically on $\chi^{(2)}$, distance z, and the input fundamental (pump) power, in the limit of negligible pump depletion.

Using dispersion data for fused silica [11] yields $n(\omega) = 1.4497$ and $n(2\omega) = 1.4607$ at 1.064 μm and 0.532 μm, respectively. The material dispersion is then $\Delta n = 0.011$; this will be increased slightly for co-doping with P or Ge. The waveguide contribution to dispersion for usual fibers for ω in the vicinity of 1.06 μm is typically at most $0.5\Delta n$, and is often much less. However, it can either add to or subtract from Δn, leading to a ΔN either larger or smaller than the Δn due to material dispersion alone. The modal propagation constant for a step-index fiber is given by

$$\beta = k\sqrt{b(n_{co}^2 - n_{cl}^2) + n_{cl}^2} \ , \tag{4}$$

where n_{co} and n_{cl} are the core and cladding indices at the frequency of interest. In evaluating $\Delta\beta$, the first term under the radical would contribute to waveguide dispersion and the second term to material dispersion. Here b is the normalized propagation constant which must be obtained numerically from the eigenvalue equation for a fiber with a normalized frequency V given by

$$V = k\rho\sqrt{n_{co}^2 - n_{cl}^2} \ , \tag{5}$$

where ρ is the core radius. Waveguide dispersion in the form of b-V plots for different fiber modes is given in [12]. Ignoring waveguide dispersion for the moment and setting $\Delta N = \Delta n = 0.011$ yields $\Delta\beta = 2\omega\Delta n/c = 1304\ \text{cm}^{-1}$ and $\mathcal{L} = 48$ μm.

From the above it is clear that the high SH powers seen in fibers demand not only a non-zero $\chi^{(2)}$, but also some sort of phase matching mechanism. One possibility is to use modal dispersion to cancel the material dispersion Δn. This is illustrated in Fig. 1 where the frequency dependence of N for the various modes of a step index fiber is shown [13]. The curves are labeled according to the LP_{mn} mode designation, where the far-field intensity pattern of the mode has m radial nodes (diametrical lines) and $n-1$ azimuthal nodes (circles). The dashed curves are the LP_{01} mode data replotted with the frequency scale doubled. The intersections with the other modal curves give the SH frequencies 2ω at which phase matching is possible with the input beam in the LP_{01} mode. For example, in the upper figure doubling into the LP_{21} mode at a frequency of ≈15,000 cm^{-1} would be possible, corresponding to an input of the fundamental (ω) at a wavelength of ≈1.33 μm.

The difficulty in using waveguide dispersion to achieve phase matching in fibers lies in the rather large value of $\Delta n \approx 0.011$, which requires a fiber with $n_{co} - n_{cl}$ of comparable magnitude. Nonetheless, *Fermann* et al. [14] accomplished phase-matched doubling of the LP_{01}

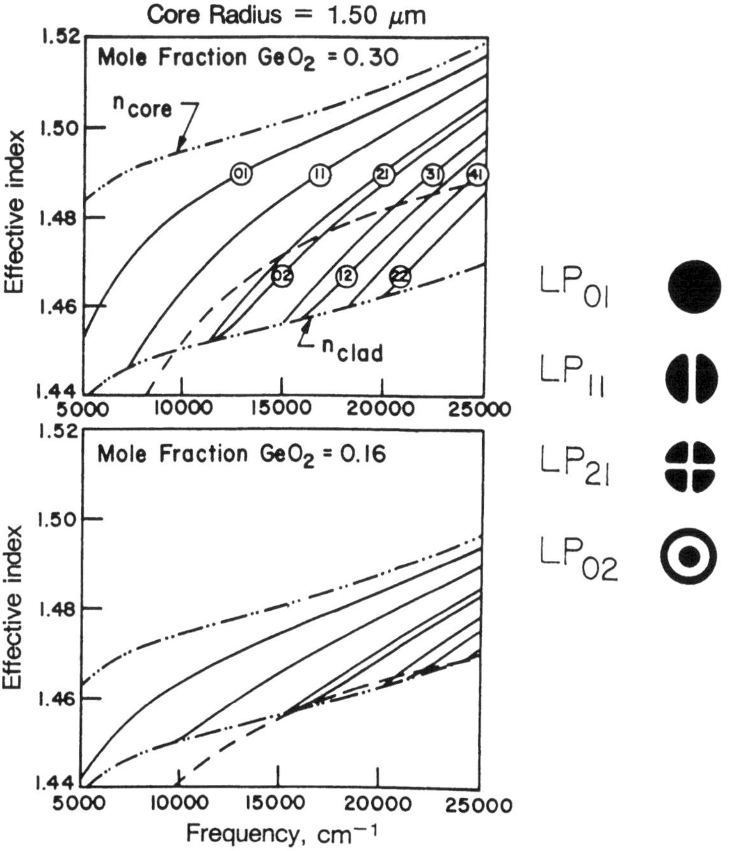

Fig. 1. Frequency dependence of the effective index of refraction N for fused silica fibers of core radius 1.5 μm with two different mole fractions of GeO_2 in the core. A few of the lowest order mode intensity profiles are indicated.

mode into the LP_{02} mode at a fundamental wavelength of 1.208 μm; and *Kashyap* [15] fabricated a fiber with $n_{co} - n_{cl} \approx 0.3$ to demonstrate phase-matched doubling for $LP_{01} \rightarrow LP_{31}$ at 1.06 μm.

Another problem occurs because of the nonuniformity of fibers, which effectively leads to a wandering phase match condition along the length of the fiber. For example, it is not unusual for the core radius to vary by as much as one part in 10^4 during the fiber drawing process. Figure 2 shows how the phase match frequency ω for doubling of the LP_{01} mode varies with core radius [13]. It is clearly evident that such variations could limit the SH power: the frequency range over which SH could be observed, typically several cm^{-1}, would be inversely proportional to the slope of the curves. *Terhune and Weinberger* [13]

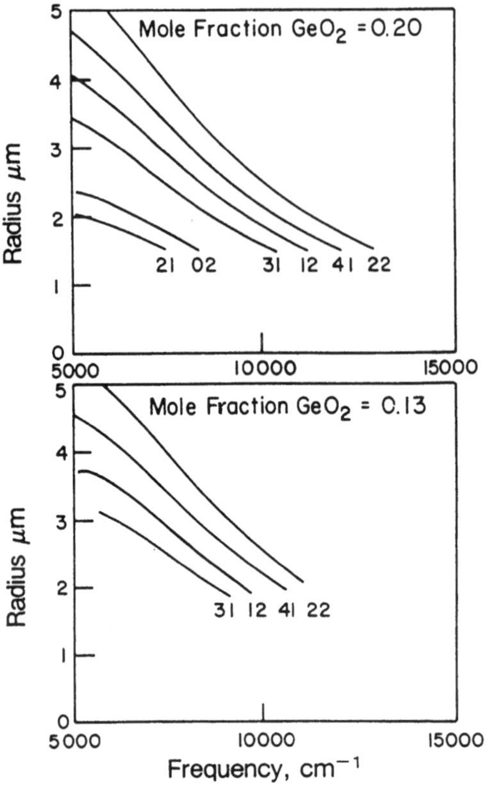

Fig. 2. Variation of the phase-matching frequency for doubling of the LP_{01} mode, as the core radius is varied, for two different mole fractions of GeO_2.

treated this problem by modeling an imperfect fiber as consisting of a number of segments characterized by some mean length. Within each segment the fiber parameters are constant, and the SH generated from different segments is incoherent. With this model, if all segments are far from phase matched, then the SH power is the average power generated in a segment times the total number of segments. Close to phase matching, the total SH power integrated over frequency is constant, independent of the segment length. Variation in the fiber parameters will thus lead to a broadening of the frequency range over which SH can be observed, with a proportional decrease in the peak intensity.

A final problem with modal phase matching is that the overlap integral of the fields taking part in the nonlinear interaction may become very small or vanish completely, so that not all mode combinations are possible. The SH power $P(2\omega) \equiv \iint |E(2\omega)|^2 \, dx\, dy$ is obtained from (2) by evaluating the integral over the fiber cross section

$$O \sim \int_{-\infty}^{+\infty} \int_{-\infty}^{+\infty} |E(\omega)|^2 \, E^*(2\omega) \, dx\, dy \,. \qquad (6)$$

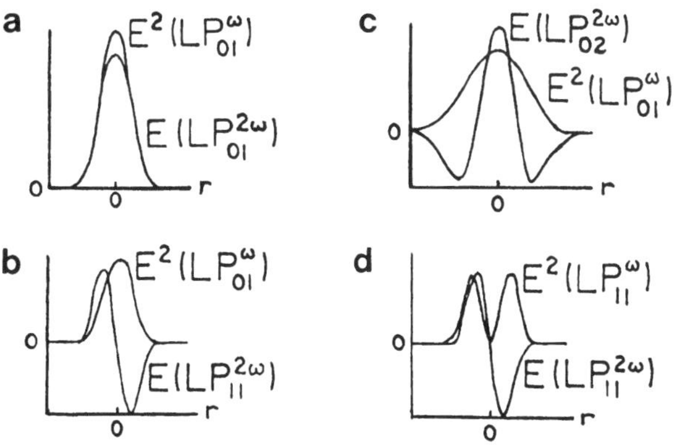

Fig. 3. Examples of mode combinations used to evaluate the overlap integrals defined in (6), showing (a) good overlap; (b) partial cancellation; (c), (d) complete cancellation.

From Fig. 3 it is evident that mode cancellations can significantly reduce P(2ω). For the $LP_{01} \to LP_{11}$ and $LP_{11} \to LP_{11}$ mode combinations the overlap integral O vanishes completely, while for $LP_{01} \to LP_{02}$ there is partial cancellation.

A completely different phase matching mechanism is provided by a $\chi^{(2)}$ which alternates sign periodically, referred to as quasi-phase matching [16]. If $\chi^{(2)} = |\chi^{(2)}| \exp(-i\Delta\beta z)$, then from (2) phase matching is accomplished. Note in this case that the period of $\chi^{(2)}$ is just $\mathscr{L}$, the usual coherence length in the absence of phase matching.

3. Electric Quadrupole and Related Nonlinearities

In an attempt to explain the earliest reports of SHG in fibers with efficiencies $\sim 10^{-3}$, *Terhune and Weinberger* [13] considered the contribution of nonlinearities at the core-cladding interface and a bulk nonlinear polarization proportional to **E** ∇**E**, which involves electric quadrupole and magnetic dipole contributions. This is the lowest order nonlinearity that is nonvanishing in centrosymmetric media; the polarization can be written as follows using three constants:

$$\mathcal{P}_k(2\omega)/\epsilon_0 = (\delta - \beta - 2\gamma)E_i(\omega)\,\nabla_i E_k(\omega) + \beta E_k(\omega)\,\nabla_i E_i(\omega) + 2\gamma E_i(\omega)\,\nabla_k E_i(\omega)\,, \qquad (7)$$

where repeated indices indicate summation. The current density **J**, the source term in Maxwell's equations, can be expanded in terms of multipole moments:

$$\mathbf{J} = \partial\mathbf{P}/\partial t + c(\nabla \times \mathbf{M}) - \partial(\nabla \cdot \mathbf{Q})/\partial t\,, \qquad (8)$$

where **P**, **M**, and **Q** are the electric dipole, magnetic dipole, and electric quadrupole moments per unit volume, respectively. To obtain the SH polarization one would then calculate **P** proportional to **E** ∇**E**, **M** proportional to $\mathbf{E} \times \partial\mathbf{E}/\partial t$, and **Q** proportional to $\mathbf{E}^2$. The total **J** for these effects can be represented by an effective induced dipole moment. This has been done in defining (7), and the coefficients $(\delta - \beta - 2\gamma)$, β, and 2γ are effective dipole polarizabilities.

Many effects can lead to a polarization at the core-cladding interface; for example, aligned molecules could produce a piezoelectric layer. These effects cannot be separated from those of the field gradients in the vicinity of the interface leading to SHG through a nonlinear polarization as in (7). Rather, all of these effects together are described phenomenologically by assuming there is an effective nonlinear polarization just inside the interface dependent on the components of the fields on the core side as follows:

$$
\begin{aligned}
\mathcal{P}_r(2\omega)/\varepsilon_0 &= A\,E_r^2(\omega) + C\,[E_r^2(\omega) + E_\phi^2(\omega) + E_z^2(\omega)]\,, \\
\mathcal{P}_\phi(2\omega)/\varepsilon_0 &= B\,E_r(\omega)\,E_\phi(\omega)\,, \\
\mathcal{P}_z(2\omega)/\varepsilon_0 &= B\,E_r(\omega)\,E_z(\omega)\,. \qquad (9)
\end{aligned}
$$

Here the coefficients A, B, and C again represent dipole polarizabilities.

The expressions for the nonlinear polarization in (7) and (9) are used to evaluate the SH power, leading this time to overlap integrals of the form

$$
O_q \sim \int_{-\infty}^{+\infty}\int_{-\infty}^{+\infty} E(\omega)\,\nabla E(\omega)\,E^*(2\omega)\,dx\,dy\,. \qquad (10)
$$

Note that this quadrupolar-type nonlinearity cannot couple the mode combination $LP_{01} \rightarrow LP_{01}$. This can be visualized from Fig. 3 , since for the LP_{01} mode ∇E resembles the LP_{11} modal field. In fact, the quadrupolar nonlinearity can only couple the even LP_{01} mode (typical fibers are usually single mode at the pump frequency ω) with a mode of odd symmetry. This was consistent with what was seen in the intial observations of SHG in long fibers, where the SH output was never observed to be in the lowest order fiber mode.

Nonetheless, *Terhune and Weinberger* concluded that this type of nonlinearities were far too small to explain the observed efficiencies, as a consequence of significant cancellations between the various terms in the nonlinear polarization. The highest SHG efficiency calculated was $\approx 10^{-5}$, for the phase-matched $LP_{01} + LP_{11} \rightarrow LP_{22}$ mode combination. For calculations attempting to duplicate the non-phase-matched experimental conditions of *Sasaki and Ohmori* [6], conversion efficiencies of 10^{-10} were obtained, nearly seven orders of magnitude too small. The experiments of *Østerberg and Margulis* [8] posed even more of a problem; in fact, on occasion they did observe the SH output to be in the LP_{01} mode. This was a clear indication that the fiber core had indeed changed symmetry during the preparation process, and that some other mechanism must be operative.

Other authors have also performed theoretical calculations of SHG in fibers. *Manassah* et al. [17] investigated SHG in birefringent fibers due to quadrupolar terms and proposed a method to combine material, modal, and geometrical dispersion to accomplish phase matching. *Payne* [18] analyzed SHG due to quadrupole and magnetic dipole sources and predicted that phase-matched SHG with 50% efficiency should be possible in a fiber only 50 cm long. This has not been supported by experiments [14,15]. Further, he concluded that the electric quadrupole contribution was negligible and so retained only magnetic dipole terms; it is likely that important cancellations of terms were thereby overlooked. *Chmela* [19] attempted to explain the self-preparation of fibers for SHG over many hours as arising due to a fifth order nonlinearity with its origin in quantum noise. Finally, *Chen* [20] performed calculations similar to those in [13], specifically for the $LP_{01} \rightarrow LP_{11}$ mode combination, and concluded that the roles of the core-cladding and bulk nonlinearities were comparable. This was in contrast to [13] where the interface contributions almost always dominated; in only a few cases of non-phase-matched doubling were the interface and bulk terms comparable.

4. The Mixing Model for Self-Organization of $\chi^{(2)}$ Gratings

It was *Stolen and Tom* [21] in 1987 who proposed the phenomenological model that was successful in explaining many of the observations of SHG in fibers. They reasoned that quasi-phase matching was a likely candidate to explain the high observed efficiencies, and that the $\chi^{(2)}$ grating probably arose due to photo-induced (charged) defects. But how did those defects, presumably created uniformly throughout the fiber, know how to alternately orient with just the right phase matching periodicity? *Stolen and Tom* realized that a dc polarization could be created in the fiber core through the third-order dipole-allowed nonlinearity:

$$\mathcal{P}_{dc} \sim |\chi^{(3)} E(\omega) E(\omega) E^*(2\omega)| \cos(\Delta\beta z) . \quad (11)$$

In other words, if a fundamental wave at frequency ω *and* a second harmonic wave at 2ω are lauched into the fiber, they can mix through the $\chi^{(3)}$ nonlinearity to automatically create a dc field which is spatially modulated at frequency $\Delta\beta$. If one further assumes that the induced $\chi^{(2)}$ amplitude is proportional in some way to $\mathcal{P}_{dc}$, then the period of the $\chi^{(2)}$ grating is just $\mathcal{L}$, and quasi-phase matching has been accomplished.

To test their theory *Stolen and Tom* simultaneously injected infrared light at 1.06 μm and green light at 532 nm into a fiber that was single mode at both wavelengths, which ruled out the possibility of SHG arising due to quadrupolar-type effects. They observed that it took only ≈5 min of this so-called seeding to permanently condition the fiber for SHG, as opposed to many hours of self-preparation. Thus the essence of the model was confirmed.

A similar macroscopic model was put forth by *Farries* et al. [22], who proposed that color centers were written periodically into the fiber at locations where the green intensity was the highest, that is where the pump and SH waves were in phase. This model works equally well, even though the $\chi^{(2)}$ grating has a dc component in this case. *Baranova and Zel'dovich* [23] obtained a similar model for the $\chi^{(2)}$ grating by considering a sum of one- and two-photon absorption processes.

Note that the *Stolen and Tom* model makes no assumptions about the nature of the defects or how they are created. As discussed later, the defects are likely excited through multi-photon absorption and there is strong evidence they are Ge-related. It is tempting to assume that the dc field is responsible for orienting the photo-induced defects. However, the strength of the internal mixing field can be calculated to be only about 1 V/cm. *Mizrahi* et al. [24] applied external electric fields of nearly 10^4 V/cm to preforms from which efficient SHG fibers had been drawn. Although defects inside the preform could indeed be oriented to produce a non-phase-matched, measurable SH signal (which could be separated from the background electric-field-induced SHG due to the $\chi^{(3)}$ nonlinearity), the effect was not permanent: the SH signal decayed within minutes of the removal of the external field. They obtained similar results in the fibers themselves [25]. Thus, it is highly unlikely that the dc field itself is responsible for orienting the defects. By what mechanism the defects are permanently oriented in a periodic fashion is a completely open question at present, although some clever (but unverified) models have been proposed [26,27].

Recall that for self-prepared fibers many hours of illumination with just the pump at frequency ω are required to condition the fiber for SHG. The question of how this self-preparation process occurs, within the context of the mixing model, is also unresolved. Assuming the same mechanism is operative, then the harmonic field at 2ω must arise spontaneously. One obvious source is the weak SH due to quadrupolar-type effects. Strong evidence in support of this is that certain self-prepared fibers always produced the SH output in a higher order fiber mode of odd symmetry; and other fibers which were single mode at 2ω could not be self-prepared [21]. Counter-evidence was provided by *Margulis and Østerberg* [28] who occasionally did observe the SH to emerge in the lowest order fiber mode. Also, *Saifi and Andrejco* [29] observed that the fiber modal structure of the SHG always duplicated that of the seed beam at 2ω, in distinct contrast to the self-preparation case. It may in fact be that the self-seeding field arises due to different mechanisms in different fibers, or even that the self-preparation process itself is fundamentally different from that of seeding.

Another question concerns the effect of the finite bandwidth of the seeding pulses, which will impose a fundamental limit on the maximum SHG attainable. One can conceptually think of the problem in terms of different frequencies in the seeding beams writing gratings of slightly different periods. If the gratings of minimum and maximum period get out of step in a distance L by some amount, say π, then L defines the maximum useful grating length for producing coherent SHG. [A comment here on terminology: this maximum useful grating length has been referred to by many authors as the "coherence length" of the grating. This usage will be avoided to prevent any confusion with the usual meaning of coherence length as defined in (3), which is just the *period* of the $\chi^{(2)}$ grating.] If the seeding pulses have bandwidths of $\Delta\beta(\omega)$ and $\Delta\beta(2\omega)$, then by differentiation (3) gives the resulting range in grating periods $\Delta\Lambda$:

$$\Delta\Lambda \;=\; 2\pi\,\Delta[\beta(2\omega) - 2\beta(\omega)]^{-1} \;=\; 2\pi\,[2\Delta\beta(\omega) - \Delta\beta(2\omega)]\,/\,(\Delta\beta)^2 \;, \qquad (12)$$

which leads to a maximum useful grating length

$$L \;=\; \pi\,/\,[2\Delta\beta(\omega) - \Delta\beta(2\omega)] \;. \qquad (13)$$

Taking as a worst-case crude estimate $\Delta\beta(\omega) \approx 6\ \text{cm}^{-1}$, which is the gain bandwidth of Nd:YAG (by comparison $\beta(\omega) \approx 90{,}000\ \text{cm}^{-1}$), and assuming $\Delta\beta(2\omega) = \sqrt{2}\,\Delta\beta(\omega)$, we find $L \approx 1$ cm. This is in good agreement with direct cut-back measurements made on fibers prepared with mode locked, Q-switched infrared pulses [9], in which the SHG signal descreased by about 70% after the first few cm of the input end of the fibers were cut off; and with other inferred grating length measurements of approximately 12 cm [22].

This limitation imposed by the bandwidth of the seeding pulses was analyzed in detail by *Tom* et al. [30], who incorporated the effect of the differential fiber dispersion in evaluating $\Delta\beta$. They pointed out that other nonlinear effects in the fiber, such as self-phase and cross-phase modulation, can further spectrally broaden the seeding pulses. They reasoned that it should therefore be advantageous to use mode locked pulses for seeding, since the higher-peak-power, mode locked, Q-switched pulses would suffer a larger spectral broadening. By measuring the spectral width of the SHG and comparing with the predictions of their analysis, they demonstrated a grating length of 35 cm for mode locked seeding. This improvement of a factor of $\approx$10 was also the factor by which their mode-locked pulse spectrum was reduced relative to that of the mode locked, Q-switched pulses. It is unclear why other mode locked seeding experiments lead to grating lengths of only a few cm, as determined by fiber cutback measurements [31]. However, a very recent analysis which explicitly incorporates the effects of self-phase and cross-phase modulation predicts that there is an intensity-dependent deviation from exact phase matching that varies across a pulse, so that a direct comparison of fibers seeded with low and high powers is not so straightforward [32].

A number of experiments have now confirmed the existence of a $\chi^{(2)}$ grating in fibers conditioned for SHG. First it was demonstrated that the SHG from a fiber was coherent with the transmitted infrared light which had been doubled in a KTP crystal, proving that the SHG was indeed phase-matched [28]. Next, several experiments with birefringent fibers provided indirect proof for the grating. In one experiment the wavelength at which the peak SHG could be obtained from a conditioned fiber was tuned by axially stretching the fiber; the new grating period led to a new wavelength at which the phase match condition (3) was satisfied [33]. In another experiment the fiber birefringence was used to obtain phase-matched SHG over an interval of ±14 nm centered on the writing wavelength at 1.064 μm [34].

Finally, the first direct confirmation of the $\chi^{(2)}$ grating was obtained using a spatially resolved Raman spectroscopic analysis, which was

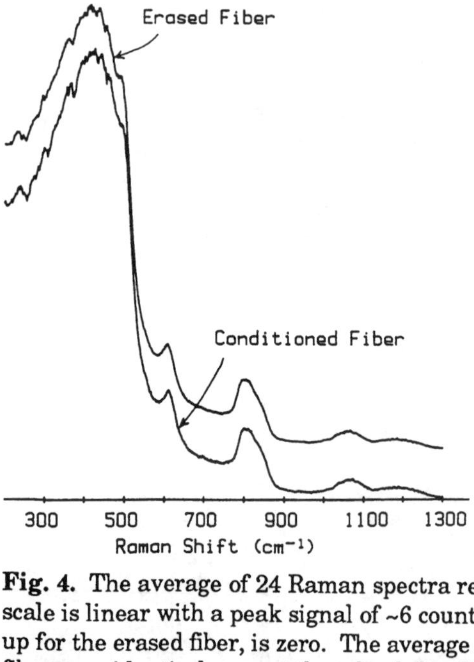

Fig. 4. The average of 24 Raman spectra recorded with 2 μm steps. The vertical scale is linear with a peak signal of ~6 counts/s, and the baseline, which is shifted up for the erased fiber, is zero. The average spectra for the erased and conditioned fibers are identical, except that the defect line at 490 cm^{-1} appears slightly more distinct in the erased fiber.

motivated by previous work in which small differences in defect lines in the Raman spectra of fibers which could and could not be conditioned for SHG were described [35]. *Kamal* et al. [36] recorded the spontaneously scattered Raman signal emitted 90° to the fiber axis from a 5-μm long section of fiber, into which was coupled the argon laser excitation at 514.5 nm. They recorded 24 such spectra displaced in steps of ≈2-3 μm along the fiber. Figure 4 displays the average of these 24 spectra for a fiber conditioned for SHG by seeding, and for the same fiber after the grating had been erased. (The erasure process is described in detail in the next section.) The broad peak at 450 cm^{-1} is attributed to the symmetric stretching of the bridging oxygens; and the weak shoulder at 490 cm^{-1} and the peak at 607 cm^{-1} are the defect lines in which the changes previously alluded to were seen by [35].

In Fig. 5(a) the average Raman signal from the conditioned fiber in the spectral interval 210-500 cm^{-1} is plotted for each of the 24 fiber posi-

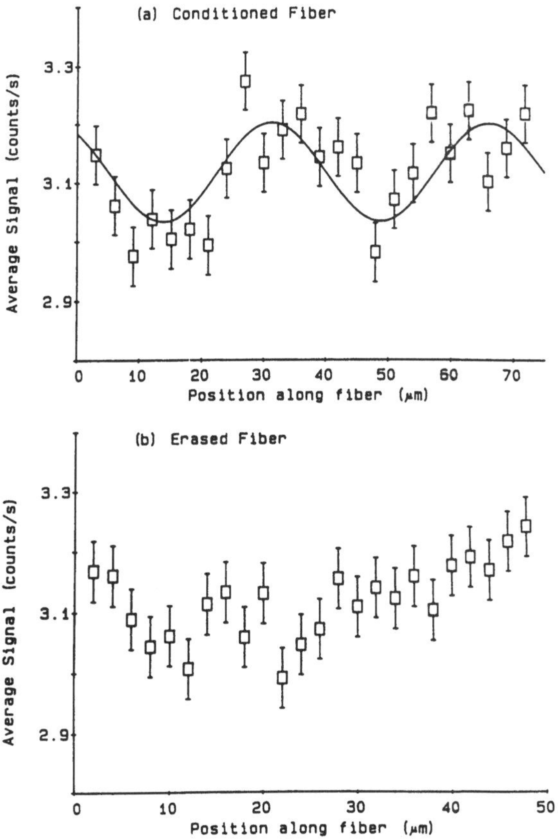

Fig. 5. Average Raman signal in the 210-500 cm^{-1} region as a function of the relative position along the fiber, for (a) conditioned fiber, and (b) the same fiber with the grating erased. Each spectrum was normalized to the 820 cm^{-1} band.

tions. A least-squares sinusoidal fit to the data gives a period of 35 μm; in fact a similar plot for *any* sub-region of the Raman spectrum below ≈600 cm^{-1} yields the same period to within a few percent. There is thus a periodic modulation of the entire low frequency part of the Raman spectrum, which is evidently a manifestation of the periodic arrangement of defects in the conditioned fiber. It is sensible that this modulation would appear in the softest modes and defect lines, both of which are at low frequency. This experiment was repeated three times on other conditioned fibers with similar results. Also, two similar experiments were performed as controls on erased fibers; the data for one is shown in Fig. 5(b). Attempts to fit this data to a sine wave gave very different results, depending on the spectral interval chosen. Further, the periods were either extremely short, approaching the spatial sampling period, or extremely long (≥140 μm).

The beauty of the self-organization process is that even in the presence of nonuniformities such as inhomogeneities, varying core size, etc., the fiber compensates to automatically arrange itself to be phase matched. A similar type of self-organized behavior in fibers was actually seen previously in 1978 by *Hill* et al. [37], who launched two single-frequency, counter-propagating argon laser beams into a Ge-doped fiber to form a standing wave, thereby writing a permanent, self-organized refractive index grating. Such Hill gratings were recently demonstrated to have negative group velocity dispersion in the visible region below 550 nm, raising the possibility of accomplishing soliton propagation at these wavelengths [38].

5. $\chi^{(2)}$ Grating Dynamics: Growth and Erasure

The mixing model describing the creation of the $\chi^{(2)}$ grating cannot address the issue of the dynamics of the grating writing process. One important question concerns saturation of $\chi^{(2)}$: does it occur and, if so, why? *Stolen and Tom* [21] predict one type of saturation mechanism: they showed that the SH generated by the induced $\chi^{(2)}$ will write an additional $\chi^{(2)}$, which is 90° out of phase with the induced $\chi^{(2)}$. As a result, when the SH generated by the fiber and the seed light at 2ω become of comparable intensity, the total writing field is decreased and the magnitude of $\chi^{(2)}$ along the fiber reaches a steady-state. This 90° phase shift has been recently confirmed experimentally [39]. Thus one can never hope to produce SHG at a power level greater than that of the SH seed.

Another obvious limit to the SHG arises when all the available defects are used up. A figure of merit characterizing conditioned fibers is $|\chi|L$, where $|\chi|$ is the peak amplitude of $\chi^{(2)}$ and L is the grating length as given in (13). The figure of merit leads to $|\chi| \approx 5 \times 10^{-12}$ esu for seeding with mode locked pulses , and to $|\chi| \approx 10^{-11}$ esu for seeding with mode locked, Q-switched pulses [30]. This rather startling result that $|\chi|$ is nearly the same (within a factor of two) for vastly different peak seeding powers is evidence in favor of saturation. Other experiments indicated saturation of the SHG at high preparation powers [40], but it was noted that this was probably an artifact of the logarithmic scales used to display the data [41].

An unexpected observation by *Ouellette* et al. [42] is that the $\chi^{(2)}$ grating can be *erased* by illumination with intense blue or green light, either from the Nd:YAG laser at 532 nm or from an argon laser. Such an erased fiber can be easily re-conditioned to its previous level of SHG, indicating that the bleaching process is probably randomizing the defect arrangement, rather than destroying the defects. The erasure can be monitored temporally by interrupting the bleaching beam periodically to measure the SH signal. Erasure data for a fiber co-doped with P and Ge at a bleaching power of P_{bl} = 125 mW (Nd:YAG average laser power at 532 nm) is shown in Fig. 6 [31].

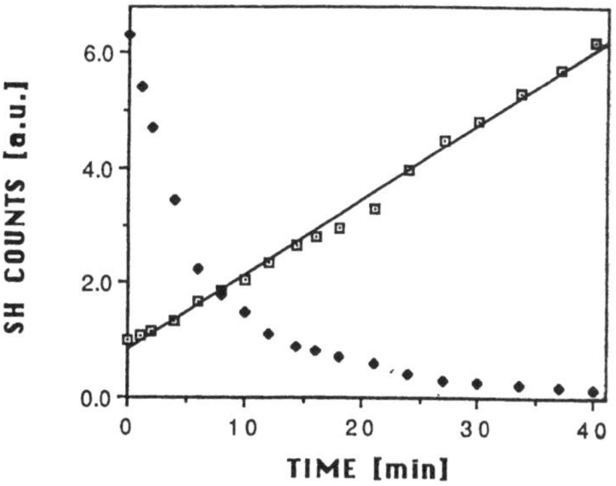

Fig. 6. Average SH power generated by the fiber from 100-ps mode locked pulses at 1.06 μm, as a function of the bleaching time. The bleaching power is 125 mW of average power at 532 nm; the fiber is continually irradiated except for a momentary interruption for the SHG measurement. The data illustrates decay according to (14); the straight line is a best fit to (15).

Ouellette et al. showed that the decay rate of the $\chi^{(2)}$ grating amplitude A has the following dependence:

$$dA/dt = -\beta A^2 , \tag{14}$$

where β is an effective time constant that depends on P_{bl}. The quadratic dependence on A is quite surprising, since bleaching of non-mutually interacting defects should yield a linear dependence on A. The erasure mechanism is therefore reminiscent of binary recombination statistics. If one further assumes that the SHG is proportional to A^2 [21], then the time dependence of the decay of the SH power has the following form:

$$P(t) = P(0) / [t/\tau + 1]^2 , \qquad (15)$$

where the time constant $\tau = \beta A(0)$. Thus a plot of $[P(0)/P(t)]^{1/2}$ vs time yields a straight line as shown in Fig. 6; for these data the time constant $\tau \approx 8$ min.

Krotkus and Margulis [43] investigated the conditioning of fibers co-doped with Ge and P in an attempt to address the question of why the preparation process terminates after a certain time. They concluded that self-erasure could lead to saturation of the SHG growth since the seeding green beam is simultaneously participating in writing and erasing the grating. They also determined that the time constant of the initial exponential growth (before saturation occurs) depended inversely on the square root of the seeding green power; thus fibers seeded with higher powers at 2ω prepare faster. These results imply that the conditioning rate scales as the seed field $E(2\omega)$; this led *Krotkus and Margulis* to consider the parametric mixing process $\chi^{(3)}(4\omega = 2\omega, \omega, \omega)$ which could lead to the generation of UV light at 0.266 μm. It is well known that UV light can induce permanent defects in fibers, and in fact the substitutional P_2 defect has an optical absorption energy which is very close to that of the 4ω quantum, so that such a third-order process could be resonantly enhanced.

Kamal and Weinberger [44] subsequently developed a rate equation model in an attempt to characterize the growth of the $\chi^{(2)}$ grating amplitude. They assumed that the induced $\chi^{(2)}$ has a time-dependent evolution in the case of seeding described by

$$d\chi^{(2)}/dt = \alpha E^2(\omega) E(2\omega) - \beta E^4(2\omega) [\chi^{(2)}]^2 - \gamma \chi^{(2)} . \qquad (16)$$

The first term represents creation of defects due to a three-photon process $(\omega, \omega, 2\omega)$ which contributes to writing of the grating; the second term represents grating erasure due to two-photon absorption at 2ω; and the last term takes into account possible disorienting mechanisms. It should be emphasized that the model makes no claims as to the microscopic physical mechanisms involved in the above processes. However, the various phenomenological terms are written in functional form consistent with experimental observations described previously.

The model predicts different behavior in fibers where grating erasure is more important and in fibers where it is less important. In the latter case, the steady-state grating amplitude is given by

$$\chi^{(2)}_{ss} = \alpha E^2(\omega) E(2\omega) / \gamma . \qquad (17)$$

Assuming the SHG is proportional to $[\chi^{(2)}_{ss}]^2$, then one expects the SH power $P(2\omega) \sim P_0^2(\omega) P_0(2\omega)$, where $P_0(\omega)$ and $P_0(2\omega)$ are the writing powers of the fundamental and second harmonic, respectively. Figure 7 shows experimental data obtained from fibers co-doped with Ge and P that approximately confirm this behavior. In the case where grating

Fig. 7

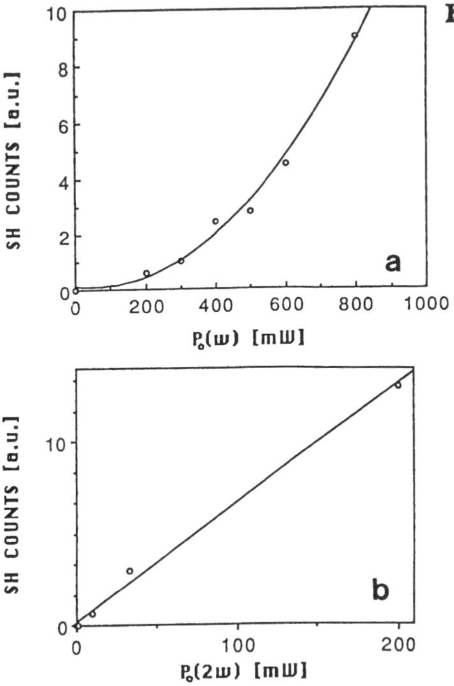

Fig. 8

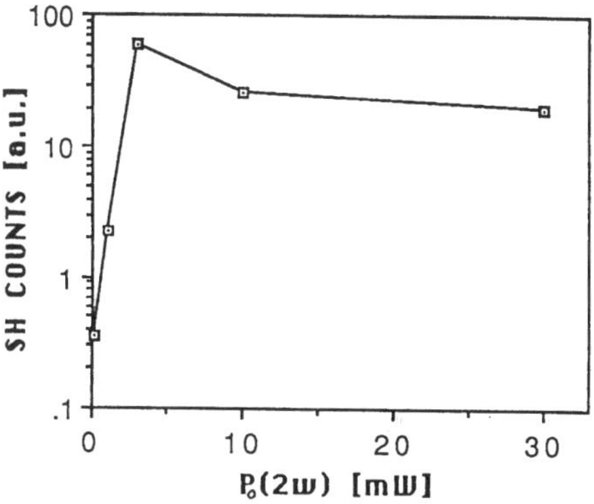

Fig. 7. Steady-state SH signal level produced by mode locked pulses at 1.06 μm, as a function of seeding powers used to write the grating: (a) quadratic dependence on $P_0(\omega)$, and (b) linear dependence on $P_0(2\omega)$. Solid curves indicate best fits to the data.

Fig. 8. Steady-state SH signal level produced by mode locked pulses at 1.06 μm, as a function of green seeding power used to write the grating, for a fiber in which self-erasure is significant. Saturation is evident at high seeding powers.

erasure during the writing process is significant, the model predicts that $\chi^{(2)}{}_{ss}$ attains a *maximum value* for an optimum value of the seeding power $P(2\omega)$. Figure 8 shows data obtained for a fiber doped solely with Ge; in addition the fiber was highly deficient in oxygen so that GeO rather than GeO_2 substitutional defects were dominant. The data display an evident turn-over in the maximum SH signal achievable after the seeding power $P(2\omega)$ is increased beyond a certain level.

This rate equation model thus predicts that grating erasure during the writing process can indeed lead to saturation of the SHG. Further, fibers with a small β coefficient should exhibit much larger conversion efficiencies. One can then speculate that in P-doped fibers, for which high conversion efficiencies have been observed, the grating erasure mechanism is somehow inhibited. This could occur, for example, due to the presence of deep traps associated with phosphorus defects: once

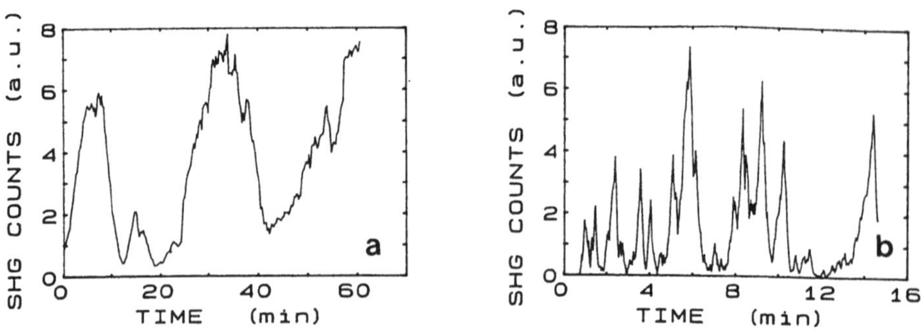

Fig. 9. Growth of the SH signal in time as the fiber is being conditioned, exhibiting (a) oscillations, and (b) chaotic behavior. The fiber is being continuously seeded with mode locked pulses at 1.06 μm and 532 nm. The seeding (writing) beams are interrupted every 1.6 s to make a rapid reading measurement of the SH signal.

photo-induced charges have been frozen into the $\chi^{(2)}$ grating, it is much more difficult for light at 2ω to liberate them and randomize the defects to wash out the grating.

There is one further comment: in certain fibers it has been very recently observed that the SHG approach to a steady-state value can exhibit large oscillations, as shown in Fig. 9 [44]. The period of these oscillations decreases with increasing $P_0(2\omega)$; for very large seeding powers the output can even appear chaotic in time. These preliminary results are currently not explicable in the context of the rate equation model. However, one possibility could be that the SH produced by the induced $\chi^{(2)}$ as the grating is being written starts to write its own grating; the SH generated from this grating can interfere with the initial SH and further write another grating, and so on. In any case, these observations indicate that the approach to steady-state can be quite complex, and that under certain conditions a "steady-state" as such may never exist. Nonetheless, the experimental results in support of the rate equation model demonstrate that self-erasure plays an important role in the ultimate saturation of the $\chi^{(2)}$ grating.

6. Evidence for Defect Structures

The mystery of the microscopic nature of the defect creation and subsequent alignment processes may prove the most difficult to unravel. One fact that is known with certainty is that doping of germanium in the fiber core is essential; no one has been able to observe SHG in fibers of pure fused silica. The first self-preparation experiments [8,9] were performed in fibers co-doped with phosphorus, and the highest efficiencies observed to date have been produced by fibers co-doped with both P and Ge [10]. So it was initially assumed that P-related defects played a crucial role. However, fibers with cores doped solely with Ge have now been observed to produce SHG with efficiencies ≈0.1-1% when seeded [29, 31], whereas fibers doped only with P produce very little or no SHG when seeded [29, 45]. In contrast, fibers doped only with Ge cannot be self-prepared at 1.06 μm [31]; this is a further indication that the seeding and self-preparation mechanisms may be different. Finally, purely Ge-doped fibers have been self-prepared for doubling of the 647.1 nm line from a mode locked krypton laser, although the output was very unstable [46]. Clearly the question of the role of defects is a complicated one!

In any case it seems that a Ge-related defect is the most likely source of the dipole allowed nonlinearity in seeded fibers, while P-related defects are also important in self-prepared fibers. The body of literature on defects in fused silica and optical fibers is extensive [47-53]. A strong candidate for involvement is the Ge E′ center, which is formed by removing an electron from the bond between neighboring Ge atoms (at the site of an oxygen vacancy), leaving behind a trapped hole. *Tsai* et al. [54] examined the electron spin resonance spectra of Ge- and Ge-P-doped fibers before and after conditioning, and showed that the SHG efficiency correlates with the concentration of Ge E′ centers formed during the conditioning process.

It is well known that short-wavelength radiation (UV, X-rays, and γ-rays) can induce numerous defects in glass [55-59]. As discussed previously, the erasure experiments point to defect creation by multi-photon absorption. There are several other experimental studies that indicate that the SHG conditioning process occurs through excitation of defects by multi-photon absorption at the longer wavelengths of 1.06 μm and 532 nm; these defects then relax (guided by the dc field in the mixing model) into states which lack inversion symmetry. *Bergot* et al. [40] used blue light from an argon laser coupled into a fiber and applied a dc field ≈10 V/μm across the fiber core to obtain large permanent enhancements of the second-order nonlinearity. Presumably the blue light greatly accelerated defect creation, while the large external poling field acted in similar fashion to the internal dc polarization in the mixing model. *Farries and Fermann* [60] seeded a Ge- and P-doped fiber at 1.319 μm to produce SH at 659 nm. (This wavelength, along with 1.064 μm and 647 nm, are the only ones to have been doubled to date.) The SH wavelength was too long to produce significant defect centers through direct one-photon absorption; the defect creation was therefore attributed to absorption of the weak third harmonic generated in the fiber at 439 nm. Also, a hydrogen and heat treatment of fibers prior to seeding has been demonstrated to enhance the conversion efficiency by four times [61]. The suggestion was that the in-diffused H_2 reacted with defect centers in the glass, perhaps to form GEC centers in which an electron is localized at a Ge site with four neighboring oxygens.

Finally, some very recent work by *Kamal* et al. [62] provides further evidence that Ge E′ centers, created through γ-ray irradiation, may play an important role. In this work, Ge-doped fiber which could not be conditioned to produce measurable SHG even after many hours of seeding (efficiency $< 10^{-13}$) was irradiated with a 10^5 rad dose of γ-rays. After irradiation the fiber could be conditioned to an efficiency of 10^{-5} in 2 min of seeding; the grating showed no measurable decay over a period of weeks. In fibers that initially showed weak SHG (efficiency $\approx 10^{-9}$ after 17 h of seeding), the γ-ray irradiation dramatically increased both the efficiency and preparation rate (efficiency $\approx 10^{-6}$ after 2.5 min of seeding). To explain these results, the authors theorized that it is the number of oxygen vacancies and the related Ge E′ centers that determine which fibers can be efficient SH producers. Ge-doped fibers which cannot be conditioned (like telecommunications-grade fibers) have very few oxygen vacancies that can lead to formation of E′ centers. Although seeding can break a Ge-Ge bond through multi-photon absorption, there is insufficient energy to knock loose an oxygen to produce a vacancy. Conditionable fibers, on the other hand, have plenty of oxygen vacancies. The role of the γ-rays is to liberate oxygen atoms from the silica/germania matrix in initially unconditionable fibers, thereby producing defects which can participate in the conditioning process.

7. Related Experiments and Questions

There is an interesting implication in the Raman experiments described earlier. In measuring the periodicity Λ of the Raman scattered signal, one might expect it to be *half* the coherence length $\mathcal{L}$, since the Raman process is not sensitive to the orientation of the defects but merely to the number of oriented defects. The Raman signal can then be viewed as arising due to a rectified $\chi^{(2)}$ grating, so that a dc offset in the Raman signal averaged over several grating periods relative to the average signal in an erased fiber would be expected. But as noted in Fig. 4, there is no difference between the two except in the defect line at 490 cm^{-1}. This seems to imply a fundamental difference between fibers which have been conditioned and subsequently erased, and fibers which have never been conditioned. Further, this would instead imply that $\Lambda = \mathcal{L}$. Unfortunately, due to uncertainties in the fiber parameters, the calculations of the expected coherence length $\mathcal{L}$ are not accurate enough to resolve this question.

Another interesting experiment by *Batdorf* et al. [63] measured the length dependence of SHG in seeded optical fibers. They found that for high seed green powers of ~1 mW, the SH rises rapidly to a maximum value in the first cm of fiber before leveling off. For low seed powers this dynamic region over which the SH rapidly grows moves farther into the fiber, becoming about 20 cm for a seed power $\lesssim 1$ nW. Further, the

SH power grows quadratically with length (as expected) only at the high powers; at low powers it grows much more rapidly (as high as the 13th power of length). These results seem to imply an intensity-dependent saturation of $\chi^{(2)}$, which is difficult to explain in the context of the mixing model.

Electric-field induced SHG in glass [64] can occur through the allowed dipole susceptibility $\chi^{(3)}$, thereby producing an induced effective $\chi^{(2)}$. Calculations of field-induced SHG in fibers predicted that phase-matched conversion effciencies $\lesssim 0.1\%$ should be possible with the right fiber geometry [13, 65]. The poling experiments discussed in Section 6 demonstrated that it is possible to orient defects with an external electric field and so produce an effective $\chi^{(2)}$ that was 100 times larger than that obtained in seeded fibers [40]. However, the SHG is of course not phase matched. More recently, *Fermann* et al. [66] used these poling techniques to produce $\chi^{(2)}$ gratings written by modal interference of blue light launched into a fiber, thereby achieving phase matching with SHG conversion efficiencies ~1%. *Kashyap* [67] performed a detailed theoretical and experimental study of field-induced SHG in fibers, where the phase matching was accomplished by using a periodic electrode structure overlaid on the fiber. Rotation of the electrode grating relative to the fiber axis could be used to phase match different mode combinations, with conversion efficiencies of nearly 10^{-5}.

In conclusion, it should be clear at this point that the phenomenon of second harmonic generation in fibers is a fascinating one. Stable, efficient fiber doublers are not yet around the corner, but conditioned fibers with an induced $\chi^{(2)}$ have been used in autocorrelators [68] and Pockels modulators [69]. Although some features of the basic mechanisms of SHG in fibers are now well understood, there are obviously many questions that remain to be answered. The dynamics of the self-organization process, which hint of temporal periodicities and chaos, are particularly intriguing. It could well be that the study of SHG in fibers leads to discovery of general concepts of nonlinear feedback that are applicable to the wide variety of self-organized systems found in nature.

References

1. R. H. Stolen, in *Optical Fiber Telecommunications*, edited by S. E. Miller and A. G. Chenowyth (Academic Press, New York, 1979),p. 125.
2. H. G. Winful, in *Optical-Fiber Transmission*, edited by E.E. Basch (Sams, Indianapolis IN,1986), p. 179.
3. G. P. Agrawal, *Nonlinear Fiber Optics* (Academic Press, San Diego, 1989).
4. L. F. Mollenauer and K. Smith, Opt. Lett **13**, 675 (1988).
5. Y. Fujii, B. S. Kawasaki, K. O. Hill, and D. C. Johnson, Opt. Lett. 5, **48** (1980).

6. Y. Sasaki and Y. Ohmori, Appl. Phys. Lett. **39**, 466 (1981);
Y. Ohmori and Y. Sasaki, IEEE J. Quantum Electron. **QE-18**, 758 (1982);
Y. Sasaki and Y. Ohmori, J. Opt. Commun. **4**, 3 (1983).
7. J. M. Gabriagues and L. Fersing, *Conference on Lasers and Electro-Optics Technical Digest* (Optical Society of America, Anaheim CA, 1984), p. 176.
8. U. Østerberg and W. Margulis, Opt. Lett. **11**, 516 (1986).
9. U. Østerberg and W. Margulis, Opt. Lett. **12**, 57 (1987).
10. M. C. Farries, *Proc. Colloquium on Nonlinear OpticalWaveguides* (IEE, London, 1988), p. 88.
11. J. W. Fleming, Appl. Opt. **23**, 4486 (1984).
12. D. Gloge, Appl. Opt. **10**, 2252 (1971).
13. R. W. Terhune and D. A. Weinberger, J. Opt. Soc. Am. B **4**, 661 (1987).
14. M. E. Fermann, L. Li, M. C. Farries, and D. N. Payne, Electron. Lett. **24**, 894 (1988).
15. R. Kashyap, *Nonlinear Guided Wave Phenomena: Physics and Applications Technical Digest, Vol. 2* (Optical Society of America, Houston TX, 1989), p. 255.
16. J. A. Armstrong, N. Bloembergen, J. Ducuing, and P. S. Pershan, Phys. Rev. **127**, 1918 (1962).
17. J. T. Manassah, R.R. Alfano, and S. A. Ahmed, Phys. Lett. A **115**, 135 (1986).
18. F. P. Payne, *Proc. of the 4th Internatl. Symposium on Optical and Optoelectronic Applied Science and Engineering: Vol. 800 Novel Optoelectronic Devices* (SPIE, The Hague, 1987), p. 132.
F. P. Payne, Electr. Lett. **23**, 1215 (1987).
19. P. Chmela, Opt. Lett. **13**, 669 (1988).
20. Y. Chen, Appl. Phys. Lett. **54**, 1195 (1989).
21. R. H. Stolen and H. W. K. Tom, Opt. Lett. 12, 585 (1987).
22. M. C. Farries, P. St. J. Russell, M. E. Fermann and D. N. Payne, Electron. Lett. 23, 322 (1987).
23. N. B. Baranova and B. Ya. Zel'dovich, JETP Lett. **45**, 717 (1987).
24. V. Mizrahi, U. Osterberg, J. E. Sipe, and G. I. Stegeman, Opt. Lett. **13**, 279 (1988).
25. V. Mizrahi, U. Osterberg, C. Krautschik, G. I. Stegeman, J. E. Sipe, and T. F. Morse, App. Phys. Lett. **53**, 557 (1988).
26. N. M. Lawandy, *Optical Society of America Annual Meeting Technical Digest* (Optical Society of America, Santa Clara CA, 1988), paper FS4.
N. M. Lawandy, *Conference on Quantum Electronics and Laser Science Technical Digest* (Optical Society of America, Baltimore MD, 1989), paper FCC2.
27. D. Z. Anderson, *Proc. SPIE Conference on Nonlinear Optical Properties of Materials Vol. 1148* (SPIE, San Diego CA, 1989).
28. W. Margulis and U. Østerberg, J. Opt. Soc. Am. B **5**, 312 (1988).
29. M. A. Saifi and M. J. Andrejco, Opt. Lett. 13, 773 (1988).
30. H. W. K. Tom, R. H. Stolen, G. D. Aumiller, and W. Pleibel, Opt. Lett. **13**, 512 (1988).
31. A. Kamal and D. A. Weinberger, unpublished.
32. F. Ouellette, Opt. Lett. **14**, 964 (1989).

33. M. C. Farries, M. E. Fermann, P. St. J. Russell, and D. N. Payne, *Digest of the Optical Fiber Communications Conference* (Optical Society of America, New Orleans LA, 1988), paper THE2.
34. M. E. Fermann, M. C. Farries, P. St. J. Russell, and L. Poyntz-Wright, Opt. Lett. **13**, 282 (1988).
35. J. M. Gabriagues and H. Fevrier, Opt. Lett. **12**, 720 (1987).
36. A. Kamal, D. A. Weinberger, and W. H. Weber, to appear in Opt. Lett., April 1989.
37. K. O. Hill, Y. Fujii, D. C. Johnson, and B. S. Kawasaki, Appl. Phys. Lett. **32**, 647 (1978).
B. S. Kawasaki, K. O. Hill, D. C. Johnson, and Y. Fujii, Opt. Lett. **3**, 66 (1978).
38. C. P. Kuo, U. Østerberg, C. T. Seaton, G. I. Stegeman, and K. O. Hill, Opt. Lett. **13**, 1032 (1988).
39. W. Margulis, I.C. S. Carvalho, and J. P. von der Weid, Opt. Lett. **14**, 700 (1989).
40. M.-V. Bergot, M. C. Farries, M. E. Fermann, L. Li, L. J. Poyntz-Wright, P. St. J. Russell, and A. Smithson, Opt. Lett. **13**, 592 (1988).
41. V. Mizrahi and J. E. Sipe, Appl. Opt. **28**, 1976 (1989).
42. F. Ouellette, K. O. Hill, and D. C. Johnson, Opt. Lett.**13**, 515 (1988).
43. A. Krotkus and W. Margulis, Appl. Phys. Lett. **52**, 1942 (1988).
44. A. Kamal and D. A. Weinberger, in preparation.
45. M. C. Farries, M. E. Fermann, and P. St. J. Russell, *Nonlinear Guided Wave Phenomena: Physics and Applications Technical Digest, Vol. 2* (Optical Society of America, Houston TX, 1989), p. 246.
46. B. Valk, E. M. Kim, and M. M. Salour, Appl. Phys. Lett. **51**, 722 (1987).
47. R. A. B. Devine, ed., *The Physics and Technology of Amorphous* SiO_2, (Plenum Press, New York, 1988).
48. E. J. Friebele, D. L. Griscom, and G. H. Sigel, Jr., J. Appl. Phys. **45**, 3424 (1974).
49. F. Hanawa, Y. Hibino, M. Shimizu, H. Suda, and M. Horiguchi, Opt. Lett. **12**, 617 (1987).
50. W. A. Sproson, K. B. Lyons and J. W. Fleming, J. of Non-Crystalline Solids **70**, 45 (1985).
51. S. K. Sharma, D. W. Matson, J. A. Philpotts and T. L. Roush, J. of Non-Crystalline Solids **68**, 99 (1984).
52. A. Chmel and S. B. Eronko, J. of Non-Crystalline Solids **70**, 45 (1985).
53. D. C. Douglass, T. M. Duncan, K. L. Walker and R. Csencsits, J. Appl. Phys. **58**, 197 (1985).
54. T. E. Tsai, M. A. Saifi, E. J. Friebele, D. L. Griscom, and U. Østerberg, Opt. Lett. **14**, 1023 (1989).
55. D. L. Griscom, E. J. Friebele, and K. J. Long, J. Appl. Phys. **54**, 3743 (1983).
56. E. J. Friebele, C. G. Askins, and M. E. Gingerich, Appl. Opt. **23**, 4202 (1984).
57. Y. Watanabe, H. Kawazoe, K. Shibuya and K. Muta, Japanese J. of Appl. Phys. **25**, 425 (1986).
58. R. A. B. Devine, J. Non-Crystalline Solids **107**, 41 (1988).
59. L. J. Poyntz-Wright, M. E. Fermann, and P. St. J. Russell, Opt. Lett. **13**, 1023 (1988).

60. M. C. Farries and M. E. Fermann, Electron. Lett. **24,** 294 (1988).
61. F. Ouellette, K. O. Hill, and D. C. Johnson, Appl. Phys. Lett. **54**, 1086 (1989).
62. A. Kamal, D. A. Weinberger, and J. H. Chu, in preparation.
63. B. Batdorf, C. Krautschik, U. Østerberg, G. Stegeman, and T. F. Morse, *Nonlinear Guided Wave Phenomena: Physics and Applications Technical Digest, Vol. 2* (Optical Society of America, Houston TX, 1989), p.259.
64. C. G. Bethea, Appl. Opt. 14, 2435 (1975).
65. D. A. Weinberger and R. W. Terhune, *Conference on Lasers and Electro-Optics Technical Digest* (Optical Society of America, Baltimore MD, 1987), p. 84.
66. M. E. Fermann, L. Li, M. C. Farries, L. J. Poyntz-Wright, and L. Dong, Opt. Lett. **14**, 748 (1989).
67. R. Kashyap, J. Opt. Soc. Am. B **6**, 313 (1989).
68. U. Østerberg and W. Margulis, IEEE J. Quantum Electron. **24**, 2127 (1988).
69. L. Li and M. E. Fermann, *Conference on Quantum Electronics and Laser Science Technical Digest* (Optical Society of America, Baltimore MD, 1989), paper FCC1.

Two-Wavelength Nonlinear Prism Coupling to Planar Waveguides in Semiconductor Doped Glasses

*M. Kull and J.-L. Coutaz**

The Royal Institute of Technology, Department of Physics II, S-10044 Stockholm, Sweden

*Permanent address: Laboratoire d'Electromagnétisme, Microondes et Optoélectronique, Ecole Nationale Supérieure d'Electronique et de Radioélectricité, Grenoble, France

Abstract. Experimental results on pump and probe nonlinear coupling to a planar semiconductor doped glass waveguide are presented. A short wavelength (532 nm) pump induces fast refractive index changes seen by a probe with longer wavelength (565-620 nm). Power-, time-, and spectral-dependences of the nonlinearity are measured. In addition the thermal nonlinearity is distinguished from the electronic one.

1. Introduction

Light by light interactions have been studied extensively during the past years. The aim is all-optical switching and all-optical logic operations. Numerous different schemes have been suggested for practical devices.

Nonlinear prism coupling to semiconductor doped glass (SDG) waveguides has so far only been demonstrated in single beam experiments, in which the self-induced nonlinearities were explored [1-2]. Here we report experimental results on two-wavelength nonlinear prism coupling to a single mode planar waveguide, fabricated in SDG by ion-exchange. We use a pump and probe version of the focused light attenuated total reflection method [3] to measure the pump power-, time-, and wavelength-dependence of the induced refractive index change. The pump and probe method also allows us to measure electronic and thermal nonlinearities separately, due to their different time evolution.

2. Experiments and results

Waveguides were fabricated in commercial glasses from Corning (C3-66). These glasses contain sodium, but only to a small extent, resulting in long ion-exchange times. Single mode guides were obtained after 71 hrs treatment in a melt of KNO_3 at 350°C. The waveguides were characterized by measurement of the m-line angles for wavelengths applicable to the nonlinear studies. From a luminescence spectrum we found the bandgap wavelength to be 560 nm.

In the nonlinear prism coupling experiment a frequency doubled, Q-switched (1.2 kHz) and mode-locked (100 MHz) Nd:YAG laser is used both to synchronously pump a dye laser (Rhodamine 6G) and to provide an intense pump, at λ = 532 nm, for the prism coupler. The FWHM duration of the pump and probe pulses are 80 ps and 40 ps respectively. Both beams are TE polarised and focused onto the base of the coupling prism. As the waveguide is pressed against the base, the pump and probe beams are coupled into the guide and in the reflected field, dark m-lines are seen at the specific coupling angles. The probe pattern is then magnified and imaged onto a detector, which is either a pin-diode or a 512 element diode-array. Interference filters ensure that no pump light is detected. The intense pump induces a refractive index change in the waveguide, so that $n = n_0 + \Delta n(P)$, where P is the pump power. This causes an angular shift of the m-line.

Using the diode array we measure a shift to smaller angles as the pump is switched on. The corresponding index change is calculated and found to be negative. In the following measurements of pump power-, time- and wavelength-dependence of Δn, the diode-array was replaced by a single pin-diode positioned at the maximum slope of the m-line and the pump beam was chopped at 36 Hz. The diode signal was then read with a lock-in amplifier and normalized with the result obtained with the diode array.

A delay line in the pump beam allows us to distinguish between thermal (slow) and electronic (fast) effects. If the pump and probe coincide temporally, we measure electronic nonlinearities superimposed on the slower thermal ones, whereas if the

Fig. 1

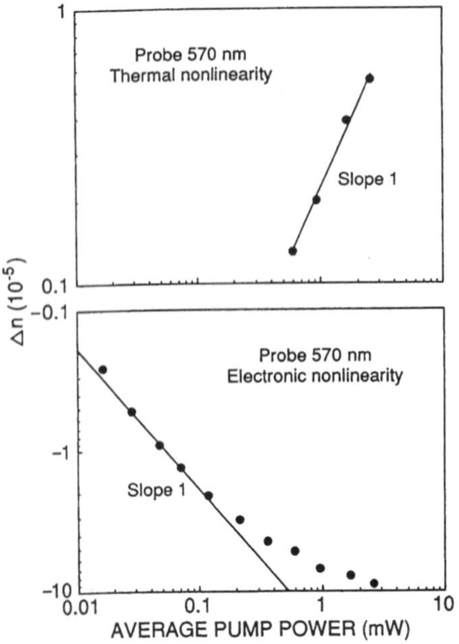

Fig. 2

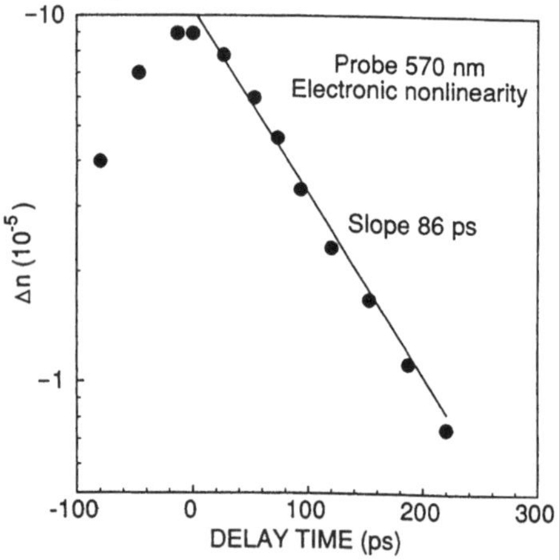

Fig. 1. Power dependence of the calculated refractive index changes. The upper curve was taken at 10 ns delay between pump and probe giving the thermal contribution to Δn and the lower one is at zero delay resulting in negative index changes. At high pump power the nonlinearity saturates.

Fig. 2. Time dependence of the electronic nonlinearity. The solid line is a least squares fit to an exponential decay with a lifetime of 86 ps.

pump pulses precede the probe pulses with several nanoseconds we measure thermal effects only. Figure 1 shows both contributions separated. The thermally induced index changes are positive and proportional to the pump power as expected. The electronic nonlinearity stems from band-filling and has a negative sign for probe wavelengths larger than the gap wavelength. As the pump power increases, conduction band states are filled and the index change saturates. With the power levels available we reach a maximum Δn on the order of -0.0001. By varying the delay we find the relaxation time of the electronic nonlinearity to be 86 ps as indicated in fig. 2.

Finally, by changing the dye laser wavelength, we measured the spectral dependence of the induced index changes. The results in fig. 3 are normalized with respect to detected probe power and spectral sensitivity of the pin-diode. The data are associated with quite large uncertainties, mainly due to differences in waveguide coupling as the probe wavelength is changed, but nevertheless there is a clear decrease of the nonlinearity close to, and far away from the gap wavelength.

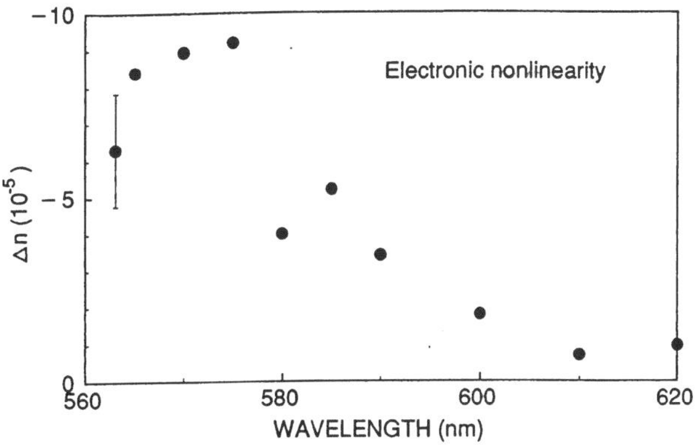

Fig. 3. The spectral characteristic of the electronic nonlinearity for probe wavelengths between 565 and 620 nm.

3. Discussion

From the data in fig. 1, at power levels below the onset of saturation, we can estimate the electronic contribution to the intensity-dependent refractive index coefficient $n_{2EL}(\lambda_{PUMP}, \lambda_{PROBE})$. Assuming a mode area of 200 μm^2 and a coupling efficiency of 10%, we find $n_{2EL}(532,570)=2\cdot 10^{-8}$ cm^2/kW. This is in close agreement with interferometric measurements of self-induced n_2-coefficients in SDG [4].

The relaxation time of the nonlinearity is a measure of the excited carrier lifetime. Large variations of the carrier lifetime in SDG can be found in the literature - from several nanoseconds to a few picoseconds. The reason is a photochemical reaction, named darkening, that permanently decreases the carrier lifetime as the sample is exposed to intense light [5]. Our measured value of 86 ps indicates that the sample is darkened to some extent.

The absolute thermal contribution to Δn is smaller than the electronic, but since Δn_{EL} saturates at high power, Δn_{TH} will become dominant at average powers exceeding 40 mW. In fact this was observed in the single beam experiments in ref. [1].

To conclude, we have in a two wavelength, pump and probe experiment measured optically induced refractive index changes in a SDG waveguide, by means of nonlinear prism coupling. As we pump the waveguide with light having shorter wavelength than the bandgap wavelength and at moderate powers, the induced fast negative index change for longer wavelength probes dominates the thermal change. The estimated n_{2EL}-coefficent is on the same order of magnitude as self-induced coefficients reported.

Acknowledgements

We gratefully acknowledge support by the Swedish Natural Science Council.

References

1. G. Assanto, J. Modern Optics 36, 305 (1989).
2. P. Dannberg, T. Possner, A. Bräuer and U. Bartuch, Phys. Stat. Sol. (b) 150, 873 (1989).
3. E. Kretschmann, Opt. Comm. 26, 41 (1978).
4. G.R. Olbright, N. Peyghambarian, Appl. Phys. Lett. 48, 1184 (1986).
5. M. Kull, J.L. Coutaz, G. Manneberg and V. Grivickas, Appl. Phys. Lett, **54**, 1830 (1989).

Part V

Semiconductors

The Nonlinear Optics of Semiconductor Quantum Wells: Physics and Devices

I.I. Abram

Centre National d'Etudes des Télécommunications,
196, Avenue Henri Ravera, F-92220 Bagneux, France

Abstract. In this lecture we review some recent experimental and theoretical investigations on the nonlinear optical properties of GaAs/$Al_xGa_{1-x}As$ Quantum Well Structures and on the use of these optical nonlinearities for the design and realization of the first few exploratory devices for all-optical signal processing.

1. Introduction

Quantum wells are artificial semiconductor structures whose microscopic spatial configuration is especially designed so as to modify significantly some of the properties of the parent semiconductor. In particular, the confinement of carriers into regions of reduced dimensionality quantizes their motion and enhances the optical properties of the semiconductor in the vicinity of the band gap. These novel features of semiconductors with reduced dimensionality elicit a lot of interest in basic physics. At the same time, Quantum Well Structures (QWS) hold great promise for the development of devices for all-optical signal processing, both because of their enhanced optical nonlinearities and because of their compatibility with semiconductor lasers which are often made of the same materials.

In this lecture, we start with some highlights of the electronic structure of GaAs/AlGaAs QWS (Section 2) and a few basic considerations on the relationship between this structure and the optical response of QWS (Section 3). Then (Section 4) we examine some manifestations of the nonlinear optical response, pointing out the features that may be relevant in the study of the basic physics of QWS and those that pertain to device considerations. In the next three Sections, we present some recent experimental and theoretical results on the optical nonlinearities of QWS and on the corresponding optical devices that have been realized. In particular, Section 5 focuses on effects observable in the transparency region of QWS, Section 6 discusses the coherent transient effects that occur at short time scales on the excitonic resonaces of QWS and Section 7 is devoted to the physics and devices that involve a real excitation of a QWS. Finally, in Section 8 we present the conclusions of this lecture.

2. The Semiconductor Quantum Wells

In this Section we discuss very briefly the electronic properties of semiconductor quantum wells that are relevant to nonlinear optics. We assume no previous knowledge of solid-state physics and we develop these preliminary notions only to the extent that is necessary for an understanding of the nonlinear optical properties of QWS. For a more thorough discussion of the electronic structure and properties of QWS, the reader is referred to textbooks on Solid State Physics, Semiconductor Physics and to recent books and reviews on semiconductor heterostructures and QWS [1].

Springer Series in Wave Phenomena, Vol. 9 **Nonlinear Optics in Solids**
Editor: O. Keller

2.1 The Electronic Properties of III-V Semiconductors

III-V semiconductors are crystalline compounds composed of an element of column III of the periodic table (Al, Ga, In) and an element of column V (P, As, Sb). Here we focus our attention more particularly to the case of GaAs, as this is the III-V semiconductor that has been studied more extensively, especially in Nonlinear Optics.

In the crystalline lattice, each atom of one element is bound tetrahedrally to four atoms of the other element, through covalent bonds. In the simplest case, each inter-atomic bond may be described in terms of linear combinations of the sp^3-hybridized atomic orbitals of the two bonded atoms, such that the symmetric combination corresponds to a bonding orbital, while the anti-symmetric combination has anti-bonding character.

Alternatively, we may take advantage of the translational symmetry of the crystal lattice to describe the electronic states of the semiconductor in terms of a basis set that is delocalized over the whole crystal. These delocalized states are linear combinations of the local orbitals and are invariant under the symmetry operations of the space group of the crystal lattice. They generally correspond to harmonic plane waves characterized by a translational quantum number, k, called the wavevector. Usually, the energy of such a delocalized state is a function of the wavevector, and thus the energies of extended states of a given type are spread over a relatively broad band. In this simple model, the bonding and the anti-bonding orbitals give rise to two different bands, called valence and conduction bands, respectively. Near an extremum of a band, the energy can be written to a first approximation as

$$E = E_i \pm \frac{\hbar^2 k^2}{2m_i^*} \tag{1}$$

in which i (= v,c) is the band index, the coefficient m_i^* is called the effective mass of the i-th band, while the sign is taken by convention as positive for a conduction band and negative for a valence band. According to this equation, within each band, the energy has the form of a kinetic energy, in which $\hbar k$ is the momentum associated with the corresponding quasi-particle. The energy separation of the conduction and valence bands

$$E_g = E_c - E_v \tag{2}$$

reflects the energy difference between the bonding and anti-bonding orbitals and is called the band gap. For GaAs $E_g = 1.52$ eV at low temperatures. There are no plane-wave electronic states whose energy lies in the band gap. For an electronic band described by eq. (1), the density of states as a function of energy is given by

$$\rho_i(E) = \frac{1}{2\pi^2}\left(\frac{2m^*}{\hbar^2}\right)^{3/2}\sqrt{E - E_i} \quad . \tag{3}$$

That is, the density is very small in the vicinity of the gap but increases with energy in the band.

When the crystal is in its ground state all bonding orbitals are filled, each with two electrons, one of spin "up" and one of spin "down", while all anti-bonding orbitals are empty. An elementary optical excitation of the semiconductor corresponds to the promotion of one electron from a bonding to an anti-bonding orbital of the same interatomic bond. In the delocalized (wavevector) viewpoint, the ground state of the crystal corresponds to the valence band full of electrons and the conduction band completely empty, while the optical excitation corresponds to the promotion of an electron from the valence to the conduction band. When an electron is removed from the conduction band it leaves in its place a hole, which we may consider as new type of Fermi quasi-particle having an effective positive charge, since it corresponds to the absence of a charge in a sea of negative charges. In the delocalized viewpoint, therefore, the optical excitation process may be viewed as the simultaneous creation of two quasi-particles in the crystal, a negatively-charged electron in the conduction band and a positively-charged hole in the valence band.

This simple picture of the optical excitation process provides, nevertheless, a relatively accurate description for the basic features of the types of states that may be accessed optically, that is, of the optical selection rules. The interaction of the semiconductor with light preserves the translational invariance and thus the wavevector remains a good quantum number for the light-plus-semiconductor system. This feature implies that the absorption of a photon by the semiconductor produces an electron-hole pair such that the sum of the wavevectors of these two quasi-particles is equal to the photon wavevector. Since for visible and for infra-red radiation the photon wavevector is small ($\sim 10^4\ cm^{-1}$) compared with the possible range of crystal wavevectors (up to $\sim 10^8\ cm^{-1}$), this means that the only electron-hole pairs that can be excited optically are such that the electron and hole wavevectors are approximately equal and opposite to each other ($k_e = - k_h$). Alternatively, description of photoexcitation in terms of local orbitals implies that the oscillator strength of a given transition (i.e. the probability of creating a given electron-hole pair satisfying the wavevector selection rule) is proportional to the probability that the electron and hole may be found on the same site. For any pair of plane-wave states of the valence and conduction bands this probability is of order 1/N, where N is the number of sites in the crystal. Because of the small density of states near the band gap (eq. 3), the overall oscillator strength in this spectral region is rather weak.

A very important feature of the photo-excited electron-hole pair is the electrostatic attraction between the positively-charged hole and the negatively-charged electron, which imposes a correlation in their relative positions and their relative motion. This correlation is best seen in the formation of bound complexes, called Wannier excitons, in which the electrons and holes stay close to each other while their relative motion resembles that of the electron and the proton in the hydrogen atom. In GaAs, the lowest-energy exciton has a spherical hydrogenic internal structure with an effective Bohr radius of approximately $a_B \sim 140$ Å. Because of the binding between the electron and the hole, the energy of the exciton is lower than that of any plane-wave electron-hole pair and is, therefore, within the band gap. The binding energy of the exciton in GaAs is 4.2 meV. The electron-hole correlation is also manifested quite strongly in the low-energy states of the valence and conduction bands. Thus, when the Coulomb interaction is taken into account, these electron-hole pair states can no longer be described as plane-waves, but rather as highly-correlated scattering states, analogous to the ionization continuum of the hydrogen atom (Sommerfeld states).

2.2 The Electronic Properties of Semiconductor Quantum Wells

Quantum Wells Structures (QWS) are stratified semiconductor microstructures that have been synthesized in the laboratory over the past decade. Their fabrication has been made possible by the development of sophisticated semiconductor growth techniques that permit to control the growth process on the atomic level, such as Molecular Beam Epitaxy. QWSs consist of stacks of ultra-thin layers of two different semiconductors,

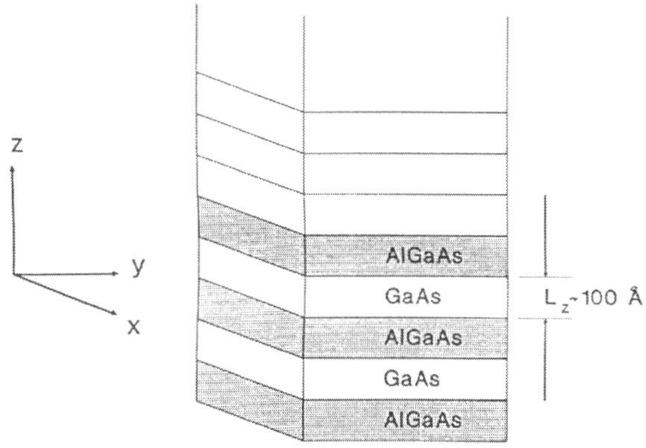

FIGURE 1: Schematic representation of a quantum well structure, composed of an alternation of thin layers of GaAs and $Al_{0.3}Ga_{0.7}As$; typical layer thickness is $L_z \sim 100$ Å.

each a few atomic monolayers thick, grown epitaxially one on top of the other, in alternation. In most studies on the Nonlinear Optics of QWS, the two semiconductors are GaAs and $Al_xGa_{1-x}As$, the thickness of each layer is typically of the order of 100 Å, and a QWS sample consists usually of 10 to 100 pairs of layers. We shall therefore examine more particularly the electronic structure of this type of QWS. To facilitate the discussion we shall designate by $\hat{z}$ the direction perpendicular to the layers and we shall take the $\hat{x}$ and $\hat{y}$ axes in the plane of the layers, as shown in Figure 1.

The ternary semiconductor $Al_xGa_{1-x}As$, with $x \sim 0.3$, has a band gap of 1.89 eV, as compared with 1.52 for GaAs. Thus, in each heterojunction between the two semiconductors of the QWS, the local value of the band-gap changes. Roughly 60% of the band-gap discontinuity (i.e. 220 meV) appears as a discontinuity in the conduction band energy across the heterojunction, while the rest (i.e. 150 meV) corresponds to a discontinuity in the valence band energy. This causes a spatial modulation of the conduction and valence band energies along the $\hat{z}$-axis, each alternating between a high and a low value. Thus, the motion of electrons and holes along $\hat{z}$ is subject to a periodic potential composed of square energy-wells corresponding to the semiconductor with the smallest gap (in our case GaAs), surrounded by energy barriers, corresponding to the semiconductor with the largest gap (in our case AlGaAs), as shown in Figure 2. An electron (or a hole) that has little kinetic energy, cannot traverse the energy barriers, and stays confined in an energy well. Its characteristics (effective mass m^*, symmetry properties) are thus similar to those of the electrons (or holes) in the well material (GaAs), however with one very important difference. Its motion is free only along the $\hat{x}$-$\hat{y}$ plane, and in these directions it can be represented in terms of two-dimensional waves characterized by a two-dimensional wavevector. Along $\hat{z}$-axis, on the other hand, it behaves like a quantum mechanical particle confined in a box: its wavefunction is localized in the well and is evanescent in the barriers, while its energy levels are discrete. If the barriers were infinitely high, the energy levels would be given by the expression

$$E_n = n^2 \frac{\hbar^2 \pi^2}{2 m_i^* L_z^2}, \tag{4}$$

where n = 1,2,3,... is the quantum number that identifies each level, and

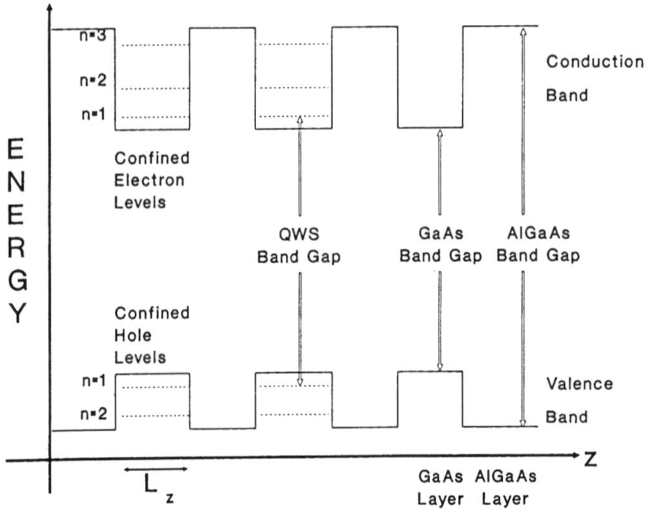

FIGURE 2: Band structure of a QWS as a function of the distance along the $\hat{z}$-axis.

L_z is the well width. Comparison of eqs. (1) and (4) shows that the confinement of the carriers in the $\hat{z}$ direction causes the $\hat{z}$-wavevector to take discrete values

$$k_z = n \frac{\pi}{L_z} , \tag{5}$$

so that the energy bands of the QWS can be written as

$$E = E_i \pm n^2 \frac{\hbar^2 \pi^2}{2m_i^* L_z^2} \pm \frac{\hbar^2 k^2}{2m_i^*} \tag{6}$$

where k is now two-dimensional. Clearly, for barriers of finite height (in our case 220 meV for the electrons and 150 meV for the holes) only a small number of localized levels may exist in the well, and this number depends essentially on the well width. Carriers whose energy is higher than the barrier energy are not confined in a well: their wavefunction is delocalized throughout the $\hat{z}$ axis and they have a mixed GaAs and AlGaAs character.

This simple analysis of the confinement of the low-energy carriers in the quasi-two-dimensional GaAs layers reveals some important consequences for the electronic structure for QWS:

1) The effective band gap of the QWS, which is the energy difference between the lowest-energy conduction state and the highest-energy valence state is, according to eq. (6)

$$E_g^{eff} = (E_c - E_v) + \left(\frac{1}{m_c^*} + \frac{1}{m_v^*}\right) \frac{\hbar^2 \pi^2}{2L_z^2} . \tag{7}$$

In other words, the effective band gap is larger than in the parent semiconductor, and its magnitude is determined by the well width L_z. This is an important feature of QWS because it permits one to custom-design their band-gap and to adjust it to an optimal value for a given application, simply by choosing the appropriate well width.

2) The valence and conduction bands are split each into a series of subbands, each subband corresponding to a confined energy level. Within each subband the electrons and holes display essentially a two-dimensional behavior. Thus, the density of states of each subband is a step function

$$\rho(E) = \begin{cases} m^*/\pi\hbar^2 & \text{for } E \geq E_n \\ 0 & \text{otherwise} . \end{cases} \qquad (8)$$

The step-like form of $\rho(E)$ implies that for QWS the density of states in the vicinity of the bandgap is relatively large, in contrast to the case of three-dimensional semiconductors in which it vanishes (eq. 3). This increases the oscillator strength in the vicinity of the bandgap and in general enhances all the optical properties of the semiconductor in this spectral range. This is particularly important for the functioning of semiconductor lasers: the finite density of states increases the gain in the vicinity of the bandgap. To a large extent, this accounts for the increased efficiency (reduced threshold) of QWS lasers.

3) As suggested in (2), the quasi-two-dimensional character of the subbands implies that the exciton structure also obeys quasi-two-dimensional physics. The spherical hydrogenic exciton becomes disk-like and this increases its binding energy and reduces its radius. For the extreme case of purely two-dimensional excitons, the binding energy increases by a factor of 4, while the radius reduces by a factor of 2 (respectively ~16 meV and a_{2D} ~ 70 Å for GaAs QWS).

These features may be observed in the absorption spectrum of QWS (Figure 3), which consists of a series of step-like plateaux, each representing

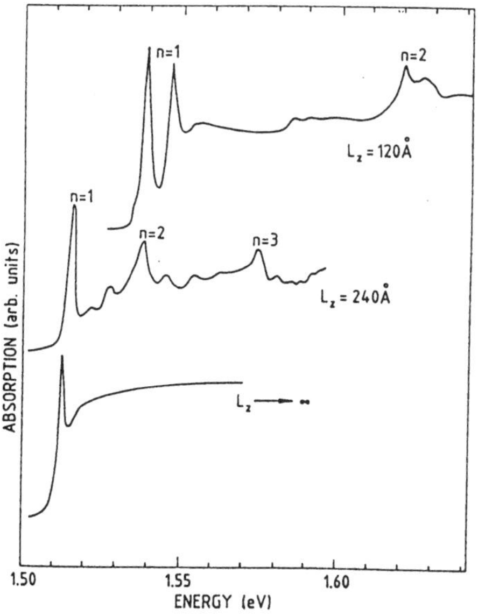

FIGURE 3: Absorption spectra of two GaAs/$Al_{0.3}Ga_{0.7}As$ QWS, compared with the spectrum of bulk GaAs. The sharp peaks are excitons involving the n-th subbands, whereas the plateaux which follow are transitions producing unbound electron-hole pairs in these subbands. The effective band-gap energy increases with decreasing layer thickness.

essentially the joint density of states of a pair of valence and conduction subbands. To the red of each plateau, there is a sharp and strong resonance corresponding to the 1s hydrogenic exciton associated with each subband of electron-hole pairs. Excitons of higher quantum numbers (2s, 3s etc.) are masked by the continuum transitions.

3. Exciton Optics

The correlation between the electron and the hole due to their mutual electrostatic attraction has an important consequence: it increases the electron-hole overlap and thus enhances considerably the oscillator strength of the excitonic electron-hole pairs. The probability that the electron and hole may be present on the same site is now of order $1/N_x$ (where N_x is the number of sites within the exciton volume) whereas it is of order $1/N$ for plane-wave electron-hole pairs. For GaAs, this represents an enhancement of the order of 10^{17} with respect to the oscillator strength of a single plane-wave electron-hole pair. Thus, the electron-hole correlation re-distributes the oscillator strength of the optical transitions of the semiconductor towards lower energies, in favor of the excitons and the highly correlated electron-hole scattering states (Sommerfeld states). It should be noted that in the absorption spectrum of the semiconductor (see Figure 3), the exciton appears only a few times more intense (and not 10^{17}) than the high-energy band-to-band transitions which to a good approximation can be described as plane-wave electron-hole pairs. The reason is that, in the spectrum, the low oscillator strength of plane-wave states is compensated by their very large number (high density of states), while the exciton peak represents a single electron-hole state.

In QWS, the reduced dimensionality increases the density of plane-wave states in the vicinity of the band-gap, while at the same time i enhances degree of the electron-hole correlation, reduces the exciton radius and further increases the oscillator strength of the exciton. The optical properties of QWS in the vicinity of the bandgap are enhanced when compared with those of the bulk, and are dominated by the strong coupling of the excitons to the electromagnetic field. In fact in many cases consideration of the excitons alone may be sufficient to describe the (linear or nonlinear) optics of QWS in the spectral region of the bandgap. In this Section, therefore, we shall examine the relationship between the structure of the excitons and their optical properties.

Excitons are "composite" quasi-particles composed of two Fermions, the electron and the hole, and are therefore Bosons. Thus, the Hamiltonian of the exciton field has the structure of a harmonic oscillator and this implies that when the exciton field is driven by an external force its response is always proportional to that force. Because of this feature, the response of the excitons to an incident light beam should be strictly linear. However, when the exciton system is driven very strongly the details of its internal structure become important because exciton-exciton interactions bring about a deviation from ideal Bose behavior. In the optical domain this deviation is manifested by the appearance of an optical nonlinearity. In QWS the confinement of electrons and holes in quasi-two-dimensional regions favors the appearance of exciton-exciton interactions and enhances the nonlinear optical properties of the semiconductor.

Following this viewpoint, we can distinguish two types of optical nonlinearities, according to the interactions that we can attribute as being at their origin. (1) The nonlinearities that result from the filling of single-particle states (phase-space filling effects) and (2) those that are due to many-body interactions. We shall examine briefly these two types of nonlinearities:

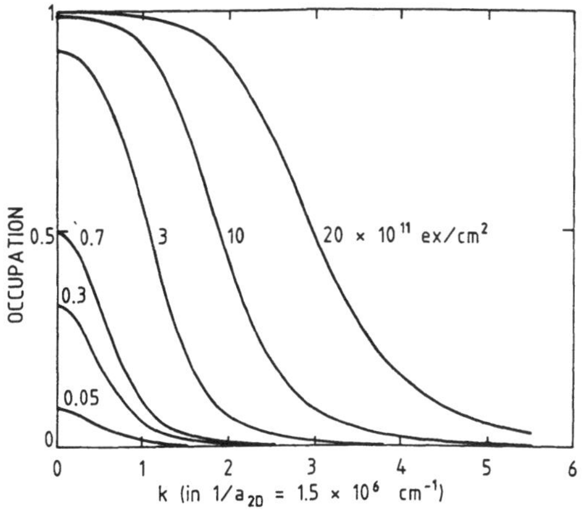

FIGURE 4: Filling of electron (or hole) phase-space by the creation of a high density of excitons in a quasi-two-dimensional layer of GaAs. The density of 1 exciton per Bohr volume is $1/(\pi a_{2D}^2) = 7\ 10^{11}\ ex/cm^2$.

(1) The phase-space filling nonlinearities arise because the electron and hole that constitute the exciton are both Fermions and are thus both subject to the Pauli exclusion principle: two electrons (or two holes) cannot both occupy the same state. When a large number of excitons is created in the QWS, the phase-space of the electrons and holes becomes progressively filled (see Figure 4) and the conditions for creating additional excitons change. For an incident light beam that can excite excitons in the semiconductor, the filling of the phase-space therefore appears as an optical nonlinearity. An example of an optical nonlinearity that can be attributed to phase-space filling is the saturation of absorption of the excitonic resonance. To understand how saturation occurs, we have to remember that the presence of a population of excitons in the semiconductor implies that the semiconductor can be de-excited by emitting photons and that the probability that such an emission can be induced is proportional to the occupation of the phase-space. When a large number of excitons accumulates in the semiconductor, the emission probability can become equal to the probability of absorption. In this case, the semiconductor appears transparent and the exciton resonance is said to be bleached: A light beam that traverses the semiconductor emerges unaffected, since the attenuation and the gain of the beam balance each other exactly.

(2) The second type of optical nonlinearities arises because the electrons and holes that constitute the excitons are charged particles and thus can interact through the Coulomb interaction. To analyze these nonlinearities, we may use the arsenal of Many-Body theory, but for our purposes it will be sufficient to examine only three many-body effects that are encountered most often in the nonlinear optics of excitons in semiconductors.

The first many-body effect we shall examine is the screening of the Coulomb interaction which arises in a highly excited semiconductor. The presence of dense exciton gas (or a dense electron-hole plasma if the excitons are dissociated) changes the dielectric constant of the semiconductor. Since the excitons (or the electron-hole plasma) are polarizable, the effective dielectric constant of the semiconductor increases upon excitation and thus the exciton binding energy decreases, implying a decrease in the electron-hole correlation. One spectral manifestation of this effect would be a shift of the exciton resonance to higher energies and a decrease of its oscillator strength, that is, of its coupling with the electromagnetic field. We note, however, that in quasi-two-dimensional structures screening is rather weak because a two-dimensional plasma is not very efficient in screening the three-dimensional force field of the Coulomb interaction between two test charges.

The second many-body effect we shall consider is <u>bandgap renormalization</u>. In a dense electron-hole plasma the electrons tend to avoid each other because of their correlation and exchange interactions. The same is true among the holes. This mechanism, whereby carriers of the same charge are kept apart from each other, lowers the energy of the electron-hole gas. This lowering of energy is often represented by a shift of the valence and conduction bands of the semiconductor towards each other upon excitation, with a concomitant shrinking of the bandgap -- hence the term "bandgap renormalization". The simplest optical manifestation of this effect would be a shift of the spectrum of the semiconductor to lower energies.

Finally, we shall mention the possibility that higher-order complexes of excitons may be possible in the semiconductor, such as the <u>biexcitons</u>. Such complexes are more common in II-VI compound semiconductors [2], but there are indications that they may also exist in GaAs QWS [3]. The existence of such complexes opens the possibility of having two-photon (or multi-photon) resonances in the semiconductor, in which excitons may serve as relay levels. Thus, the high-intensity spectrum of the semiconductor in the vicinity of the band-gap may display additional resonances, light-induced gaps or light-induced shifts of the exciton frequency.

In most optical experiments, all sources of optical nonlinearity contribute simultaneously, and the individual contributions discussed here cannot be easily disentangled. It is possible, however, to design specific experiments which isolate one type of nonlinearity from the others and thus permit the investigation of the physics that underly this nonlinearity.

4. The Nonlinear Optical Response

Before looking into the particular manifestations of the nonlinear optical processes in the vicinity of the band gap of QWS, we shall review some general considerations on the nonlinear optical properties of a material from the points of view of basic and of applied physics. However, we shall restrict this review only to those considerations that pertain to recent experiments and exploratory devices using QWS.

For the purposes of our discussion, we can make a very rough distinction between two types of optical nonlinearity. The nonlinear optical response that is observed in the transparency region of a material and the nonlinear optical processes that occur when the incident light beam is resonant with an absorption band. Traditionally, the term "Nonlinear Optics" is reserved for the former type of optical response. In this lecture, however, we use the enlarged definition, since the two types of optical response have a common origin but are differentiated essentially by the relative magnitudes of the characteristic times for the material-light interaction and for the observation process.

4.1 Optical Response in the Transparency Region

When a light beam traverses a transparent medium it cannot exchange any energy with the material system. However, its oscillating electric field can polarize the electronic cloud of the material and thus induce an oscillating macroscopic polarization which may, in turn, act as a classical dipolar antenna and emit light according to Maxwell's equations. In the simplest case, the induced polarization is proportional to the instantaneous value of the electric field of the incident radiation,

$$P = \chi E, \tag{9}$$

the proportionality constant χ being the optical susceptibility of the material. In this case, the polarization oscillates at the same frequency and has the same phase and same spatial characteristics as the incident electric field. Therefore the wave that it emits interferes with the incident radiation to produce all the effects of refraction that we know in classical optics. In fact, the quantity $n = \sqrt{1 + \chi}$ (in MKS units) is the refractive index of the material.

In general, however, the induced polarization is a non-linear function of the incident electric field and for this reason the optical response of the material has a more complicated dependance on the incident field. To make contact with the traditonal way of developing the theory of nonlinear optics (although this viewpoint is not essential for our subsequent discussion on the excitonic nonlinearities) we may expand the polarization into a perturbation series in increasing powers of the electric field,

$$P = \chi^{(1)}E + \chi^{(2)}E^2 + \chi^{(3)}E^3 + \ldots, \tag{10}$$

the coefficients $\chi^{(1)}$, $\chi^{(2)}$, $\chi^{(3)}$, ... are called the susceptibilities of the first, second, third order and so on. The linear term in the induced polarization gives rise to the index of refraction, as we saw above. All higher order terms, on the other hand, constitute the nonlinear polarization, whose interaction with the electromagnetic field gives rise to the different nonlinear optical effects discussed throughout this volume. In all the cases that we know, the linear polarization dominates the optical response of the material and thus it is necessary to have strong incident electric fields (corresponding to high incident beam intensities) so that the nonlinear polarization terms can give a measurable effect, comparable to that of the linear refractive index.

The induced polarization of eq. (10) has the form of an apparent interaction of the electromagnetic field with itself, in which the material simply enters parametrically, in the determination of the susceptibilities that characterize this interaction. The different orders of the susceptibility are directly related to the electronic structure of the material, as they represent the coefficients that result in the light-matter Hamiltonian after adiabatic elimination of the material degrees of freedom. Each order of the susceptibility thus exhibits a strong enhancement in the vicinity of the electronic resonances of the material that have the proper symmetry characteristics. This enhancement is often exploited in the laboratory in the different techniques of nonlinear spectroscopy, to explore the electronic states of the material by measuring the nonlinear optical response as a function of the wavelength of one or more incident beams.

This brief analysis of the nonresonant nonlinear optical response brings to the fore several features that are important to keep in mind when examining the implementation of these optical nonlinearities for the design of devices that could be used for optical signal processing.

The transparency of the material implies that the light-matter interaction is characterized by a detuning frequency Δ (i.e. the frequency difference between the material resonances and the light beam) that is relatively large. Thus, no real states of the material are excited and the induced polarization must be described in terms of transitions to a virtual state whose lifetime is of the order of $1/\Delta$. For materials that are transparent in the visible or near infrared, this time-scale is usually in the femtosecond range, implying that the macroscopic polarization follows very faithfully all the variations in intensity, frequency or phase of the incident light beam and decays as soon as the incident field is removed. Thus, these nonlinearities (1) present an effectively instantaneous optical response and (2) preserve the phase information of a coherent incident optical signal. These two points are favorable considerations in the design of devices for very fast optical signal processing. However, the feasibility of such devices is compromised by the fact that the intensities that are necessary to obtain a measurable nonlinear optical response in the transparency region are often in the MW/cm^2 or even GW/cm^2 range, necessitating very big laser sources. This requirement precludes the implementation of non-resonant optical nonlinearities in practical devices that can operate for purposes other than carefully-controlled laboratory demonstrations.

4.2 The Resonant Optical Response

When the frequency of the incident light beam is resonant with an absorption band of the material, the beam deposits part or all of its energy in the material. More precisely, the electric field of the incident beam induces an oscillating coherent polarization, which can remain active over a relatively long time since the de-tuning frequency is practically zero ($\Delta \sim 0$), in contrast to the case of the non-resonant response. However, because of the random processes that exist in real materials (such as collisions, inhomogeneities or interaction with thermal vibrations) the macroscopic polarization may lose its phase memory. Consequently, on average its instantaneous value may become zero, while the energy that excites it is still stored in the material.

This brief discussion of the photoexcitation process indicates that two types of resonant optical response can be expected, depending on the time-scales at which the observation occurs. For observation times shorter than (or of the order of) the dephasing time, the optical response of the material is dominated by the electromagnetic interaction of the coherent polarization and gives rise to <u>optical coherent transient</u> effects. At much longer times, such that energy becomes stored in the material even though the macroscopic polarization has decayed, the material response is often referred to as a <u>dynamic optical nonlinearity</u>. We shall examine briefly these two types of response.

4.2.1 Coherent Transients

The dephasing time-scale associated with an electronic transition is given by the inverse of the linewidth of that transition. For example, a decay-time of 1 ps corresponds to a linewidth of approximately 5 cm^{-1} (0.6 meV). Thus, in order to observe coherent transient effects, two requirements must be met by the incident light beam: (1) it must be resonant with a relatively sharp and isolated transition of the medium and (2) it must consist of pulses whose duration is shorter than (or of the order of) the material dephasing time. In semiconductors, generally, sub-picosecond pulses are necessary for the observation of coherent transient phenomena, since most spectral features are usually broad and merge into absorption continua. For the particular case of excitons in QWS, however, which appear in the spectrum as sharp and intense lines, pulses as long as a few picoseconds may be adequate for some types of coherent transient experiments, at least at low temperatures.

In a resonant situation, the induced polarization is described as a coherent superposition of the two states involved in the transition. Thus, for the description of the resonant optical response, the electronic structure of the material system can be reduced to these two states. Following this simplification, coherent transient experiments are usually treated through an elegant formalism that takes advantage of the analogy between two-level systems and a collection of spins [4]. Within this formalism, the coherent photo-excitation process is visualized through a geometrical model based on a formal vector (the "Bloch vector") that behaves as the optical analog of a spin: its longitudinal component gives the optical inversion of the material system, while its transverse component corresponds to the induced oscillating coherent polarization. Thus, optical coherent transient effects can be understood as direct analogs of the corresponding magnetic phenomena, through the precession, nutation, or decay of the Bloch vector and are described through the equations of motion of this vector, called the optical Bloch equations.

To date, the interest in coherent transients has been confined essentially to basic physics: many types of experiments (such as free-induction decay or photon echoes) have been designed to provide information on the dynamics of the dephasing process, each experimental technique bringing to the fore a different contribution to the overall dephasing process. From the point of view of applied physics, however, coherent transients on semiconductor QWS have not yet produced any viable application. Their implementation in practical devices, however, seems rather unlikely, at least in the short term, since these phenomena usually occur at very short time-scales and their observation requires intricate experimental setups and, generally, liquid Helium temperatures.

4.2.2 Dynamic Optical Nonlinearities

At times longer than the dephasing time, the coherent polarization decays and, according to Maxwell's equations, the material medium can no longer emit radiation. The energy that induced the polarization cannot be restored to the electromagnetic field and is therefore stored in the medium in the form of an incoherent population of excited states. The types of excited states that come into play depend on the time-scale in which the material medium is observed. In semiconductors, in the picosecond and nanosecond range the stored energy is most often in the form of an excited electron-hole plasma, or a gas of excitons (electronic response). Such a population can be dissipated either through spontaneous emission (which does not require a macroscopic polarization) or through nonradiative decay processes. In the microsecond to millisecond time-scales, the dynamic nonlinearities may involve transport and trapping of the electronic excitation (photorefractive response) [5], or relaxation of the electronic energy into the lattice vibrations (photo-thermal response) [6]. Here, we shall focus our attention on the picosecond and nanosecond electronic response.

The presence of the excited-state population modifies the optical properties of the material. For example, it may block further absorption and thus produce a bleaching of the optical transition, or it may make accessible transitions to new states and thus produce an additional induced absorption, usually at a different wavelength. Clearly, this modification of the optical properties is as an optical nonlinearity that modifies the propagation of the incident light beam or of a second "probe" beam.

In contrast to the nonlinearities in the transparency region, the dynamic nonlinear optical response does not depend on the instantaneous value of the incident electric field, but is rather a function of the integral of the incident intensity over some characteristic time, since it depends on the energy accumulated and stored in the excited material:

$$I_{out} = F(\int I_{in}(t)dt) . \tag{11}$$

The study of this nonlinearity gives information on the structure and on the dynamics of the excited states of the material that are populated by the photoexcitation process. In particular, since the dynamic nonlinearities persist as long as there are excited species in the material, a study of their time-dependance gives access to the characteristic relaxation times to which is subject the excited-state population.

At this point, it is interesting to draw a few comparisons between the dynamic nonlinearities and the nonlinear optical response of the material in the transparency region, regarding their implementation in nonlinear optical devices. Clearly, dynamic nonlinearities are relatively slow. In particular, they do not preserve the phase memory of the incident radiation and constitute therefore an incoherent optical response, since they occur after the dephasing of the induced macroscopic polarization. At the same time, since they depend on the presence of an excited-state population, their response time is conditioned by the characteristic times of that population. First, there is a finite "incubation" time before dynamic nonlinearities can become effective, corresponding to the time necessary to accumulate a sufficient excited-state population in the material. Second, they present a relatively long recycling time since, in a given sequence of operations in one device, the energy has to be evacuated after each operation so that the device can be re-set for the next operation. This latter step is usually the limiting factor for the speed of devices based on dynamic nonlinearities.

However, in contrast to the non-resonant nonlinearities, the incident light intensity necessary for the observation of dynamic nonlinearities is relatively low, since in this case it is not the instantaneous value of the intensity that counts, but rather its integral over time. Because of this feature, dynamic nonlinearities have an advantage of five or six orders of magnitude in the incident intensity (when compared with the non-resonant optical response) and can thus operate in the range under $1kW/cm^2$. In this intensity range compact and efficient light sources already exist for optoelectronic applications, namely the semiconductor lasers. This last point makes the implementation of nonlinear optical devices based on dynamic nonlinearities feasible, even if their recycling time and lack of coherence may be a handicap for some ultra-fast applications.

5. The Non-Resonant Response of Excitons

5.1 Physics: The Optical Stark Effect

Nonlinear optical effects occurring in the transparency region of QWS and involving excitons were explored in a series of experiments undertaken simultaneously by two different groups [7]. In these experiments a light pulse (called "pump") of short duration (~150 fs), very high intensity (a few GW/cm^2) and sharp spectrum is incident along the $\hat{z}$-axis of a QWS; its wavelength is tuned a few tens of meV to the red of the lowest exciton resonance. A second, relatively weak pulse (called "probe") can arrive on the sample with an adjustable delay with respect to the pump pulse. The probe has a broad spectrum and thus measures essentially the transmission spectrum of the QWS at differerent instants of time. When the pump pulse traverses the QWS sample, the probe measures a displacement of the exciton resonance to higher energies. However, the exciton spectrum comes back to its original position as soon as the pump pulse leaves the sample, indicating that there is no residual population of excitons and that no energy is deposited in the sample. The spectral displacement observed in this experiment is an optical nonlinearity that can be understood as a manifestation of the Optical Stark Effect, by analogy to what is observed in atomic two-level-like transitions. That is, in the very intense field of the pump pulse, the exciton energy levels are "dressed" by a cloud of photons and this causes a shift of the resonance frequency in a manner analogous to the "dressed atom" model of atomic optics.

The dressing of the exciton by photons at low incident field is a well-established concept in solid state physics, known as the polariton effect [8]. A polariton is a coupled mode of the exciton and electromagnetic fields that arises because of their very strong mutual interaction. A quantitative measure of this interaction is, of course, the very strong oscillator strength of the exciton. In the conventional theory of the polariton, however, the exciton is treated as an ideal Boson with no internal structure, and thus its optical response is strictly linear: the conventional polariton is an effect of linear optics. The Optical Stark Effect, and more generally the nonlinear optical response of the excitons, constitute a deviation from ideal Bose behavior that arises because of interactions among the excitons due to their internal structure. The exciton-photon interaction is sensitive to this internal structure, even when excitons are accessed off-resonance as virtual particles.

Several theoretical papers [9] have appeared since the first experiments were announced, each author relying on differerent features of the exciton to obtain a two-level-like system on which the ideas of the dressed two-level atom model can be implemented. Thus, even though the basic interpretation of this phenomenon as a manifestation of the "dressed exciton" is widely accepted, the details of the "dressing" process are still subject to controveresy.

More recently, in a series of very elegant experiments, the Optical Stark Effect has been used to induce a splitting between two different types of excitons that are normally degenerate but differ from each other by their symmetry properties [10].

5.2 Devices: An Optical Gate

The dispersive optical nonlinearity associated with the Optical Stark Effect has been used to demonstrate the possibility of producing an ultra-fast optical gate. The large spectral displacement of the exciton resonance frequency observed in the Optical Stark Effect is accompanied by a large change in the refractive index of the material, as required by the Kramers-Kronig relations. If the QWS is placed inside a Fabry-Perot cavity, this change in refractive index can be translated into a change of the transmittance (or reflectance) of the Fabry-Perot at a given wavelength. Thus, the presence of an intense ultra-short pulse that induces the non-linear index change in the QWS can be used to control the switching of the Fabry-Perot from a transmitting to a reflecting state. Soon after the first observation of the Optical Stark Effect, a nonlinear Fabry-Perot optical gate which could switch a light beam "on" or "off" with a switching time of the order of 0.7 ps [11]. In spite of its remarkable temporal characteristics, however, such a gate cannot be used as a functional device, as it requires for its operation a large, high-power femtosecond laser which is not amenable to miniaturization.

6. Excitonic Coherent Transients

QWS excitons are ideal candidates for the observation of optical coherent transient effects since they appear in the spectrum as sharp and relatively isolated lines with a high oscillator strength. Two such effects have been observed to date: (1) arrested free-induction decay [12] and (2) photon echoes [13].

Arrested free-induction decay is observed in a pump-and-probe configuration with ultra-short light pulses (~100 fs) in which the probe monitors the sample a few hundred femtoseconds before the pump arrives. The transmission spectrum measured by the probe pulse displays an oscillatory structure superimposed on the exciton peak, even though the pump has not yet excited the sample. This seemingly paradoxical result is due to the fact that in ordinary transmission spectroscopy the macroscopic polarization that is induced by the probe pulse emits radiation as it decays; this phenomenon is known as free-induction decay. The Fourier-transform of the time-dependance of the emitted radiation corresponds essentially to the observed spectrum. For example, a 1-ps exponential free-induction decay corresponds to a 5 cm^{-1} wide Lorentzian spectrum. However, if an intense pump pulse arrives soon after the probe pulse, it suddenly introduces scattering centers (excitons or electron-hole pairs) that cause an abrupt interruption of the free decay of the induced polarization. The spectrum that is obtained, then, is no longer Lorentzian but involves an oscillatory structure corresponding to the Fourier transform of the interrupted exponential decay.

Photon echo experiments are understood traditionally [4] by modeling the material as a statistical ensemble of two-level systems, in which the coherent polarization can decay through two types of processes:
(1) "homogeneous" processes, in which the phase of the macroscopic polarization is randomized because of the dynamics of the excited species (e.g. because of collisions or other brief sudden and random fluctuations) and
(2) "inhomogeneous" processes, in which the statistical distribution of the resonance frequencies of the individual two-level systems causes the microscopic polarization of each two-level system to oscillate at a slightly different frequency from the others; thus, even though locally the microscopic polarization has a non-zero oscillating value, the ensemble-averaged (experimentally measured) macroscopic polarization decays as the local phases evolve at different rates. Clearly, this latter decay process is non-ergodic, while "homogeneous" decay is ergodic.

In an echo experiment, a short light pulse excites coherently the ensemble of two-level systems and induces a macroscopic polarization which is allowed to evolve freely during a time-interval T. At the end of this time-interval a second light pulse inverts the polarizations of the individual two-level systems. Thus, at time T after the second pulse, any statistical spread that may have existed in the temporal evolution of the phases of the individual microscopic polarizations is compensated and the microscopic polarizations re-phase to reconstitute the macroscopic polarization. This reconstituted polarization emits, in turn, a coherent pulse (the "echo") like a classical dipolar antenna. The photon echo experiment permits to probe the ergodic (dynamical) decay processes of the coherent polarization, since it can circumvent its non-ergodic (inhomogeneous) decay. In the excitons of QWS dynamical dephasing times of the order of 5 to 20 picoseconds have thus been measured, at liquid Helium temperatures.

The observation coherent transients on Wannier excitons has raised the problem of developing the appropriate theoretical framework for describing such phenomena. The traditional two-level model may be adequate for excitonic coherent transients at very low excitation densities, but certainly breaks down in the high-intensity regime because it neglects the hydrogenic internal structure of the excitons and considers them effectively as rigid spheres (or rigid disks, in two-dimensional QWS). Since excitons are extended quasi-particles, at high densities they can "interpenetrate" and, thus, the details of their internal structure become important in the determination of their high-intensity resonant optical properties.

One way of accounting properly for high-intensity excitonic coherent transients [14] is to adopt the reciprocal-space description of excitons as collective excitations of the semiconductor crystal, involving all plane-wave electron-hole pairs, each weighted according to its amplitude in the momentum-space hydrogenic wavefunction. The wavefunction of a two-dimensional hydrogenic exciton can be written in momentum-space as

$$\phi(k) = \frac{\sqrt{2\pi}\, a_B}{(1 + (a_B k/2)^2)^{3/2}}, \tag{12}$$

where a_B is the effective Bohr unit of length. A plot of this wavefunction is given in the low-density curves of Figure 4. Within this viewpoint, the multi-exciton state that is driven coherently by a short laser pulse may be expressed as the product of the wavefunctions of all the plane-wave electron-hole pairs composing the exciton, each pair excited into a coherent superposition of its ground and excited states

$$|\alpha\rangle = \prod_k \{\cos[\tfrac{\alpha}{2}\phi(k)]\,|0\rangle_k + \sin[\tfrac{\alpha}{2}\phi(k)]\,|1\rangle_k\}, \tag{13}$$

where $|i\rangle_k$ (i=0,1) is the i-th state of the plane-wave electron-hole pair of relative wavevector k, and α is a parameter that gives the excitation amplitude the Wannier exciton system. It is related to μ, the transition dipole of the exciton resonance, and to E(t), the envelope function of the electric field of the incident pulse, by

$$\alpha = \int_{\infty}^{\infty} \frac{\mu E(t)}{\hbar} dt \quad . \tag{14}$$

The coherent wavefunction (13) accounts for the phase-space filling interactions among excitons, due to their extended hydrogenic structure. Many-body interactions may be included by adjusting variationally the single-exciton wavefunction $\phi(k)$.

All experimental observables pertaining to the coherently excited semiconductor may be calculated by taking the expectation value of the corresponding operator over the coherent exciton state (13). Thus, the coherent polarization, which gives access to the description of all coherent transient experiments, can be calculated as

$$P = \frac{\mu}{2} \sum_k \phi(k) \sin[\alpha\phi(k)] \quad . \tag{15}$$

The coherent polarization induced in a hydrogenic Wannier exciton system by a short laser pulse is plotted in Figure 5, as a function of the excitation parameter α. The coherent polarization induced in an ensemble of two-level systems is also plotted for comparison. Clearly, at high excitation densities, the coherent transient optical response of hydrogenic excitons deviates significantly from the response of the two-level model used traditionally to describe such transients. The deviation of the excitonic polarization from the two-level model reflects the phase-space filling interactions due to the hydrogenic structure of the excitons.

This theoretical analysis suggests that coherent transient experiments, such as photon echoes, which monitor directly the macroscopic polarization induced in a Wannier exciton system by a short incident light pulse, can give information on the internal structure of the excitons and on the exci-

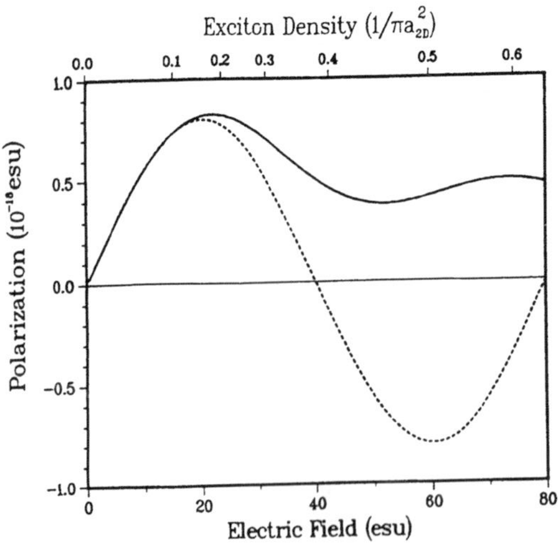

FIGURE 5: Resonantly induced coherent polarization of an exciton in a 2D GaAs layer, as a function of the electric field of the incident light pulse. Dashed line gives the response of a two-level system, for comparison. Upper horizontal axis gives exciton density attained by exciting pulse. (From Ref. 14)

ton interactions in the dense exciton gas, in addition to the information on the dephasing processes that they have already produced.

7. Dynamic Nonlinearities

Dynamic optical nonlinearities have been studied quite extensively in QWS, both in basic and in applied physics, essentially because of the relative simplicity in the experimental apparatus that is required. The time-scale of observation (10 ps to 10 ns) as well as the light intensities necessary (~1 kW/cm^2) are quite easily accessible with the lasers commonly available in the laboratory. In addition, many of the manifestations of dynamic nonlinearities can be observed with a minimal preparation of the QWS samples, directly at room temperature.

7.1 The Physics of Dense Electron-Hole Populations

From the point of view of basic physics, the study of dynamic nonlinearities interfaces quite well with our understanding of dense carrier populations in semiconductors, since such nonlinearities involve the presence of a significant population of excitons or electron-hole pairs. Thus, the analysis of dynamic optical nonlinearities provides a powerful tool for probing the structure and the dynamics of dense semiconductor plasmas.

We shall describe here a series of simple pump-and-probe type experiments performed in our laboratory on several QWS. In these experiments we investigated optically the phase-space filling and many-body interactions among electrons and holes in the dense photo-excited exciton population [15], and we followed the dynamics of the relaxation of the carriers between different subbands of the QWS [16]. Similar experiments have also been performed elsewhere [17] and there is a vast theoretical literature dealing with the dense exciton gas or electron-hole plasma [18].

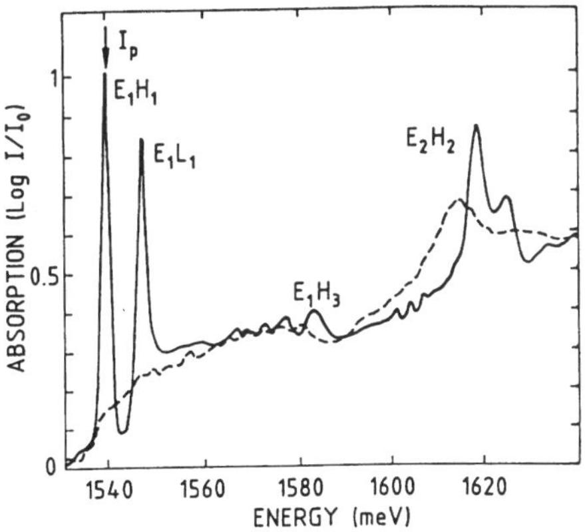

FIGURE 6: Solid line: absorption spectrum of a GaAs/$Al_{0.3}Ga_{0.7}As$ QWS with L_z = 120 Å. Dashed line: spectrum after excitation by short pulse resonant with n = 1 exciton (shown by arrow). Note bleaching of n = 1 excitons and shift of n = 2 exciton. (From Ref. 15)

In the first series of experiments [15], the pump pulse creates excitons in the lowest-energy subband of a QWS, while the probe pulse measures the spectrum of the excited sample when the pump pulse has deposited all its energy. A typical spectrum is displayed in Figure 6. We observe that the spectral modifications that occur on the excitons of the first two subbands of the QWS are quite different: the excitons of the first subband saturate and disappear, whereas the exciton of the second subband broadens and shifts to lower energies. Clearly, the presence of excitons in the first subband fills the phase-space of this subband. This causes the corresponding excitonic resonances to bleach and masks any other effects that there may have been due to many-body interactions. The second subband, however, is completely unoccupied and thus all the spectral modifications that are observed in it are due exclusively to many-body interactions. Comparison of the spectral modification of the two subbands shows that the optical nonlinearities of QWS are dominated by phase-space filling effects while the optical manifestations of many-body interactions are relatively weak, at least in the exciton densities that can be attained through direct optical excitation.

This simple pump-and-probe technique permits to separate in the spectrum the manifestations of the different types of nonlinearity and gives access to the study of many-body interactions through the experimental techniques of Nonlinear Optics. For example, one manifestation of many-body interactions that can be observed directly in the spectrum of Figure 6, is the slight shift to lower energies of the excitonic resonance of the second subband: it may be attributed to a large extent to the band-gap renormalization of the second subband. This effect arises from the exchange and correlation interactions of a test exciton (injected in the second subband by the probe pulse when measuring the spectrum) as it interacts with the dense electron-hole population that is already present in the first subband. The magnitude of the spectral shift observed in this experiment (a few meV for an exciton density of ~4 10^{11} cm^{-2}) contrasts sharply with the magnitude of the spectral shift observed in luminescence or gain experiments which give the band-gap renormalization of the electron-hole population of the first subband itself (a few tens of meV). This result shows that band-gap renormalization for an unoccupied subband is much smaller (by an order of magnitude) than the band-gap renormalization of a subband that

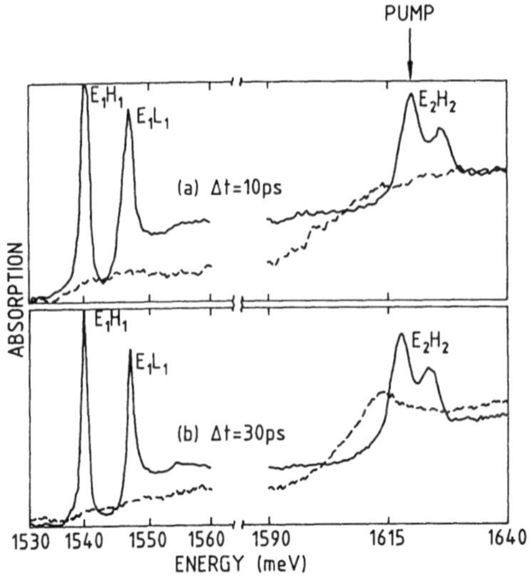

FIGURE 7: Solid line: absorption spectrum of a $GaAs/Al_{0.3}Ga_{0.7}As$ QWS with L_z = 120 Å. Dashed line: (a) spectrum after excitation by short pulse resonant with n = 2 exciton (shown by arrow). Note bleaching of all excitons. (b) 20 ps later, the n = 1 excitons are still bleached but the n = 2 exciton re-appears, indicating relaxation of the exciton population from the n = 2 to the n = 1 subbands. (From Ref. 16)

contains a large population of carriers, and underscores the idea that the naive model that commonly views renormalization as a rigid shift of the whole band structure is not a good approximation when states far above the Fermi level are under consideration.

In the second series of experiments [16], the pump pulse is tuned to the exciton of the second subband. In this frequency range, the absorption spectrum of the unexcited QWS is a superposition of the second subband exciton spectrum and the high-energy part of the first subband. The pump, therefore, creates excitons in the second subband and high-energy electron-hole pairs in the first subband. The transmission spectrum measured by the probe pulse immediately after the pump pulse has deposited all its energy in the sample indicates that the excitons of both the first and second subbands are saturated, because of phase-space occupation effects (see Figure 7). The second subband excitons are occupied directly, while those of the first subband are occupied after relaxation of the high-energy electron-hole pairs to the bottom of the band. This latter process, which was investigated through separate experiments involving a 100-fs laser [19,17a], is due to carrier-carrier (binary) collisions and therefore occurs in very fast (sub-picosecond) time-scales. As can be seen in Figure 7, transmission spectra measured a few tens of picoseconds later, however, indicate that the exciton resonance of the second subband re-appears but is broadened and shifted to lower energies, as it was under pumping of the first subband. This observation indicates that the phase-space filling effects of the second subband disappear but the many-body interactions due to the presence of a dense population in the first subband persist: the excitons of the second subband relax down to the first subband in time-scales ranging from 10 to 20 ps, depending on the relative energy spacing of the two subbands.

Clearly, there are two very different time-scales for the relaxation processes of carriers (or excitons), even when the carriers are created initially at the same energy. The relaxation process that involves a change in subband is two or three orders of magnitude slower than relaxation within the same subband. This slowing down of the relaxation rate is due essentially to the quantization of the $\hat{z}$-wavevector of the carriers because of their quasi-two-dimensional confinement in QWS. The discrete values of the $\hat{z}$-wavevector and the sparsity of the energy levels that this entails, implies that $\hat{z}$-wavevector and energy conservation cannot be satisfied simultaneously in a collision between two carriers, unless a large number of phonons can participate in the collision. This slows down considerably the inter-subband relaxation process, when compared with intra-subband relaxation in which the $\hat{z}$-wavevector is always conserved, since it retains the same value after a binary collision.

7.2 Devices: Monolithic Optical Bistable Etalons

From the point of view of applied physics, the relative simplicity of the experimental setup involving dynamic optical nonlinearities and the possibility of working at room temperature have permitted the design and fabrication of the first viable optical bistable devices for all-optical signal processing. These devices are based on the nonlinear refractive index change that is produced in the vicinity of the bandgap by the bleaching of the exciton resonance. As in the optical gate discussed in Section 5.2, the nonlinear QWS is introduced in the Fabry-Perot cavity and thus the nonlinear index change is translated into a change in the transmission and reflection characteristics of the Fabry-Perot. When an intense light beam of the appropriate frequency is incident on the Fabry-Perot etalon, it excites the QWS and creates a dense electron-hole population which causes a change in the refractive index of the QWS. This in turn modifies the transmission characteristics of the Fabry-Perot and acts as a feedback mechanism on the excitation of the QWS inside the cavity. As discussed in the chapter on Optical Bistability in this volume, the nonlinear Fabry-Perot has two stable states, one with low and one with high transmittance, or respectively one with high and one with low reflectance. Passage from one state to the other is abrupt and involves a hysterisis loop. This device is essentially the optical equivalent of the transistor and permits the optical emulation of all the functions that the transistor has in electronics, such as switching, gating, logical operations, memory etc.

Several laboratories have been engaged recently in the development of such Optical Bistable devices. At CNET, a team is working on the design, fabrication and optimization of Optical Bistable etalons based on the dynamic optical nonlinearities of QWS in the vicinity of the band-gap. This team has produced an Optical Bistable Fabry-Perot etalon whose two mirrors are integrated on the QWS by fabricating the whole structure in the same Vapor-Phase Epitaxy growth process [20]: The "mirrors" that are deposited directly on the QWS are actually multi-layer Bragg reflectors, each consisting of an alternation of quarter-wave GaAs and AlAs layers (approximately 600 Å and 700 Å thick respectively). A stack of approximately 10 such layer pairs displays 95% reflectivity at its resonance frequency, and thus the two such stacks that surround the QWS produce a nonlinear Fabry-Perot cavity. A switching power threshold of the order of a few mW for a spot of 6 μm diameter at 850 nm and switching times of the order of 20 ns were measured for such a device. These preliminary figures can conceivably be improved, but are already quite compatible with the present performance of semiconductor lasers. This Optical Bistable device is monolithic, very compact (~4 μm thick) and is fabricated by the same processes that already exist for the fabrication of electronic devices. It can therefore be potentially integrated in an optical (or optoelectronic) circuit through the methods that have been developed for electronic integration. Similar devices have also been produced by other groups [21].

At present, the CNET team is working on the pixelization of nonlinear Fabry-Perot QWS by photolithography and reactive-ion etching to produce two-dimensional arrays of 1 to 10 μm-size optical bistable elements [22]. The fabrication of nonlinear Fabry-Perot etalons with small lateral dimensions is expected to enhance their efficiency and lower the switching power threshold. The reason is that such a micrometer-size semiconductor mesa can act as a waveguide for the incident light signals, thus increasing the effective light intensity, while at the same time it can confine laterally the electron-hole population, thus preventing carrier diffusion from diminishing the high excitation density. In addition to the advantages obtained from the delimitation of micrometer-size elements, the fabrication of arrays with a large number of elements can take advantage of the possibilities for massive parallelism that are offered in optics: such arrays could be used for processing in parallel a large number of optical signals, simultaneously.

8. Conclusion

In this lecture we reviewed the Nonlinear Optical properties of semiconductor QWS and we illustrated the interest that these properties present, both in basic and in applied physics, through a survey of recent experimental and theoretical work.

The Nonlinear Optics of QWS involve very rich physics. Recent investigations have revealed a large number of novel and interesting phenomena, such as the excitonic Optical Stark effect, arrested free-induction decay, excitonic photon echoes, the non-rigid band-gap renormalization or the slowing-down of carrier relaxation because of quantum confinement.

In applied physics, the Nonlinear Optical properties of QWS have permitted the design and the realization of new exploratory all-optical devices, such as ultra-fast gates with sub-picosecond switching times or monolithic bistable etalons whose fabrication and operation is compatible with today's optoelectronic technology. Even though the performance characteristics of such devices are still far from being optimal, they illustrate the promise that QWS-based optical devices hold for optical signal processing.

Acknowledgement: The work presented in this lecture is the result of a close collaboration with G. Dolique, J.A. Levenson, J.L. Oudar, and R. Raj. It is a pleasure to acknowledge their contribution to this work.

References

1. A partial list of recent books and compilations of articles on QWS, their structural, electronic and optical properties, is:
Heterojunctions and Semiconductor Superlattices, Ed. by G. Allan, G. Bastard, N. Boccara, M. Lannoo, and M.Voos, Springer-Verlag (Berlin, 1986);
Semiconductors and Semimetals, Vol. 24: Applications of Multiquantum Wells, Selective Doping, and Superlattices, Ed. by R. Dingle, Academic Press (New York, 1987);
Physics and Applications of Quantum Wells and Superlattices, Ed. by E.E. Mendez and K. von Klitzing, NATO ASI series B: Physics; Vol. 170, Plenum Press (New York, 1987);

Interfaces, Quantum Wells, and Superlattices, Ed. by C.R. Leavens and R. Taylor, NATO ASI series B: Physics; Vol. 179, Plenum Press (New York, 1988);
Optical Nonlinearities and Instabilities in Semiconductors, Ed. by H. Haug, Academic Press (New York, 1988);
G. Bastard, Wave Mechanics Applied to Semiconductor Heterostructures, Editions de Physique (Paris, 1989).

2. I. Abram, J. Opt. Soc. Am. B 5, 1204 (1985).

3. S. Charbonneau, T. Steiner, M.L.W. Thewalt, E.S. Koteles, J.Y. Chi, and B. Elman, Phys. Rev. B 38, 3583 (1988).

4. For a review of coherent transients and the quasi-spin formalism, see: L. Allen and J.H. Eberly, Optical Resonance and Two-Level Atoms Wiley (New York, 1975).

5. A comprehensive review can be found in:
Photorefractive Materials and Their Applications I: Fundamental Phenomena, Ed. by P. Gunter and J.-P. Huignard, Springer-Verlag (Berlin, 1988);
Photorefractive Materials and Their Applications II: Applications Ed. by P. Gunter and J.-P. Huignard, Springer-Verlag (Berlin, 1989).

6. Photothermal nonlinearities are reviewed in:
I. Janossy, M.R. Taghizadeh, J.G.H. Mathew, abd S.D. Smith, IEEE J. Quantum Electron. QE-21, 1447 (1985);
M. Dagenais, A. Surkis, W.F. Sharfin, and H.G. Winful, IEEE J. Quantum Electron. QE-21, 1458 (1985).

7. A. Mysyrowicz, D. Hulin, A. Antonetti, A. Migus, W.T. Masselink, and H. Morkoc, Phys. Rev. Lett. 56, 2748 (1986);
A. von Lehmen, D.S. Chemla, J.E. Zucker, and J.P. Heritage, Opt. Lett. 11, 609 (1986).

8. J.J. Hopfield, Phys. Rev. 112, 1555 (1958).

9. Some of the early papers are:
C. Comte and G. Mahler, Phys. Rev. B 34, 7164 (1986);
S. Schmitt-Rink and D.S. Chemla, Phys. Rev. Lett. 57, 2752 (1986);
S. Schmitt-Rink, D.S. Chemla, and H. Haug, Phys. Rev. B 37, 941 (1988);
M. Combescot and R. Combescot, Phys. Rev. Lett. 61, 117 (1988);
I. Balslev and A. Stahl, Solid State Comm. 67, 85 (1988).

10. M. Joffre, D. Hulin, A. Migus, and M. Combescot, Phys. Rev. Lett. 62, 74 (1989).

11. D. Hulin, A. Mysyrowicz, A. Antonetti, A. Migus, W.T. Masselink, H. Morkoc, H.M. Gibbs, and N. Peyghambarian, Appl. Phys. Lett. 49, 749 (1986).

12. B. Flugel, N. Peyghambarian, G. Olbright, M. Lindberg, S.W. Koch M. Joffre, D. Hulin, A. Migus, and A. Antonetti, Phys. Rev. Lett. 59, 2588 (1987);
J.P. Sokoloff, M. Joffre, B. Fluegel, D. Hulin, M. Lindberg, S.W. Koch, A. Migus, A. Antonetti, and N. Peyghambarian, Phys. Rev. B 38, 7615 (1988).

13. L. Schultheis, M.D. Sturge, and J. Hegarty, Appl. Phys. Lett. 47, 995 (1985);

L. Schultheis, J. Kuhl, A. Honold, and C.W. Tu, Phys. Rev. Lett. 57, 1635 (1986); ibid. 57, 1797 (1986);
L. Schultheis, A. Honold, J. Kuhl, K. Kohler, and C.W. Tu, Phys. Rev. B 34, 9027 (1986);
R. Raj, to be published.

14. I. Abram, Phys. Rev. B 40, (15 Sept. 1989 issue).

15. J.A. Levenson, I. Abram, R. Raj, G. Dolique, J.L. Oudar, and F. Alexandre, Phys. Rev. B 38, 13443 (1988);
J.A. Levenson, I.I. Abram, R. Raj, and G. Dolique, J. Phys (Paris) 49, C2-251 (1988);
I. Abram and J.A. Levenson, Superlattices and Microstructures 5, 181 (1989).

16. J.A. Levenson, G. Dolique, J.L. Oudar, and I. Abram, Surface Science, (in press); J.A. Levenson, G. Dolique, J.L. Oudar, and I. Abram, Solid State Electronics, (in press).

17. W.H. Knox, C. Hirlimann, D.A.B. Miller, J. Shah, D.S. Chemla, and C.V. Shank, Phys. Rev. Lett. 56, 1191 (1986);
C. Weber, C. Klingshirn, D.S. Chemla, D.A.B. Miller, J. Cunningham, and C. Ell, Phys. Rev. B 38, 12748 (1989);
D.Y. Oberli, D.R. Wake, M.V. Klein, J. Klem, T. Henderson, and H. Morkoc, Phys. Rev. Lett. 59, 696 (1987);
A. Seilmeier, H.-J. Hubner, G. Abstreiter, G. Weimann, and W. Schlapp, Phys. Rev. Lett. 59, 1345 (1987).

18. The optical properties of a dense electron-hole plasma are reviewed in:
H. Haug and S. Schmitt-Rink, Prog. Quantum Electron. 9, 3 (1984).
More recent work includes:
S. Schmitt-Rink, D.S. Chemla, and D.A.B. Miller, Phys. Rev. B 32, 6601 (1985);
R. Zimmermann, Phys. Stat. Sol. (b) 146, 371 (1988);
W. Schaefer, K.H. Schuldt and J. Treusch, Phys. Stat. Solidi (b) 147, 699 (1988).

19. J.L. Oudar, J. Dubard, F. Alexandre, D. Hulin, A. Migus, A. Antonetti, J. Phys. (Paris) Colloque 48, C5-511 (1987).

20. R. Kuszelewicz, J.L. Oudar, J.C. Michel, and R. Azoulay, Appl. Phys. Lett. 53, 2138 (1988).

21. J.L. Jewell, H.M. Gibbs, S.S. Tarng, A.C. Gossard, and W. Wiegmann, Appl. Phys. Lett. 40, 291 (1982);
O. Sahlen, U. Olin, E. Masseboeuf, G. Landgren, and M. Rask, Appl. Phys. Lett. 50, 1559 (1987).

22. R. Kuszelewicz, J.L. Oudar, R. Azoulay, J.C. Michel, J. Brandon, and O. Emile, Phys. Stat. Sol. (b) 150, 465 (1988).

Transient Nonlinear Optics in Semiconductors

J.M. Hvam

Fysisk Institut, Odense Universitet, Campusvej 55,
DK-5230 Odense M, Denmark

Abstract. The transient nonlinear optical properties of direct gap semiconductors are discussed in relation to picosecond time resolved experiments of the excite-and-probe type. In particular, degenerate four-wave mixing and transient laser induced grating spectroscopies, in two-beam and three-beam configurations, are being used to investigate the nonlinear optical resonances associated with exciton interactions. The coherence and dephasing of these resonances are also related to the exciton interactions and relaxations. Specific results from the binary compound CdSe and from mixed crystals of $CdSe_xS_{1-x}$ are presented, and some possible applications of the transient nonlinear optical response are discussed.

1. Introduction

There is an increasing demand for optical devices for light modulation and optical switching in present-day and future technologies of optical transmission and signal processing. This has triggered an intense search for materials with large nonlinear optical coefficients and a fast response time [1]. In this context, direct gap semiconductors constitute an interesting group of materials for basically two reasons: i) They exhibit very strong resonance enhancements of the linear as well as the nonlinear optical coefficients near the band gap and ii) in resonance, the response time is governed by carrier relaxations and recombination lifetimes that inherently lie in the femtosecond to nanosecond range, and can even, to some extent, be manipulated.

The linear optical response is governed by the large oscillator strength for the dipole allowed direct electron transitions across the band gap giving rise to the creation of an electron-hole (e-h) pair, strongly correlated by the Coulomb interaction (exciton effect) [2]. The nonlinearity is brought about by the final state (exciton) interactions, that occur even at fairly moderate exciton densities ($n_x \geq 10^{14} cm^{-3}$). The observed response times are also governed by final state interactions in the form of impurity, phonon and carrier-carrier scattering and recombination, depending on purity, temperature and excitation density [2,3].

The strong interest in the nonlinear optics in semiconductors is twofold. In the first place, transient nonlinear optical spectroscopies have turned out to be extremely powerful for the investigation of carrier interaction and relaxation phenomena in semiconductors [4-6]. Secondly, the detailed knowledge of the nonlinear optical properties of semiconductor materials and structures may lead to the development of new electro-optical or all-optical devices [7,8]. The present paper is mainly concerned with the first aspect, but shall also try to point out the application (device) aspects of some of the observed nonlinear optical phenomena.

In the following section will be presented the nonlinear optics formalisms employed in this work, and its application to a simple medium, in the form of a two-level system, is described in Sect.3. In Sect.4, relevant optical properties of bulk semiconductors are briefly reviewed, followed, in Sect.5, by a discussion of some general aspects of the experimental observation of transient nonlinear optics in resonant media. Section 6 is devoted to the presentation of some degenerate four-wave mixing (DFWM) experiments in CdSe and $CdSe_xS_{1-x}$, and Sect.7 contains the conclusions.

2. Nonlinear optics

The linear, as well as the nonlinear, optical response of materials is described by expressing the relation between the electric field **E** of the electromagnetic radiation and the resulting polarization **P** of the medium. In the nonlinear case, $\mathbf{P}_{NL}$ is normally expanded in powers of **E** [9,10]

$$\begin{aligned}\mathbf{P}_{NL}(\mathbf{r},t) &= \chi^{(1)}\cdot \mathbf{E} + \chi^{(2)}\cdot \mathbf{EE} + \chi^{(3)}\cdot\mathbf{EEE} + \cdots \\ &= \mathbf{P}^{(1)}(\mathbf{r},t) + \mathbf{P}^{(2)}(\mathbf{r},t) + \mathbf{P}^{(3)}(\mathbf{r},t) + \cdots \end{aligned} \quad (1)$$

where the terms on the right hand side are in a short-hand vector notation and in reality are multiple space and time integrals of the form, e.g.

$$\begin{aligned}\mathbf{P}^{(2)}(\mathbf{r},t) = \int \chi^{(2)}&(\mathbf{r}-\mathbf{r}_1,t-t_1;\mathbf{r}-\mathbf{r}_2,t-t_2) \\ &\cdot\mathbf{E}(\mathbf{r}_1,t_1)\mathbf{E}(\mathbf{r}_2,t_2)d\mathbf{r}_1 d\mathbf{r}_2 dt_1 dt_2\end{aligned} \quad (2)$$

If the electromagnetic field is composed of a group of monochromatic plane waves

$$\mathbf{E}(\mathbf{r},t) = \sum_i \mathbf{E}_i(\mathbf{k}_i,\omega_i) = \sum_i (\mathbf{E}_{0i} e^{i(\mathbf{k}_i \cdot \mathbf{r} - \omega_i t)} + \text{c.c.}) \tag{3}$$

then the polarization can likewise be expanded in plane waves

$$\mathbf{P}^{(n)}(\mathbf{r},t) = \sum_j \mathbf{P}_j^{(n)}(\mathbf{k}_j,\omega_j) = \sum_j (\mathbf{P}_{0j}^{(n)} e^{i(\mathbf{k}_j \cdot \mathbf{r} - \omega_j t)} + \text{c.c.}) \tag{4}$$

In steady-state, the different components can be expressed by

$$\mathbf{P}^{(n)}(\mathbf{k},\omega) = \chi^{(n)}(\mathbf{k},\omega) \cdot \mathbf{E}_1(\mathbf{k}_1,\omega_1)\mathbf{E}_2(\mathbf{k}_2,\omega_2) \cdots \mathbf{E}_n(\mathbf{k}_n,\omega_n) \tag{5}$$

where the n'th order nonlinear susceptibility $\chi^{(n)}(\mathbf{k},\omega)$ is the Fourier transform of the corresponding response function $\chi^{(n)}(\mathbf{r}-\mathbf{r}_1,t-t_1,\dots.\mathbf{r}-\mathbf{r}_n,t-t_n)$. Furthermore, the only components that survive the space and time integrations are those that conserve energy (for times $t-t_i \gg 2\pi/\omega_i$) and momentum:

$$\omega = \sum_i \pm\, \omega_i \qquad \text{and} \qquad \mathbf{k} = \sum_i \pm\, \mathbf{k}_i \tag{6}$$

where +/- are entered for absorbed/emitted waves, respectively. Due to the fixed dispersion relation $\omega = \omega(\mathbf{k})$ in any medium, the simultaneous fulfillment of the two Eqs.(6) is very restrictive. It is only met under special conditions, called phase-matching, and then usually only for one particular polarization wave. Phase-matching insures that the interacting waves and the nonlinear polarization propagate with the same speed through the medium, so that an appreciable energy transfer between waves can take place over a certain length.

For steady-state conditions the form in Eq.(5) is very convenient. In the usual dipole approximation, or local response theory, the susceptibility is, furthermore, independent of wave vector, so that the medium is fully described by the frequency dependent nonlinear susceptibilities

$$\chi^{(n)}(\mathbf{k},\omega) = \chi^{(n)}(\omega) = \chi^{(n)}(\omega;\pm\omega_1,\pm\omega_2,\dots,\pm\omega_n) \tag{7}$$

where in the last expression we have adopted the usual notation for nonlinear susceptibilities [9], indicating the applied frequencies and the resulting frequency from Eq.(6).

In the dynamic situation of ultra short light pulses, one might expect the time dependent response functions in Eq.(2) to be the most appropriate representation. However, except for the most extreme cases, the time variations in the field amplitudes are slow compared to the optical periods $2\pi/\omega_i$, so that the energy conservation in Eq.(6) is still valid. In this case it is convenient to work in a mixed frequency and time representation of the form

$$\mathbf{P}^{(n)}(\mathbf{k},\omega,t) = \chi^{(n)}(\omega,t)\cdot\mathbf{E}_1(\mathbf{k}_1,\omega_1)\cdots\mathbf{E}_n(\mathbf{k}_n,\omega_n) \tag{8}$$

where the normalized time dependencies of the field amplitudes $f_l(t_l)$ are included in a time dependent susceptibility:

$$\chi^{(n)}(\omega,t) = \int_{-\infty}^{0} \chi^{(n)}(\tau_1,\tau_2,\cdots,\tau_n) f_1(\tau_1)\cdots f_n(\tau_n) \cdot e^{i(\omega_1\tau_1+\cdots+\omega_n\tau_n)} d\tau_1\cdots d\tau_n \tag{9}$$

with $\tau_l = t - t_l$.

Physically, the response functions $\chi^{(n)}(\tau_1,\tau_2,\cdots,\tau_n)$ express the polarization in the medium at time t following a series of δ-function pulses at times $t_1,t_2,\cdots,t_n$. They have to be calculated quantum mechanically, knowing the electronic structure of the medium, e.g. a semiconductor. They will thus contain the resonance enhancements as well as the coherence and dephasing properties of the optical transitions in the semiconductor in question.

The nonlinear polarization induced by the driving fields (Eqs.(1),(8)) will in turn serve as a source for an electromagnetic wave, as described by the wave equation [10]

$$\left(\nabla^2 + \frac{\varepsilon}{c^2}\frac{\partial^2}{\partial t^2}\right)\mathbf{E}(\mathbf{r},t) = -\frac{4\pi}{c^2}\frac{\partial^2}{\partial t^2}\mathbf{P}_{NL}(\mathbf{r},t) \tag{10}$$

where ε is the dielectric function and c is the velocity of light. The wave equation has to be solved self-consistently with Eq.(1) to find the total nonlinear field $\mathbf{E}_{NL}(\mathbf{r},t)$.

2.1 Third order optical nonlinearities

In the present paper, we shall concentrate on the third order nonlinearity, being the lowest order in materials with inversion symmetry ($\chi^{(2)} \equiv 0$):

$$\mathbf{P}^{(3)}(\mathbf{k},\omega) = \chi^{(3)}(\omega;\omega_1,\omega_2,\omega_3)\mathbf{E}_1(\mathbf{k}_1,\omega_1)\mathbf{E}_2(\mathbf{k}_2,\omega_2)\mathbf{E}_3(\mathbf{k}_3,\omega) \tag{11}$$

One light beam. The simplest case to consider is the one where only one monochromatic light beam is present in the medium. Then in Eq.(11), $\omega_1=\omega_2=\omega_3=\omega'$ and $\mathbf{k}_1=\mathbf{k}_2=\mathbf{k}_3=\mathbf{k}'$, which from Eq.-(6) offers two possibilities:

1) Third harmonic generation with $\omega=3\omega'$ and $\mathbf{k}=3\mathbf{k}'$. In most media, however, phase matching will be impossible to achieve, since normally $\omega(3\mathbf{k}')\neq 3\omega'$.

2) Degenerate "four-wave-mixing" (DFWM) within the same beam with $\omega=\omega'$ and $\mathbf{k}=\mathbf{k}'$. This situation is always phase-matched and the nonlinear polarization

$$\mathbf{P}^{(3)}(\mathbf{k}.\omega) = \chi^{(3)}(\omega;\omega,-\omega,\omega)\left|\mathbf{E}(\mathbf{k},\omega)\right|^2 \mathbf{E}(\mathbf{k},\omega) \tag{12}$$

clearly expresses the situation of intensity dependent refractive index n(I) and absorption coefficient $\alpha(I)$. With $n(I)=n_1+n_2I+\cdots$ and $\alpha(I)=\alpha_1+\alpha_2I+\cdots$, we find from Eq.(12) that $n_2=\mathrm{Re}\chi^{(3)}/\varepsilon_0n^2c$, and $\alpha_2=\omega\cdot\mathrm{Im}\chi^{(3)}/\varepsilon_0n^2c^2$. The intensity dependent refractive index is responsible for the important phenomena of self-focusing and self-defocusing, and the intensity dependent absorption coefficient describes bleaching, induced absorption, two-photon absorption, etc.

Two light beams. We shall limit ourselves to the degenerate case with $\omega_1=\omega_2=\omega$, $k_1=k_2=k$, and $\cos\theta=\mathbf{k}_1\cdot\mathbf{k}_2/k^2<1$, as in Fig.1.

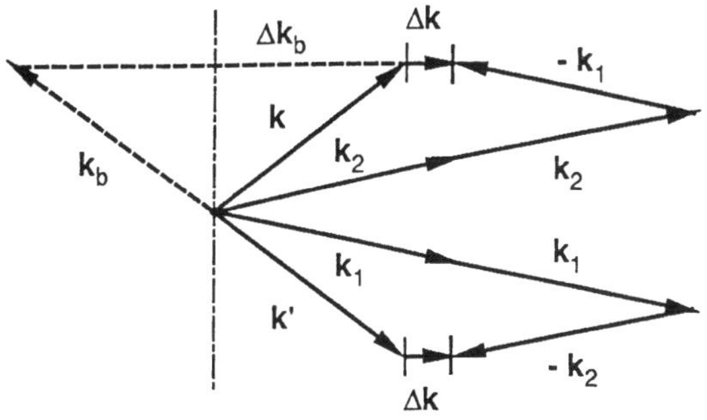

Fig.1. Wave vector conservation for DFWM with two light beams, $\mathbf{k}_1,\mathbf{k}_2$.

There are two possibilities with perfect phase-match: $\mathbf{k}=\mathbf{k}_1$ and $\mathbf{k}=\mathbf{k}_2$. As in the case with only one beam, the nonlinear signals are generated in a direction with a large linear signal and may therefore be difficult to detect at moderate intensities. There are also two possibilities with near phase-match: $\mathbf{k} = 2\mathbf{k}_2-\mathbf{k}_1-\Delta\mathbf{k}$ and $\mathbf{k}' = 2\mathbf{k}_1-\mathbf{k}_2-\Delta\mathbf{k}$, where $\Delta\mathbf{k}$ is the wave vector mismatch perpendicular to the sample plane. This geometry (see Fig.1) has the advantage that the nonlinear signal is generated in a direction where there is no linear signal. It is therefore very well suited for the detection of even very small nonlinear signals. The linear background is to a high degree eliminated by simple spatial filtering.

This DFWM can also be viewed upon as a case of light induced gratings [6,11,12] The two incident beams set up stationary polarization gratings, in the nonlinear medium, with scattering vectors $\pm(\mathbf{k}_2-\mathbf{k}_1)$ and a grating constant $\Lambda=2\pi/\left|\mathbf{k}_2-\mathbf{k}_1\right|$. These gratings in turn scatter the incident beams $\mathbf{k}_1$ and $\mathbf{k}_2$ into the directions $\mathbf{k}'$ and $\mathbf{k}$, respectively, as also discussed above (see Fig.1).

The signal intensity in the scattered direction is [6,12]

$$I_s \propto d^2 \frac{\sin^2(\Delta k \cdot d/2)}{(\Delta k \cdot d/2)^2} \tag{13}$$

where d is the sample thickness, or nonlinear interaction length in general. Near phase-match requires $\Delta k \cdot d \leq 1$, i.e. thin sample geometry. If the sample is very thin ($k \cdot d \leq 1$), the grating is essentially two-dimensional and back scattering (with $\Delta k_b \cong 2k$) will occur with about the same intensity as in the forward direction (see Fig.1). Note, however, that in both directions the signal will be weak, because $I_s \propto d^2$. Therefore strong resonance enhancement will normally be necessary to observe a back scattered signal. On the other hand, it does open up the possibility to observe the nonlinear interaction in a resonance where strong linear absorption prevents transmission even through a thin sample [6,13]. We shall later see back scattering in a semiconductor exciton resonance.

Three light beams. With three incident light beams, more freedom is at hand to perform different types of four-wave mixing experiments. One advantage is that two coherent light beams at an angle is sufficient to set up a light-induced grating in the medium, which can then be probed independently by the third beam that needs no coherent relation to the first

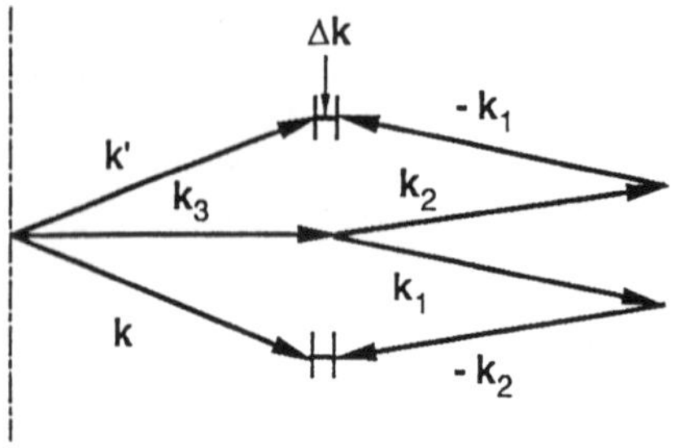

Fig.2. Wave vector conservation for DFWM with three light beams $\mathbf{k}_1, \mathbf{k}_2, \mathbf{k}_3$.

two beams. One can thus set up a nonlinear grating by exciting resonantly by the two beams and then probe the grating in the transparent region. In the degenerate case $\omega_1 = \omega_2 = \omega_3$, and with three different beam directions, $\mathbf{k}_1$, $\mathbf{k}_2$ and $\mathbf{k}_3$, a certain phase mismatch $\Delta\mathbf{k}$ again requires thin samples, $d \leq 1/\Delta k$, and gives rise to first order scattering ($\chi^{(3)}$) in the three directions $\mathbf{k} = \mathbf{k}_1 - \mathbf{k}_2 + \mathbf{k}_3 - \Delta\mathbf{k}$, $\mathbf{k}' = -\mathbf{k}_1 + \mathbf{k}_2 + \mathbf{k}_3 - \Delta\mathbf{k}$ and $\mathbf{k}'' = \mathbf{k}_1 + \mathbf{k}_2 - \mathbf{k}_3 - \Delta\mathbf{k}$. These scattering directions are shown in Fig.2, where accidentally $\mathbf{k}'' = \mathbf{k}_3$, because $\mathbf{k}_3$ is bisecting the angle between $\mathbf{k}_1$ and $\mathbf{k}_2$. Even in the degenerate situation, there are some differences between four-wave mixing with two beams and with three beams, particularly considering dynamic effects as shall be discussed in the next section.

2.2 Time resolved degenerate four-wave mixing

With two beams, the situation is sketched in Fig.3a. The two incident laser pulses are split off the same laser pulse (pulse length τ_L), and are impinging on the sample with a variable optical delay between them. In order for the two laser pulses to interact coherently in the nonlinear medium, for example by setting up a polarization grating, the delay between them should not exceed the dephasing time of the nonlinear polarization in the medium, caused by the first laser pulse [6]. The nonlinear DFWM signal is then self-diffraction of the second pulse in the grating set up by the coherent overlap between the polarizations from the first and the second pulse. For pulse #1 arriving first ($\tau_{12}>0$) as in Fig.3, a signal will thus be emitted in the direction $2\mathbf{k}_2-\mathbf{k}_1$ as indicated. For pulse #2 arriving first ($\tau_{12}<0$), the signal would be emitted in the direction $2\mathbf{k}_1-\mathbf{k}_2$ (dashed arrow in Fig.3a).

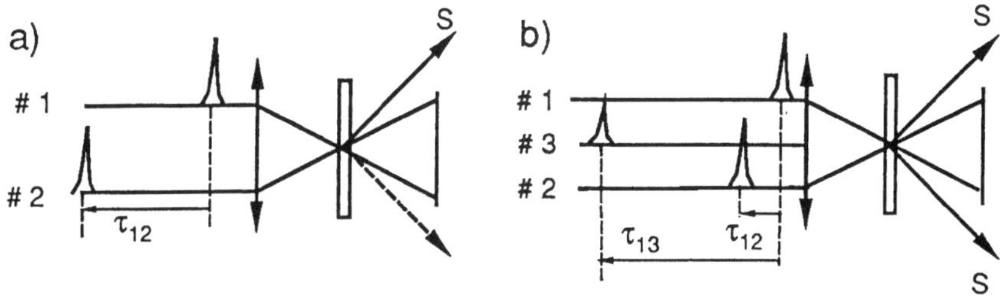

Fig.3. Time resolved DFWM with two beams (a) and with three beams (b), respectively.

With a slow detector in the signal beam, the time integral of the signal pulse is detected. As we shall see later, this integrated intensity will for delays $\tau_{12}\gg\tau_L$ vary as $I_s\propto \exp\{-a\tau_{12}/T_2\}$, [6,14] where a is a constant and T_2 is the dephasing time, or transverse relaxation time, of the optical excitation (polarization) in the medium. This two-beam configuration is therefore well suited to measure dephasing times, provided of course that they exceed the laser pulse length, $T_2>\tau_L$.

With three beams, the situation is sketched in Fig.3b. The two first pulses arrive simultaneously, or well within the dephasing time of the material ($\tau_{12}\ll T_2$), and interfere coherently to set up a nonlinear grating in the medium. This grating can then be detected at variable time delays, τ_{13}, by diffraction of the third pulse. If also $\tau_{13} \ll T_2$, then pulse #3 will diffract off a coherent polarization grating set up by pulse #1 and pulse #2, as in the self-diffraction case above. If, however, ω is in resonance with an electronic excitation in the material a real excitation density grating may persist in the material long after the coherent polarization grating has disappeared by dephasing. This type of experiment is therefore well suited to separate the purely coherent contribution to the optical nonlinearities from the more long-lived incoherent contributions from a high density of excited carriers in the medium.

The lifetime τ_G of the incoherent excitation density grating is determined by the lifetime τ_R of the excited carriers as well as by carrier diffusion [15], since the latter will wash out the spatial modulation of the carrier density.

$$\tau_G^{-1} = \tau_R^{-1} + 4\pi^2 D/\Lambda^2 \qquad (14)$$

where D is the carrier diffusion coefficient and Λ is the grating constant, as determined by the wavelength λ of the exciting light and the angle θ between the two interfering beams ($\mathbf{k}_1$ and $\mathbf{k}_2$)

$$\Lambda = 2\pi/|\mathbf{k}_1 - \mathbf{k}_2| = \lambda \;/\; 2\sin(\theta/2) \qquad (15)$$

By detecting the integrated intensity of the scattered test signal as a function of the delay τ_{13} of the test pulse (#3) one can determine the grating lifetime. By performing such transient grating experiments at different angles θ, one can determine the carrier lifetime and the diffusion coefficient separately from Eq.(14).

2.3 Measuring nonlinear coefficients

It is of course only the coherent contribution to the nonlinear signal that is well described by a nonlinear susceptibility as in Eq.(5). When using DFWM experiments to measure the magnitude (in resonance) of the latter, it is therefore essential to identify the coherent contribution as for example in a transient experiment with a time resolution better than the dephasing time of the resonance. In a resonant c.w. experiment, the incoherent contribution integrates up over the grating lifetime, and may thus exceed by several orders of magnitude the coherent contribution.

The intensity I_s of the coherent self-diffraction in a two-beam DFWM experiment, with the input intensities I_1 and

I_2, can be calculated in steady-state as [12]

$$I_s = \frac{1}{4} I_1 I_2^2 \frac{|K|^2 l_a^2}{(1 + K_2 I_2 l_a)^3} \tag{16}$$

where $K=K_1+iK_2=\omega\chi^{(3)}/\varepsilon_o n^2 c^2$ is the third order nonlinear coupling coefficient and $l_a=(1-e^{-\alpha d})/\alpha$ is the nonlinear interaction length, as determined by the linear absorption length $1/\alpha$ or the sample thickness, whichever is the smaller. It is here assumed that $l_a \ll 1/\Delta k$, where Δk is the wave vector mismatch. If nonlinear absorption of the pump beam is negligible compared to linear absorption ($K_2 I_2 l_a \ll 1$), a very simple expression for the nonlinear susceptibility is obtained in terms of the ratio between the transmitted signal and pump intensities $R_s \equiv I_s/I_2 e^{-\alpha d}$:

$$|\chi^{(3)}(\hbar\omega)|^2 = \frac{4\varepsilon_0 n^2 c^4}{\omega^2 I_1 I_2 l_a^2} R_s(\hbar\omega) \tag{17}$$

Thus, knowing the input intensities and the linear absorption coefficient, a good estimate of $\chi^{(3)}$ can be obtained from measuring the ratio R_s [7,12].

3. Simple medium: Two-level system

The simplest medium one can think of in the present context only contains two electronic levels, a and b, with energies $E_a < E_b$ interacting with electromagnetic radiation of frequency $\omega_0=(E_b-E_a)/\hbar$. In the density matrix formalism this interaction is described by the Schrödinger equation $(\partial\rho/\partial t)=(i/\hbar)[\rho,\mathbb{H}]$ with the Hamiltonian $\mathbb{H}=\mathbb{H}_0+\mu\cdot\mathbf{E}(t)$, where μ is the dipole operator and $\mathbf{E}(t)$ is the radiation electric field [10,14].

Introducing the longitudinal and transverse relaxation times T_1 and T_2 of the medium results in the Bloch equations:

$$\frac{\partial\rho_D}{\partial t} = -\frac{2i\mu E}{\hbar}(\rho_{ba} - \rho_{ab}) - \frac{\rho_D - \rho_{D0}}{T_1} \tag{18}$$

$$\frac{\partial\rho_{ba}}{\partial t} = \frac{\partial\rho_{ab}^*}{\partial t} = \frac{i\mu E}{\hbar}\rho_D - \left(\frac{1}{T_2} + i\omega_0\right)\rho_{ba} \tag{19}$$

where $\rho_D = \rho_{aa} - \rho_{bb}$ with the thermal equilibrium value ρ_{D0}. Solving for ρ_{ba} yields the polarization as a nonlinear function of E:

$$P_{NL}(t) = N\cdot Tr(\mu\rho) = N\mu(\rho_{ab}(t)+\rho_{ba}(t)) = N\mu\rho_{ab}(t)+c.c. \tag{20}$$

where N is the density of non-interacting two-level atoms in the system. If the latter is inhomogeneously broadened with the normalized line shape $g(\omega_0)$, the polarization is obtained by integration over this line shape [6,14]:

$$P_{NL}(t) = N\cdot\int_0^\infty \mu\rho_{ab}(t)g(\omega_0)d\omega_0 \tag{21}$$

For such a simple inhomogeneously broadened two-level system, one can calculate the intensity I_{sc} of self-diffracted light in a two-beam DFWM experiment as in Fig.3a , with slowly varying pulse amplitudes $E_i = E_{i0}f(t)$. In the rotating wave and perturbation approximations, one obtains [6,14]

$$I_{sc} \propto \left|\rho_{ab}^{(3)}(\mathbf{r},t,\tau)\right|^2 \tag{22}$$

where τ is the delay of pulse #2 (wave vector $\mathbf{k}_2$) with respect to pulse #1 (wave vector $\mathbf{k}_1$), and

$$\rho_{ab}^{(3)}(\mathbf{r},t,\tau) = \rho_{D0}\; E_{10}E_{20}^2\; \mu^3\; G(t,\tau,\Gamma)\; e^{i\{(2\mathbf{k}_2-\mathbf{k}_1)\cdot\mathbf{r}-\omega t\}} \tag{23}$$

where Γ is the the line width of $g(\omega_o)$, and

$$G(t,\tau,\Gamma) = \int_{-\infty}^{t}\int^{t_1}\int^{t_2} f(t_1)[f(t_2)f(t_3+\tau)F(t+t_1+t_2-t_3) + f(t_2+\tau)f(t_3)F(t+t_1-t_2+t_3)] \cdot\exp\{(t_1-t_2+t_3-t)/T_2-(t_1-t_2)/T_1\}dt_3dt_2dt_1 \tag{24}$$

$F(t)$ being the Fourier transform of $g(\omega_0)$.

In Fig.4 are shown calculated scattered intensities $I_{sc}(t,\tau)$ as a function of time t after the arrival of pulse #2 for different delays τ [6]. Note that the diffracted signal appears as a photon echo [10] with the delay τ. With an inhomogeneous broadening $\Gamma^{-1}<\tau$, the macroscopic polarization induced by pulse #1 dies out due to destructive interference between the different oscillators (free induction decay) before the arrival of pulse #2. However, the second pulse will start a rephasing process, that after a time τ will restore the macroscopic polarization, provided the individual oscillators are still coherent with the exciting laser pulses.

In experiments, correlation traces are usually recorded [6], i.e. the time integrated signal as a function of the delay $I_{sc}(\tau)=\int I_{sc}(t,\tau)dt$. The correlation trace is also shown in Fig.3.2. As mentioned already in Sect.2.2, it has a characteristic asymmetric shape with a tail towards positive delays when looking in the direction $2\mathbf{k}_2-\mathbf{k}_1$. Looking in the direction $2\mathbf{k}_1-\mathbf{k}_2$ the tail extends towards negative τ. For $\tau \gg \tau_L$, $I_{sc}(\tau)\propto\exp\{-4\tau/T_2\}$ from which the dephasing time T_2 is easily determined. Note that when comparing the correlation traces in

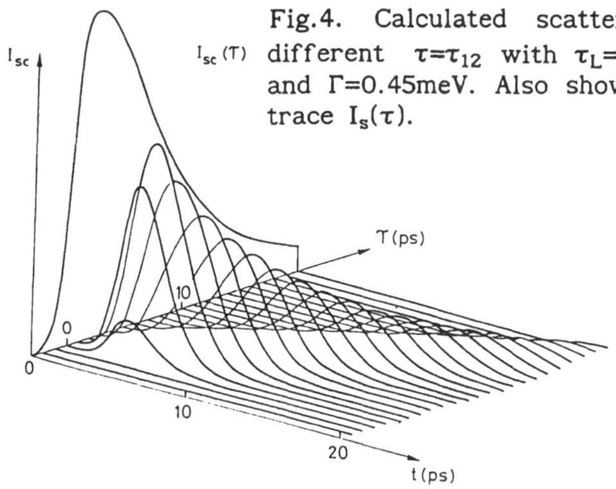

Fig.4. Calculated scattered intensity for different $\tau=\tau_{12}$ with τ_L=1.4ps, $T_1=T_2$=20ps and Γ=0.45meV. Also shown is the correlation trace $I_s(\tau)$.

the two above mentioned symmetric directions, τ=0 can be found with high precision.

4. Semiconductors

Semiconductors cannot in general be described by a two-level model, but rather by a two-band model [16] including the electronic states near the valence and conduction band edges, as sketched in Fig.5a. In direct gap semiconductors the band edges are at the same point in crystal momentum space (**k**=0), which again facilitates strong optical interaction, since photons carry negligible momentum compared to electrons and holes. For photon energies larger than the band gap, $\hbar\omega \geq E_g$, electron-hole (e-h) pairs are created by direct absorption of photons, resulting in a strong linear absorption edge (see Fig.5b).

Initially, the electrons and the holes are created in well-defined momentum states $\mathbf{k}_e, \mathbf{k}_h$ and the resulting polarization is in phase with the electromagnetic field. Scattering (even

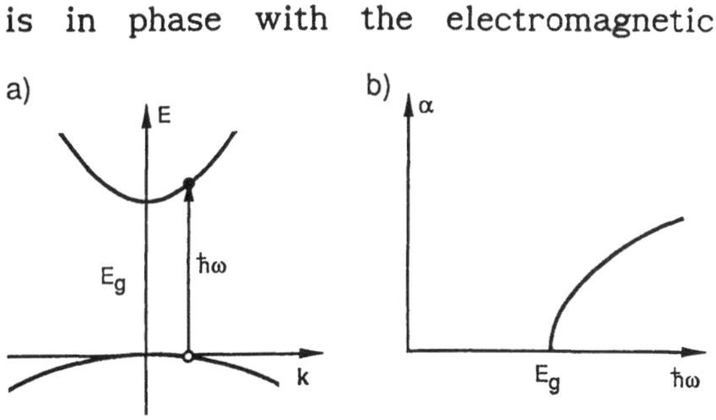

Fig.5. Semiconductor band edges (a) and absorption edge (b).

elastic) out of the initial momentum states will cause dephasing of the polarization, as characterized by the dephasing, or transverse relaxation, time T_2. Further relaxation in the form of inelastic intra band or inter band
scattering gives rise to several longitudinal relaxation and recombination times, T_1, τ_r. The predominant scattering mechanisms are impurity, phonon and carrier-carrier scattering depending on purity, temperature and excitation density of the semiconductor, and the typical scattering times cover the range from femtoseconds to microseconds [3,5,6,15,17].

If the carriers are created with kinetic energy exceeding the LO-phonon energy in the material ($\hbar\omega - E_g > \hbar\omega_{LO}$), LO-phonon scattering dominates with scattering times $\tau_{LO} \cong 150fs$ [17], except at very high carrier densities. For $0 < \hbar\omega - E_g < \hbar\omega_{LO}$, scattering times are in the sub-picosecond range, caused by acoustic phonon and impurity scattering. Inter band recombination lifetimes are typically $\tau_r \cong 1ns$ [18], except if the carriers are being trapped in deeper localized gap states (traps), where they can live for microseconds or longer depending on localization depth [19].

4.1 Exciton effects

In the above, we have neglected the Coulomb interaction between the e-h pair involved in the optical transition. Including this exciton effect [2,16] leads to a hydrogen-like series of bound states just below the band gap as sketched in Fig.6a with the exciton energies given by

$$E_{x,n}(\mathbf{K}) = E_g - \frac{E_x^b}{n^2} + \frac{\hbar^2 K^2}{2m_x} \quad ; \quad n = 1, 2, 3, \ldots \tag{25}$$

where $E_x^b = e^2/2\varepsilon a_x$ and $a_x = \hbar^2\varepsilon/\mu e^2$ are the exciton binding energy and Bohr radius, respectively. $m_x = m_e + m_h$ and $\mu = m_e m_h / m_x$ are the

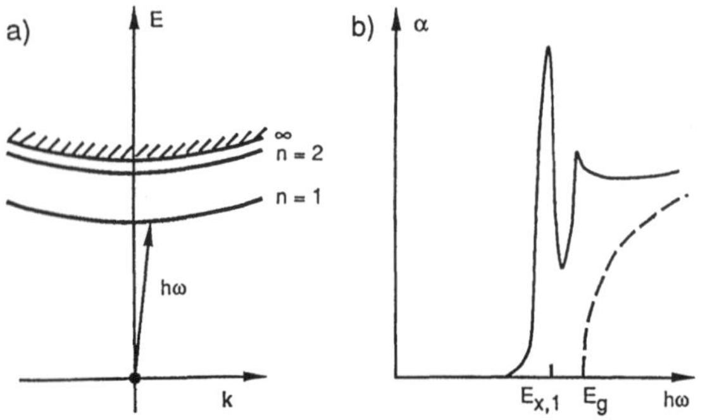

Fig.6. Exciton energy bands (a) and absorption spectrum (b).

translational and reduced masses of the exciton and $\mathbf{k}=\mathbf{k}_e+\mathbf{k}_h$. Due to the large dielectric constant $\varepsilon \cong 10\varepsilon_0$ and small reduced mass $\mu \cong 0.1m_0$ in typical semiconductors, the exciton binding energy is small, $E_x^b \cong 10$meV, and the exciton Bohr radius correspondingly large, $a_x \cong 50$Å. This has two implications: Sharp exciton resonances are normally only observed at low temperatures and exciton interactions start to occur at relatively low exciton densities. The latter will prove important, when it comes to the discussion of optical nonlinearities. Bound exciton states interact very strongly with light, because the electron and hole wave functions are correlated in space. Thus a series of sharp and intense absorption resonances are observed at low temperatures, as indicated in Fig.6b, where it is also shown that even up in the continuum states (above the band gap) there is a strong enhancement of the absorption from the exciton effect [16].

Biexcitons. In many semiconductors, exciton interactions may lead to the formation of new bound complexes like excitonic molecules, biexcitons, the ground state of which form a new energy band in the semiconductor [2,4,6]:

$$E_m(\mathbf{K}) = 2E_{x,1} - E_m^b + \frac{\hbar^2 K^2}{4m_x} \tag{26}$$

where the biexciton binding energy, E_m^b, is typically 10-20% of the corresponding exciton binding energy E_x^b.

As sketched in Fig.7, the formation of biexcitons open up new optical transitions. Besides the fundamental transitions from the ground state to the exciton state (g-x) for $\hbar\omega \geq E_x$, direct two-photon transitions (TPA) between the ground state and the biexciton state (g-m) are allowed for $\hbar\omega \geq E_x - E_m^b/2$, and induced transitions (IA) between the exciton and the biexciton states (x-m) for $\hbar\omega \geq E_x - E_m^b$. The two latter transitions are inherently nonlinear processes and therefore play a signifi-

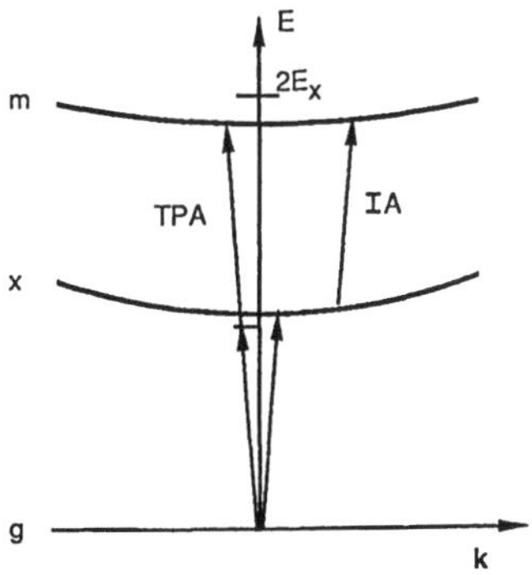

Fig.7. Exciton (x) and biexciton (m) bands with corresponding optical transitions.

cant role when discussing the nonlinear optical properties of semiconductors near the band gap [2,4,6], as we shall see in the following.

A proper theoretical model for the exciton system, as sketched in Fig.7, should thus be a three-band model. For the one-photon transitions (g-x) and (x-m), however, a two-band model could be appropriate. As we shall see later for some purposes a two-band model can even be approximated with a simple two-level model, including inhomogeneous broadening to take into account phenomenologically the exciton and biexciton dispersions as well as any potential fluctuations due to real crystal inhomogeneities [5,6].

4.2 Optical nonlinearities

Near the band edge of a semiconductor there are thus a number of optical nonlinearities that are resonantly enhanced due to real electronic, or excitonic, excitations. These nonlinearities are basically due to many-body effects in the excited semiconductor [2]. Near the fundamental absorption edge (see Fig.5b), state filling due to the Fermi nature of the excited carriers will shift the absorption edge towards higher energies and give rise to absorption saturation, or bleaching, at a given wavelength near the edge. A counteracting effect, giving rise to induced absorption, is a rigid shrinkage of the band gap with increasing carrier density due to many-body exchange and correlation effects. At low temperatures, the density of states rises very quickly with $E-E_g$, so that the latter effect dominates over state filling resulting in a net red-shift of the absorption edge (induced absorption). At higher temperatures, on the other hand, the density of states rises more slowly and state filling will dominate over band gap renormalization. This leads to a net blue shift of the absorption edge (bleaching).

In the above discussion exciton effects were excluded. This may be valid at high temperatures and/or very high excitation densities. At low temperatures and low to medium excitation densities, however, excitons dominate the linear, and even more so the nonlinear, optical properties [2,4]. As mentioned above in Sect.4.1, two inherently nonlinear absorption resonances appear in the energy range $E_x-E_m^b<E<E_x$ with corresponding nonlinear refractive indices. In the exciton resonance itself, nonlinearities in the form of broadening (bleaching) and shift of the absorption peak appear due to exciton interactions (collisions and phase-space filling [20,21]) and at higher densities by screening of the excitons by free carriers

[22]. In the high density e-h plasma limit, the band gap renormalization discussed above takes over.

5. Experimental observations - general aspects

The standard technique for observing experimentally the transient optical nonlinearities in a semiconductor sample is to excite the sample with an intense laser pulse (pump) of picosecond or sub-picosecond duration (τ_L) and then probe the inferred nonlinear changes in the optical properties of the sample by reflection, transmission, scattering, or diffraction of a weak test beam, as sketched in Fig.8.

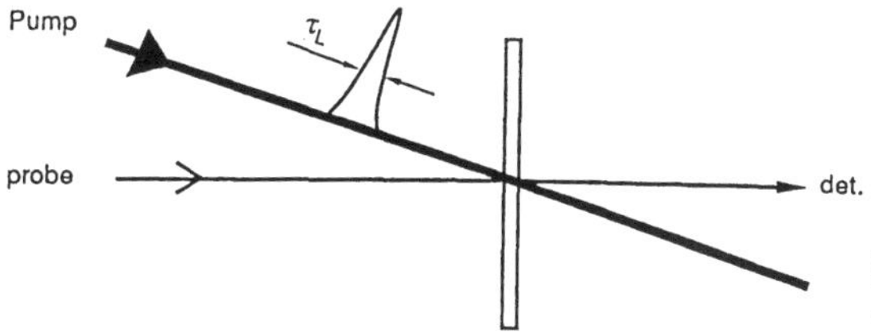

Fig.8. Pump-and-probe experiment

The time resolution, limited by τ_L, can be obtained in different ways. With a c.w. probe beam, an optical detector with a time resolution matching τ_L is needed. This can be either an optical gate, driven by the pump pulse, or a streak-camera with picosecond time resolution. A simpler, less expensive, and quite often satisfactory solution is to apply as probe a weak test pulse of the same duration as the pump pulse. With a slow detector, the integrated test signal is then recorded as a function of the optical delay between the pump and test pulses. The time resolved DFWM previously discussed (Sects.2.2,3) is one version of such a dynamic pump-probe or correlation experiment.

An idealized transient nonlinear optical response may look as sketched in Fig.9, assuming that the laser pulse length is shorter than the dephasing time of the material, which again is shorter than the recombination lifetime, i.e. $\tau_L<T_2<\tau_R$.

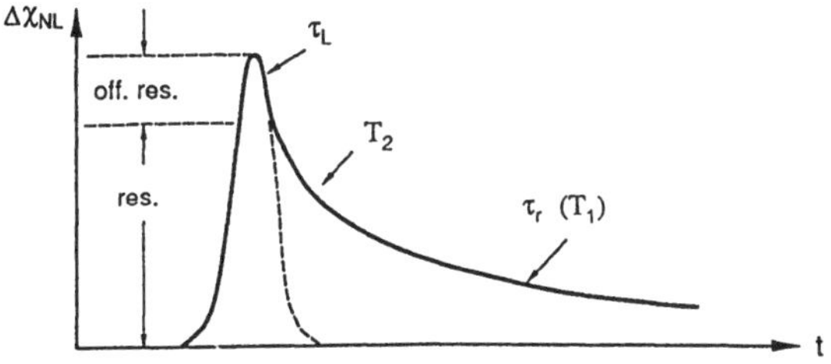

Fig.9. Transient nonlinear optical response in a semiconductor.

Initially, there is a fast response, basically following the laser pump pulse (dashed in Fig.9). This is due to a, normally small, nonresonant contribution to the optical nonlinearity. This response is fast, because it only involves virtually excited states that are extremely short-lived. If the laser light is resonant with some real excitations in the medium, a resonant contribution to the optical nonlinearity will show up with a tail, which for the purely coherent contribution will decay with the dephasing time of the optical resonance. In experiments that are sensitive to an incoherent contribution to the changes in the optical constants, the final decay will be governed by carrier recombination (or possibly diffusion, as in laser induced grating experiments, discussed in Sect.2.2). From this it should be clear that transient nonlinear optical spectroscopy is potentially very powerful to identify the different contributions to the optical nonlinearities, and to determine a number of important electronic interaction and relaxation mechanisms in semiconductors in a very direct way [6,10,11].

The nonresonant contributions are useful for a number of applications, where fast time resolution is needed, as in optical auto- or cross-correlators and in optical gating. For general purpose optical switching, however, the optical power is far too high. For such purposes the strong resonance enhancements of the optical nonlinearities must be employed to bring down the power threshold for optical switching. The price to be paid for this is a slow-down in response time due to the lifetime of the real carriers created. One can compensate for that by artificially reducing the carrier lifetime, or one can make use of the purely coherent nonlinearity with response time T_2, while keeping the carrier density sufficiently low not to interfere with the switching process [6,7].

6. DFWM experiments in CdSe and $CdSe_xS_{1-x}$

In this section I shall illustrate transient nonlinear optical spectroscopy in semiconductors with some recent results in CdSe [6,7] and $CdSe_xS_{1-x}$ [23] obtained by picosecond time resolved DFWM experiments. The experimental set-up is illustrated in Fig.10. The light source is a c.w. picosecond dye laser, synchronously pumped by a mode-locked Ar-ion laser. The dyes employed are DCM, Rh6G and Rh110, depending on the spectral range.

The pulse length (≥1.4ps) and the spectral width (≅2meV) are near transform limited and the pulse energy is moderate (≤0.15nJ, focused to $\leq 10\mu J/cm^2$).The laser beam is split into

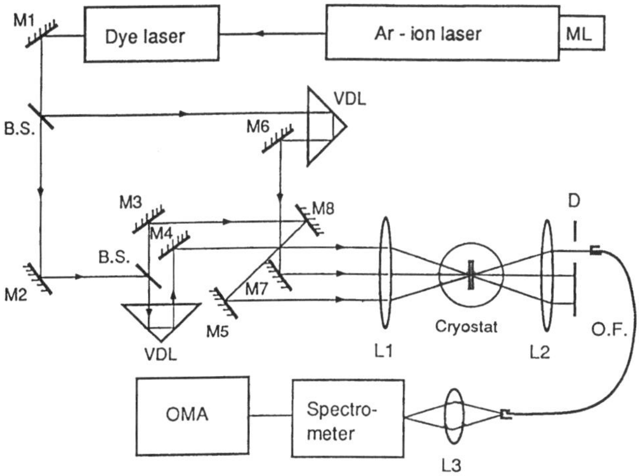

Fig.10. Experimental set-up. BS: beam-splitter, D: diaphragm, L1-L3: lenses, M1-M8: mirrors, OF: optical fiber, VDL: variable delay line, ML: mode-locker.

three beams with variable delays in two of them. The samples are mounted in a liquid helium cryostat, and the light signal is led by an optical fiber into the spectrometer, mounted with an optical multi channel analyzer (OMA). We can thus obtain excitation spectra of the nonlinear optical signal by scanning the laser wavelength while accumulating the signal on the OMA. Correlation traces are recorded by scanning the optical delay while accumulating the signal in a set spectral window on the OMA.

6.1 Nonlinear resonances

An example of the nonlinear resonances associated with exciton interactions in a direct gap semiconductor is shown in Fig.11. The self-diffracted signal in the back scattering direction [6] in a two-beam experiment (see inset in Fig.11) is shown as a function of photon energy (solid curve). For comparison is also shown the linear reflection (dashed curve), from which we can identify the free exciton resonance (E_x) at $\hbar\omega$=1.825eV. The other two resonances, at $\hbar\omega$=1.8225eV and $\hbar\omega$=1.820eV, are associated with the two-photon transition to the biexciton state (TPA) and the induced exciton-biexciton transition (M-band), respectively. These identifications are consistent with a biexciton binding energy E_m^b=5meV. The resonances in Fig.11 are precisely the nonlinear transitions discussed in Sects.4.1-4.2 (see Fig.7).

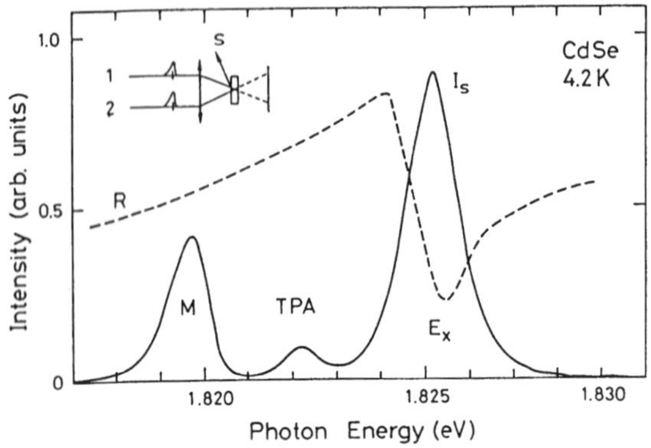

Fig.11. Spectral dependence of back scattered signal (solid curve) and linear reflection (dashed curve).

The two lowest lying resonances are also observed in forward scattering, whereas the samples are too thick (≥10μm) to transmit in the exciton resonance itself. From forward scattering experiments the magnitudes of the nonlinear susceptibility are estimated, as discussed in Sect.2.3 (Eq.(17)). In the resonances, we find maximum values $\chi^{(3)} \cong 10^{-5}$ esu, which is a significant nonlinearity corresponding to a nonlinear refraction index (Sect.2.1) $n_2 \cong 5 \cdot 10^{-8} cm^2/W$.

Similar magnitudes of $\chi^{(3)}$ have been found in the mixed crystals $CdSe_xS_{1-x}$ [23]. Here the unavoidable random compositional disorder gives rise to a fluctuating potential that tends to weakly localize the excitons. The DFWM signal shows a strong resonance in the localized exciton band tail, suggesting that state filling in this band tail is causing the nonlinearity. Localization problems in mixed crystals, or semiconductor alloys, will together with quantum confinements in semiconductor micro structures, or quantum wells, prove to have important implications on the nonlinear optical properties of these materials. The latter is discussed in a separate paper by I. Abram in this volume [24].

6.2 Transient behavior

The transient behavior of the above nonlinear resonances is found in a time resolved DFWM experiment as discussed in Sect-2.2. In a three-beam configuration, the coherent as well as the incoherent gratings contribute. This is illustrated in Fig.12, where the scattered intensity is plotted as a function of the delay of the test pulse with respect to the two pump pulses setting up the nonlinear grating. All three curves clearly show the initial coherence spike followed by the slow-

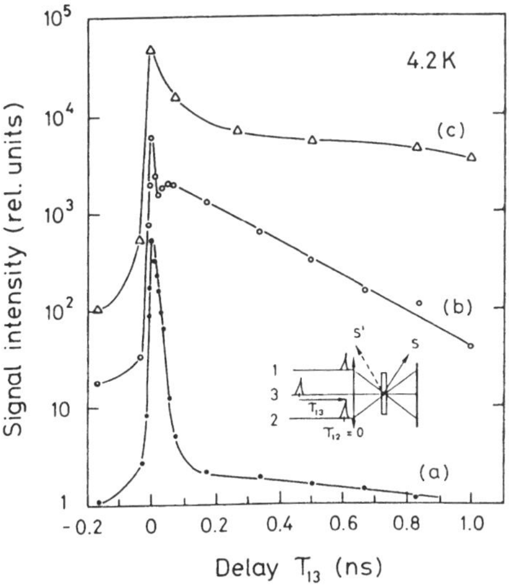

Fig.12. Transient DFWM in the resonances M (a) and E_x (b) of CdSe and in the localized exciton tail of $CdSe_{0.35}S_{0.65}$ (c).

er decay of the incoherent exciton density grating. The lower curve (a) is for the M-band resonance (x-m transition) in CdSe, where the coherent contribution is more than two orders of magnitude larger than the coherent contribution. This resonance features a strong nonlinearity even at low exciton densities ($n_x < 10^{15} cm^{-3}$). In the exciton resonance itself (curve b), the nonlinearity occurs at a higher exciton density, resulting in a larger incoherent contribution which, on the other hand, decays faster due to exciton-exciton collisions. The upper curve (c) is for the localized exciton resonance in $CdSe_{0.35}S_{0.65}$, where both the coherent and the incoherent contribution tend to decay slower than in the pure compound semiconductor. As discussed in Sect.2.2, a detailed analysis of these decays under varying experimental conditions (excitation geometry, intensity, sample temperature, etc.) can give valuable information about exciton dephasing times, diffusion coefficients and recombination times [3,5,6,11,23].

As indicated in Fig.12, the decay tends to be slower in the mixed crystals than in the pure compounds. The diffusion coefficients are one to two orders of magnitudes smaller for the localized excitons ($\leq 1 cm^2/s$, depending on the localization energy) than for the free excitons ($\geq 10 cm^2/s$). Similarly, the recombination lifetime is larger in the mixed crystal than in the pure compounds.

The dephasing times are most easily studied in a two-beam configuration where only the coherent nonlinearity gives rise to a self-diffracted signal [6]. Examples of such a correlation traces in the M-band resonance of CdSe at 4.2K and dif-

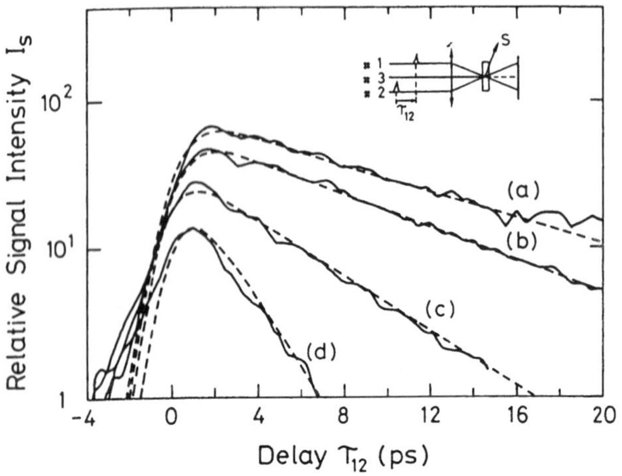

Fig.13. Experimental (full curves) and calculated (dashed curves) correlation traces in CdSe at 4.2K at different excitation intensities, resulting in different dephasing times T_2 and inhomogeneous broadening Γ (from Ref.6).

ferent intensities are shown in Fig.13. The correlation traces have the characteristic asymmetric shape with an extended tail towards positive delay as discussed in Sect.2.2 and calculated in Sect.3 for a two-level system. The dashed curves in Fig.13 are in fact calculated from a two-level model following the lines laid-out in Sect.3 with τ_L=1.4ps and different values for the dephasing time T_2 and inhomogeneous broadening Γ. From the nice fits, it can be argued that the simple two-level model brings out the essential features of this particular resonance, with the inhomogeneous broadening accounting for the exciton kinetic energy (kT_x) and the dephasing time representing the the initial scattering out of the coherent momentum states. It should be noted, however, that in more complex resonances, like the TPA resonance to the biexciton state, such a simple picture does not hold [6].

From a series of coherent light scattering experiments, the dephasing can be studied as a function of intensity, temperature, localization, etc.. It has been found that in the pure compounds, the dephasing is limited by impurity scattering and exciton collisions at low temperature and phonon scattering and thermal dissociation at higher temperatures [6]. In the mixed crystals, on the other hand, the (disorder induced) phonon scattering seem to dominate at all temperatures [23].

7. Conclusions

Time resolved nonlinear optical spectroscopies have proven very successful in studying the fast electronic interactions and relaxation mechanisms in semiconductors. In particular, the transient DFWM experiments have revealed very strong excitonic resonance enhancements of the nonlinear coefficients. At the same time, the dephasing times of these resonances are relatively long (10-100ps) at low temperature and moderate excitation densities. The purely coherent part of the resonant optical nonlinearities may prove useful for future optical switching devices.

Acknowledgements. The author is indebted to I. Balslev, C. Dörnfeld, B. Hönerlage and H. Schwab for their contributions to the present work. The work was supported by the Danish Natural Science Research Council.

References

1. For recent reviews on optical switching, see e.g.:
   H. Gibbs, *Optical Bistability: Controlling Light by Light,* (Orlando Academic, Orlando, 1985), and
   P. Mandel, S.D. Smith and B.S. Wherret, *From Optical Bistability Towards Optical Computing* (North Holland, Amsterdam, 1987).
2. C. Klingshirn and H. Haug, Phys. Rep. **70**, 315 (1981).
3. A. Miller, D.A.B. Miller and S.D. Smith, Adv. in Phys. **30**, 697 (1981).
4. B. Hönerlage, R. Lévy, J.B. Grun, C. Klingshirn and K. Bohnert, Phys. Rep. **124**, 161 (1985).
5. L. Schultheis, J. Kuhl, A. Honold and C.W. Tu, Phys. Rev. Lett. **57**, 1635 (1986).
6. C. Dörnfeld and J.M. Hvam, IEEE J. Quantum Electron. **25**, 904 (1989).
7. C. Dörnfeld and J.M. Hvam, J. de Physique, **49 suppl. 6**, C2-205 (1988).
8. C.R. Paton, Z. Xie and J.M. Hvam, to be published.
9. P.G. Harper and B.S. Wherret, eds., *Nonlinear Optics* (Academic Press, London, 1977).
10. Y.R. Shen, *The Principles of NONLINEAR OPTICS*, (Wiley & Sons, New York, 1984).
11. H.J. Eichler, P. Günter and D.W. Pohl, *Laser Induced Dynamical Gratings*, Springer Ser. Opt. Sci. **50** (Springer, Berlin, 1986).

12. A. Maruani and D.S. Chemla, Phys. Rev. **B23**, 841 (1981).
13. J.M. Hvam and C. Dörnfeld, *Optical Switching in Low-Dimensional Systems*, eds. H. Haug and L. Banyai (Plenum Publishing Corp., New York, 1989) pp. 233-241.
14. T. Yajima and Y. Taira, J. Phys. Soc. Japan **47**, 1620 (1979).
15. S.C. Moss, J.R. Lindle, H.J. Mackey and A. Smirl, Appl. Phys. Lett. **39**, 227 (1981).
16. A. Stahl and I. Balslev, *Electrodynamics of the Semiconductor Band Edge*, Springer Tracts in Modern Physics **110** (Springer, Berlin, 1987).
17. J.A. Kash, J.C. Tsang and J.M. Hvam, Phys. Rev. Lett. **54**, 2151 (1985).
18. M. Jørgensen and J.M. Hvam, Appl. Phys. Lett. **43**, 460 (1983).
19. J.M. Hvam and M.H. Brodsky, Phys. Rev. Lett. **46**, 371 (1981).
20. W.H. Knox, R.L. Fork, M.C. Downer, D.A.B. Miller, D.S. Chemla, C.V. Shank, A.C. Gossard and A. Wiegmann, Phys. Rev. Lett. **54**, 1306 (1985).
21. R. Leonelli, J.C. Mathae, J.M. Hvam, F. Tomasini, and J.B. Grun, Phys. Rev. Lett. **58**, 1363 (1987).
22. H. Haug and S. Schmitt-Rink, J. Opt. Soc. Am. B **2**, 1135 (1985).
23. J.M. Hvam, C. Dörnfeld and H. Schwab, phys. stat. sol. (b) **150**, 387 (1988).
24. I. Abram, this volume.

Phase Conjugation

T.J. Hall and A.K. Powell

Physics Department, King's College London, Strand, London, WC2R 2LS, UK

Abstract. This chapter describes the principles of optical phase conjugation and examines the nonlinear processes of four-wave mixing and stimulated scattering that give rise to phase conjugated wavefronts. Applications for good phase conjugating systems are considered along with the optical requirements for the materials. Possible future developments in the field are discussed.

1. Introduction

A phase conjugate mirror is a non-linear optical device which not only reverses the direction of propagation of an incident wave but conjugates its complex amplitude. This can be considered as equivalent to reversing time so that the phase conjugate wave will exactly retrace its original path. In particular any distortions of the wave caused by its passage through a phase aberrator will be removed on the repassage of the wave. This is unlike the case of reflection by an ordinary mirror where distortions are exacerbated [1]. These remarkable properties of phase conjugate mirrors are illustrated in Fig. 1. Phase conjugate waves may be generated either by degenerate four-wave mixing (Fig. 2a) or by stimulated scattering processes (Fig. 2b) in materials with a third order optical non-linearity. Materials with a third order optical non-linearity may be loosely thought of as possessing a refractive index which depends on light intensity [1,2]. This chapter considers the nonlinear optical processes of four-wave mixing and stimulated scattering. The discussion includes the topics of Stimulated Photorefractive Scattering (SPS) and Stimulated Brillouin Scattering (SBS). Applications for phase conjugating systems are then considered along with the material requirements for optimum performance.

2. Four-wave Mixing and Stimulated Scattering

In degenerate four-wave mixing (DFWM) two counter-propagating plane pump beams are introduced into a material with an intensity dependent refractive index and the material is probed by a less intense probe beam.

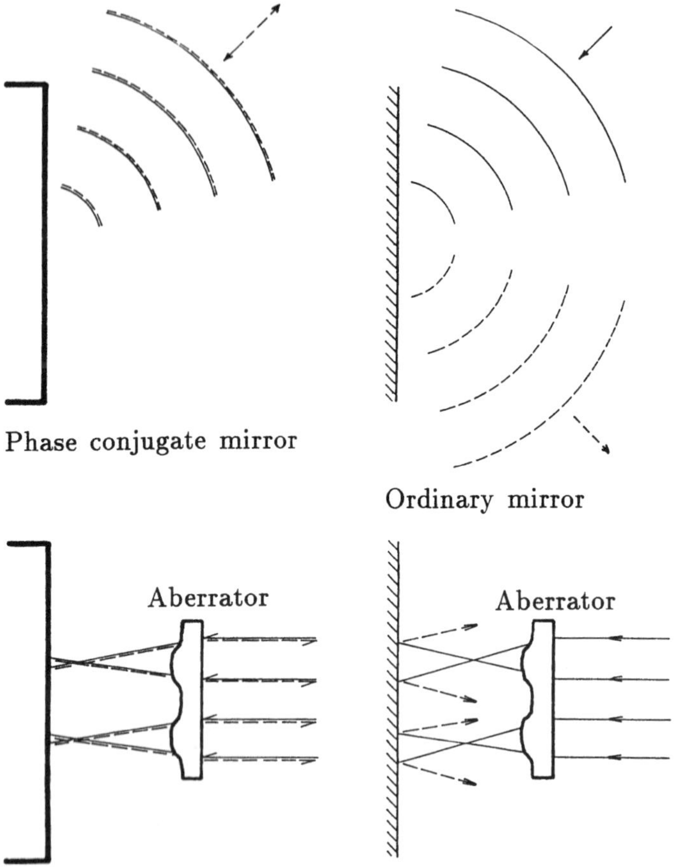

Figure 1 Comparing the properties of ordinary mirrors to those of phase conjugate mirrors.

A fourth beam is generated within the material which is the phase conjugate of the probe beam. This may be understood by analogy to real time holography. There are two gratings of importance in this picture. The first is due to the intensity fringes caused by the interference of the probe and the forward pump. This writes a refractive index grating as a result of the intensity dependence of the material refractive index. The backward pump is then diffracted by this grating into the conjugate wave (Fig. 3a). The intensity fringes caused by the interference of the probe and backward pump also write a grating. The forward pump is diffracted by this grating into the conjugate wave (Fig. 3b). A third process involving parametric mixing of the waves may also generate a conjugate wave but as it involves a nonlinear induced polarisation at twice the optical frequency it requires an ultra-fast optical nonlinearity. The magnitude of a non-linearity tends

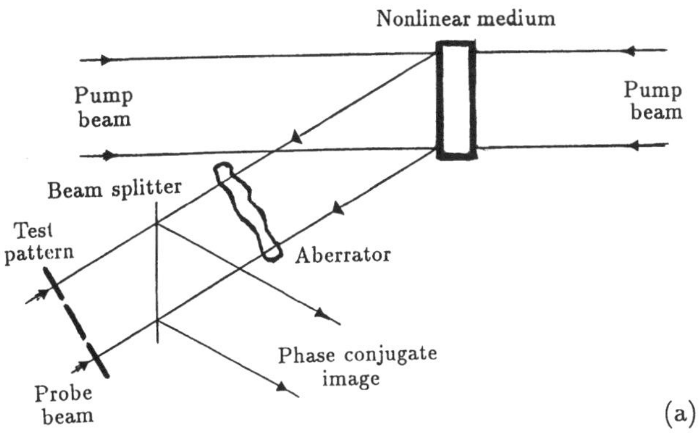

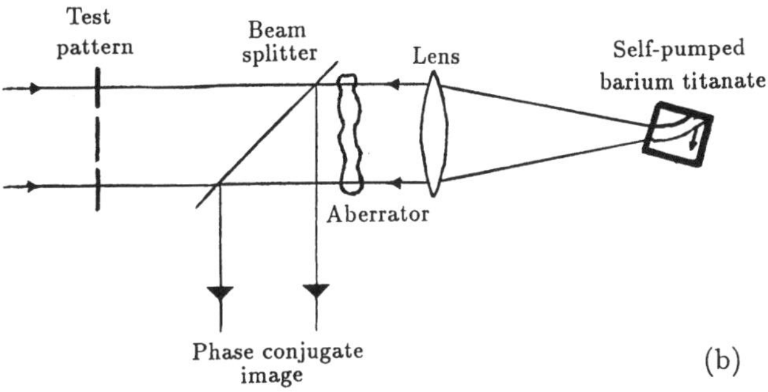

Figure 2 Methods of producing phase conjugated wavefronts.
a) Four-wave mixing.
b) Stimulated photorefractive scattering.

to scale inversely with speed and so this third process is typically negligible compared to the real time holographic processes.

In stimulated scattering processes phase conjugation is still caused by scattering by gratings but all the interacting waves are derived from the incident probe [3,4]. How this self-pumping can arise is discussed later. The main distinction between phase conjugation by degenerate four-wave mixing and by stimulated scattering is that whereas the former may have greater than unity reflectivity since external energy is supplied by the pumps, the latter can have at most unity reflectivity as all the energy is derived from the pump. To form a phase conjugate the two pumps of the four-wave mixing process must also be phase conjugates. Counter-propagating plane waves have this property but this property is lost as soon as the pump energy is depleted by absorption or efficient diffraction into the conjugate. In self-pumping the internal mechanism responsible for self-pumping can maintain the conjugate property of all interacting waves. It is observed, as a consequence, that the quality of phase conjugation by self-pumping can be superior to that of phase conjugation by four-wave mixing. To obtain high

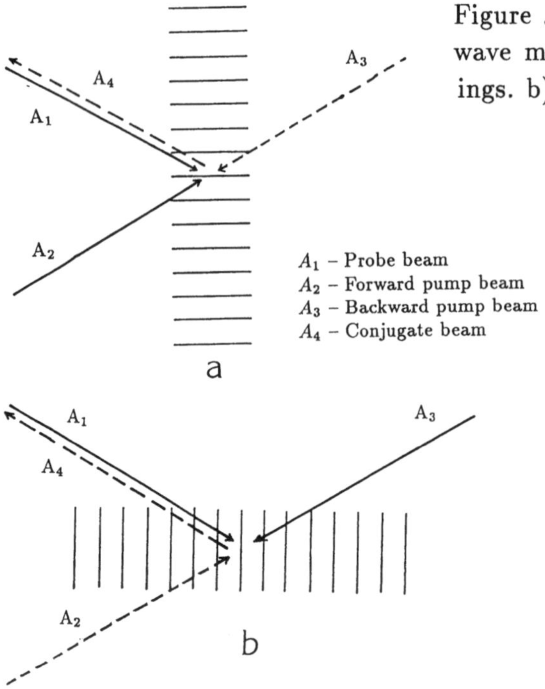

Figure 3 Grating pictures of four-wave mixing. a) Transverse gratings. b) Longitudinal gratings.

quality phase-conjugation with reflectivities greater than unity it is possible to combine self-pumping to derive phase conjugate pumps for a degenerate four-wave mixing interaction as is done, for example, in Brillouin enhanced four-wave mixing [5].

3. The Photorefractive Effect

There are a variety of non-linear optical mechanisms and hence materials capable of phase conjugation. One of the most widely used non-linear optical effects is the photorefractive effect [6]. The photorefractive effect is a term used to describe photoconductive electro-optic media and typically includes materials such as ferroelectrics ($BaTiO_3$, SBN, $LiNbO_3$, $KNbO_3$), the sillenites (BSO, BGO, BTO), the semiconductors (GaAs, InP, CdTe) and more recently organic crystals [7]. These materials operate at visible or near infra-red wavelengths.

All photorefractive materials are semi-insulating in the dark. When illuminated, charges are photoexcited from deep defect centres within the energy gap into the conduction band (electrons) or valence band (holes). The mobile charges diffuse or drift in an electric field until they recombine with an empty defect centre (Fig. 4). Charge is continuously ejected from brightly illuminated regions and will have a tendency to collect in the darkly illuminated regions. This distribution of charge gives rise to

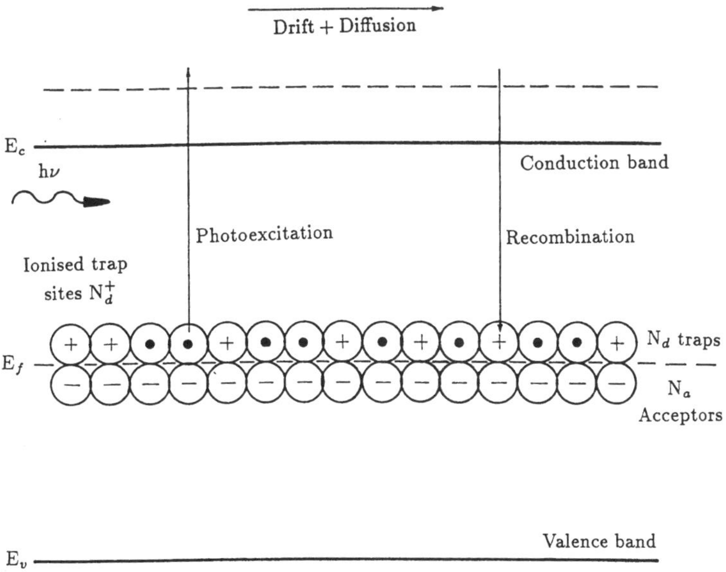

Figure 4 Photorefractive material's simplified energy band diagram.

an internal space charge electric field. This electric field then modulates the refractive index of the material via the linear electro-optic effect (Fig. 5). Since the charge distribution only depends upon relative intensities, the effect is observed at all intensities greater than an equivalent dark illumination corresponding to thermal excitations. This can be considered analogous to a saturable non-linearity at intensities well above the saturation intensity. However, unlike other non-linear mechanisms, the saturation intensity is of the order of mW cm^{-2}. Photorefractive materials are therefore compatible with low power c.w. lasers, which explains their popularity. Their speed of response does however scale with intensity and corresponds to an absorbed energy of $\simeq 100\mu$ J cm^{-2}. This is quite competitive with other non-linear optical mechanisms. In the case of ferroelectrics such as $BaTiO_3$, the induced refractive index changes can be of the order of 10^{-3} resulting in sometimes bizarre non-linear effects not yet observed in other materials. These interesting effects are a second reason for the popularity of photorefractive materials. The scope for improving the performance of these materials still further is discussed later.

From the description of the physical mechanism it is clear that the photorefractive effect is non-local. This can lead under certain circumstances to a spatial shift of up to a quarter period between the refractive index grating and the intensity fringes that induce it. This is essentially a consequence of Gauss's law linking the space charge and fields. The spatial phase shift has profound consequences for the optical interactions.

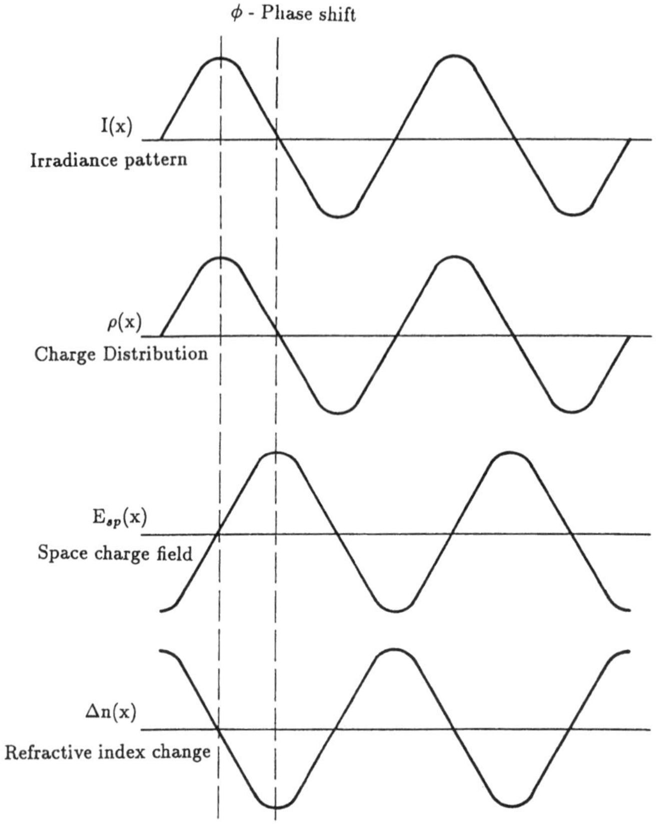

Figure 5 Spatial relationships of photoinduced charge distribution, space charge field and refractive index change with respect to cosinusoidal illumination of a photorefractive material.

Consider a wave propagating through a thin weakly absorbing inhomogeneity (Fig. 6a). It is clear that the loss of amplitude can be described as the superposition of the unperturbed incident wave and a wave in antiphase with the incident wave scattered by the inhomogeneity. In the case of a refractive inhomogeneity (Fig. 6b) the amplitude of the wave is unaffected and only the phase changes. This is equivalent to the scattering by the inhomogeneity of a wave in quadrature with the incident wave. Now consider two beams which write and are simultaneously diffracted into each other by a grating in a photoconductive material. If the grating is unshifted (Fig 7), the waves scattered by the grating are in quadrature with the respective incident waves. The phase of the waves is altered by the interaction but there is no change in their intensities compared with the case where no grating is present. If the grating is shifted by a quarter period (Fig. 7) one of

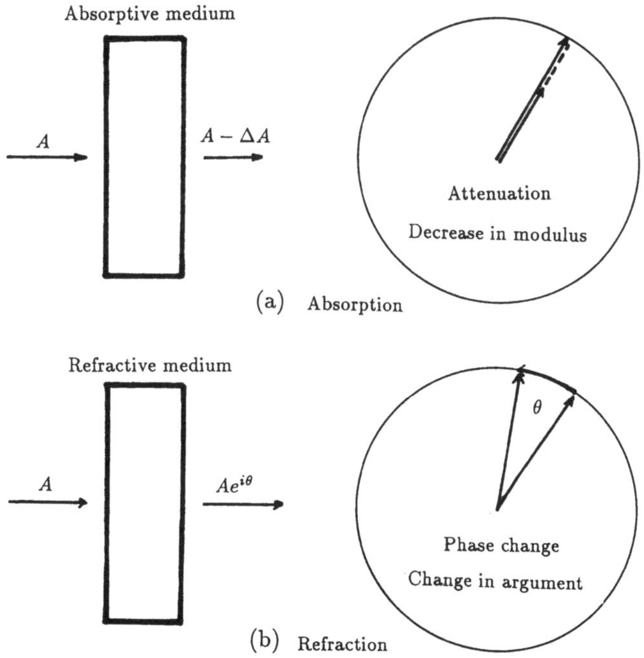

Figure 6. Phase change associated with (a) absorption, where the amplitude is diminished in the beam and energy is transferred into a scattered beam, and (b) refraction, where there is only a change in phase with passage through the medium.

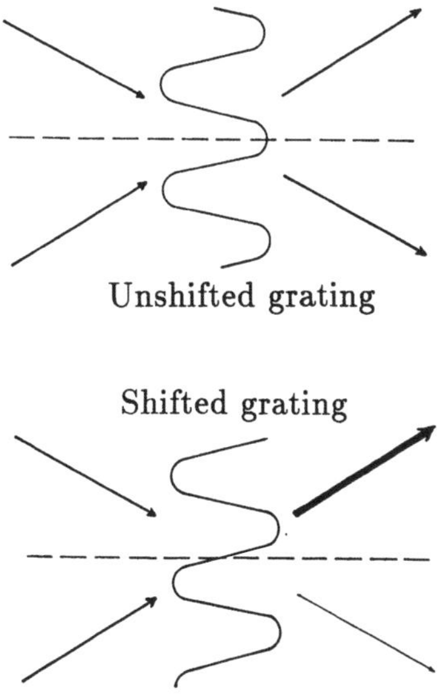

Figure 7
Energy exchanged may be established in a refractive grating by cancelling the acquired refractive phase shift by introducing a $\frac{\pi}{2}$ phase relationship between the illuminating intensity and the refractive grating. This is the principle of two-wave mixing.

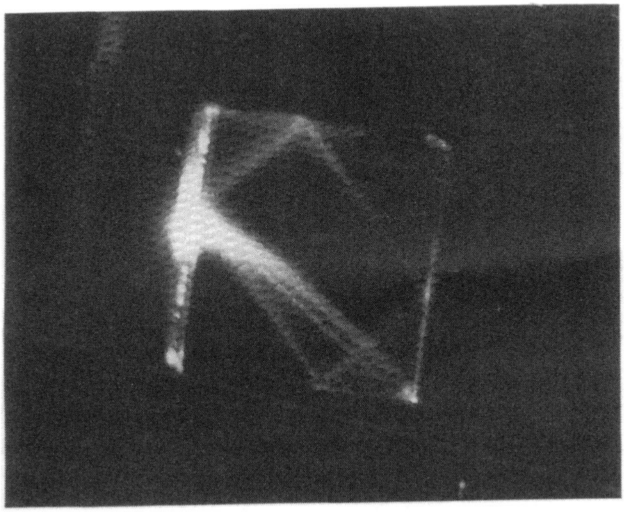

Figure 8 Beam fanning in barium titanate. Energy is also being coupled from the beam fan into a ring oscillator, the cavity being formed by internal reflections off the crystal faces.

the diffracted waves gains an additional phase of $\pi/2$ and the other loses a phase of $\pi/2$. One diffracted wave is therefore in phase and the other in anti-phase with the respective incident waves. One beam is therefore observed to grow in intensity at the expense of the depletion of the other. Which beam gains and which is depleted depends on the orientation and symmetry of the material and the polarisation of the waves. The gain via this two-wave mixing interaction can be considerable. Gain coefficients of 10 - 100cm^{-1} are easily possible in ferroelectric materials with their large electro-optic coefficients.

If a beam of coherent light is passed through a photorefractive $BaTiO_3$ crystal the initially straight beam is observed to fan into a corner [8] and the phase conjugate of the incident beam begins to grow (Fig. 8). This mechanism of self-pumped phase conjugation is known as stimulated scattering [9]. The beam fanning phenomenon may be explained by the amplification by two-wave mixing of light scattered from inhomogeneities within the crystal (Fig. 9). A detailed study [10] of two-wave mixing shows a directional dependence of the gain seen by a weak signal wave propagating at a small angle to a strong pump (Fig. 10). It turns out that the direction of the greatest gain for a signal travelling in the same direction as the strong pump coincides with the direction of the greatest gain for a signal travelling in the opposite direction to the strong pump. This direction is towards the bot-

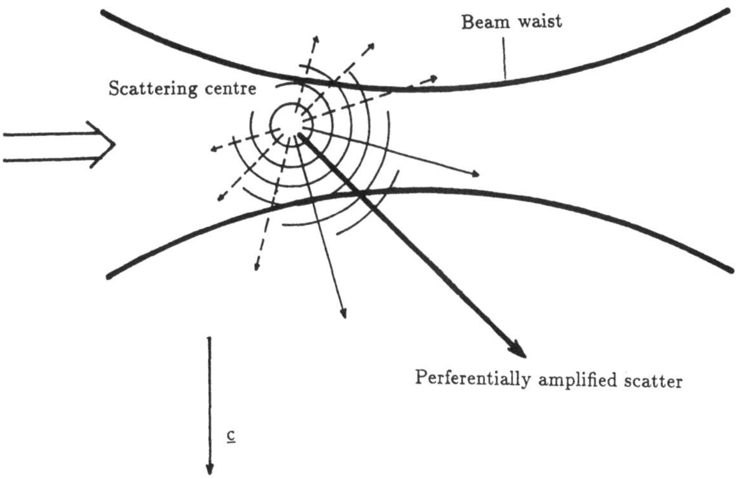

Figure 9. Beam fanning due to amplified light scatter from a random scattering centre.

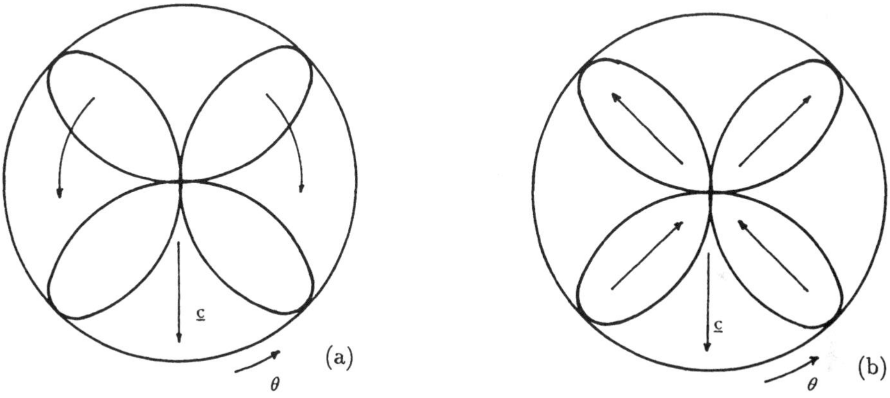

Figure 10. Polar diagrams showing the strength of the electro-optic effect at angles to the c-axis. The arrows represent the preferred direction of energy flow for (a) transverse gratings and (b) longitudinal gratings

tom right hand corner of the crystal in Fig. 8. Scattered light thus grows in intensity in the direction of the corner and in doing so depletes the incident pump. The scattered light in turn acts as a source for more scattering and is in turn depleted. The outcome of this repeated process has the appearance of a curvature of the light towards the corner. It must be emphasised that it is in fact a complicated succession of amplified scattering events.

On reaching the corner of the crystal the light will be reflected back into the fan thus providing a back pump for degenerate four-wave mixing

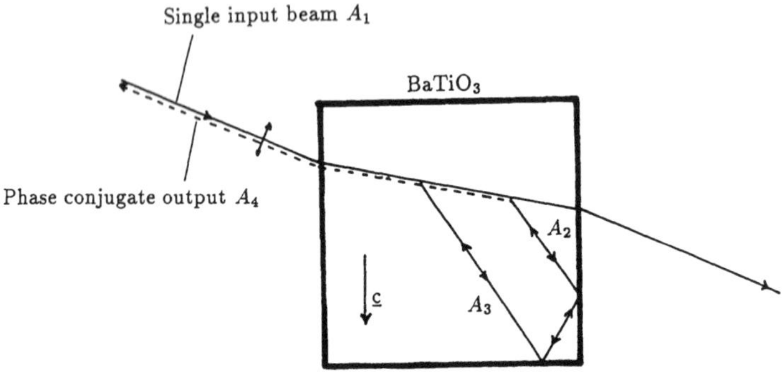

Figure 11. Corner cube reflection seeding the SPS interaction in Barium Titanate

interactions (Fig. 11), which seed the formation of a phase conjugate wave. A vast variety of competing two-wave and four-wave mixing interactions will co-exist in such a high gain environment but only the phase-conjugate of the incident wave will accurately retrace the complicated incident wave. Out of all the backward going waves it is therefore the conjugate that experiences the highest gain. This property together with the depletion of the incident wave by the conjugate tends to suppress other waves and leads to the eventual domination of the return by the phase-conjugate wave. The insensitivity of the stimulated photorefractive scattering processes on the exact nature of the back reflection has been established in experiments using incoherent erase beams to suppress fanning and index matching to suppress corner reflections. Back seeding using uncritical reflections of the transmitted light back into the crystal resulted in phase-conjugation [11]. Recently back seeding has been proposed as a means of self-pumped phase conjugation in photorefractive crystals not found to self-pump of their own accord [12].

4. Stimulated Brillouin Scattering

The stimulated photorefractive scattering (SPS) process is closely related to Brillouin scattering (SBS) illustrated in Fig. 12. An intense beam is focussed into a cell containing a liquid or a gas at a high pressure. Spontaneously scattered light interacts with the incident light to produce interference fringes. These interference fringes cause corresponding pressure variations in the fluid by electrostriction. The interference fringes are caused by light scattered at frequencies which can differ from the frequency of the incident wave and are therefore moving. Those that are moving at the acoustic velocity in the fluid resonantly drive corresponding acoustic waves. The resultant density variations and hence refractive index variations rapidly

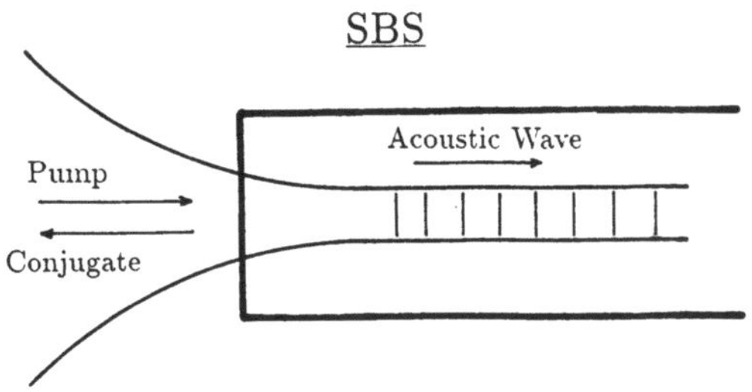

Figure 12 Stimulated Brillouin scattering. The backscattering process is likened to SPS in barium titanate that drives the self-pumped phase conjugation process.

dominate the scattering. Since the acoustic waves are resonantly driven the induced refractive index gratings are in quadrature with the driving intensity fringes leading to amplification analogous to photorefractive two-wave mixing. If the incident wave is sufficiently complex, for example by use of an aberrator, it is again the phase conjugate wave that sees the most gain and dominates the return. Self-pumped phase conjugation again arises as a consequence of high gain and competition. Both SBS and SPS consequently share the property that the more complex the structure of the incident wave the greater the phase conjugate fidelity. The main difference between the two processes is that SBS uses an unsaturated nonlinearity and so the absolute intensity is important. The nonlinearity requires the use of ns duration $MWcm^{-2}$ intensity laser pulses. The use of travelling acoustic waves results in a significant GHz down shift in frequency of the conjugate. This frequency shift however can be an advantage since scattering within the pump that is amplified in SPS is not amplified in an analogous SBS interaction since its frequency does not coincide with the Brillouin shift. SBS is intrinsically a low noise process (the use of high power lasers tends to mitigate against this but the noise then has other causes).

5. Application of Phase Conjugation

The principal application of phase conjugation to laser systems is its use for phase conjugate resonator mirrors, beam clean-up, coherent beam combination and the phase locking of lasers. Analogous techniques but at generally lower powers find application to the manipulation of coherent image carrying beams in optical image processing.

In laser mirror applications the idea is that the aberrations sampled by the incident wave on its journey to the phase conjugate mirror, particularly those induced by the gain medium, are undone as the phase conjugate wave propagates back along the original path. The result is effectively a high quality resonator even in the presence of the inevitable distortions found in high power laser gain media [13]. This application requires that all the light needed for the phase conjugate process is derived from the incident beam. This is termed self-pumping and is observed to occur naturally in two systems: SBS in liquids and gases and SPS in photorefractive crystals with large nonlinearities, such as SBN and $BaTiO_3$.

In beam clean-up applications the goal is to convert the optical energy in a high-power beam of low spatial quality to a beam that has both high quality and high power. This has been demonstrated primarily with the stimulated Raman effect but it can be accomplished also with SBS, and two-wave mixing in photorefractive materials [14]. A low power diffraction limited probe beam is first derived by spatial filtering of a small fraction of the pump beam. This is then amplified via the chosen non-linear process in the presence of the remainder of the pump (Fig. 13). In all three processes the gain only depends on the intensity of the pump and not on its aberrated phase and in favourable conditions a significant fraction of the pump's energy may be transferred to the probe. Intensity structure on the pump can be transferred but a geometry involving crossed pump beams can be used to average this out.

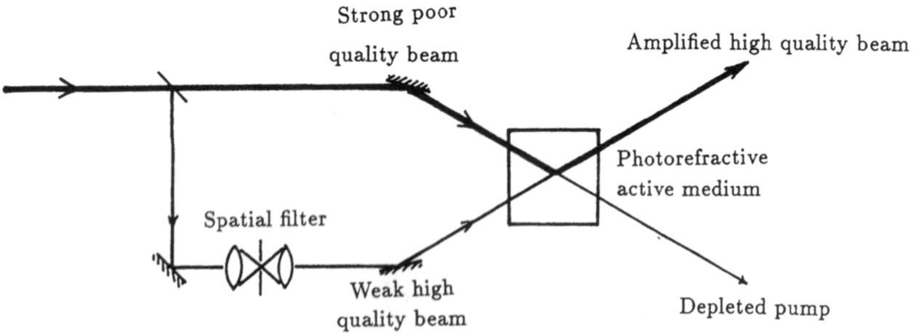

Figure 13. Beam clean-up by photorefractive gain.

In beam combination applications the goal is to combine the radiation from two or more lasers into one beam coherently so that the net result is a single diffraction-limited beam containing the total power from all the individual lasers. One method is to use the individual lasers as multiple pumps of a Raman or two-wave mixing amplifier [15, 16]. In the amplification process the probe beam takes power from each of the pump beams but preserves its original phase. Self-pumped phase conjugation can be used to combine the outputs of several amplifiers. A master output is broken into several paths each containing an amplifier. The output of each amplifier is combined and simultaneously conjugated in a phase conjugate mirror. The phase conjugate beams return through the amplifiers compensating for aberrations and path length differences and hence coherently recombining into a single high output power beam at an output coupler placed in front of the master oscillator. The output of a low power, single transverse and longitudinal mode solid state laser has been amplified in this way by an array of semiconductor laser amplifiers to form a high power single transverse and longitudinal mode output using SPS in $BaTiO_3$ [17].

The locking of the relative phase of several beams by simultaneous phase conjugation can also be used to lock the frequency of oscillation of several lasers [18] and allow the coherent subtraction and other operations on image carrying beams [19]. This technology can be injected directly into image/data processing techniques.

The phase locking and distortion correction properties of phase conjugation can be directly applied to the coherent subtraction of two images. One system capable of this operation is shown in Fig. 14a. An incident beam is split by a beam splitter and the two halves passed through two image transparencies and then simultaneously phase conjugated. The simultaneous phase conjugation of two mutually coherent incident waves is equivalent to the phase conjugate of one more complex incident wave. Thought of separately, therefore, the two phase conjugate returns must have a fixed phase relationship. The two conjugates return undoing any phase distortions until they meet at the beam splitter. The properties of a dielectric beam splitter then ensure that the two conjugates combine in anti-phase in the lower arm leading to the subtraction of the two images (Figs. 14b,14c).

The real time holographic properties of photorefractive materials have application as a programmable Fourier plane filter. For example, Fig. 15 shows how a thin photorefractive material may be used as a two input optical correlator by combining the Fourier transformation properties of a lens with the multiplication of field amplitudes that occurs in the four-wave mixing process [20]. The figure (Fig. 15) illustrates how the correlation process can be used for pattern recognition. Real time holography can be used as a reconfigurable interconnect where gratings are written to diffract light from transmitter sources onto receiver detectors. In this view the optical correlator can be considered to be a shift invariant interconnection. The use of a thick material leads to greater efficiency and shift variance due to Bragg matching considerations. A variety of schemes have been proposed [21, 22]. In all cases however it is important to realise that it is only the writing of gratings which is limited by the material response time. Diffraction by a grating is virtually instantaneous. Schemes for reconfigurable interconnections need not necessarily therefore place restrictions on the data rate,only the reconfiguration times. Similar considerations apply to one input of an optical correlator [23].

Photorefractive two-wave mixing also has application to coherent image amplification and iterative optical processing. The idea is to develop a two-dimensional analogue to the operational amplifier of electronics. Critical here is a high space-bandwidth low noise amplifier. Until recently only $BaTiO_3$ was capable of amplification considerably exceeding absorptive loss. The problem with $BaTiO_3$ is the susceptibility to competing processes. This one expects would be a feature of all high-gain nonlinear optical materials.

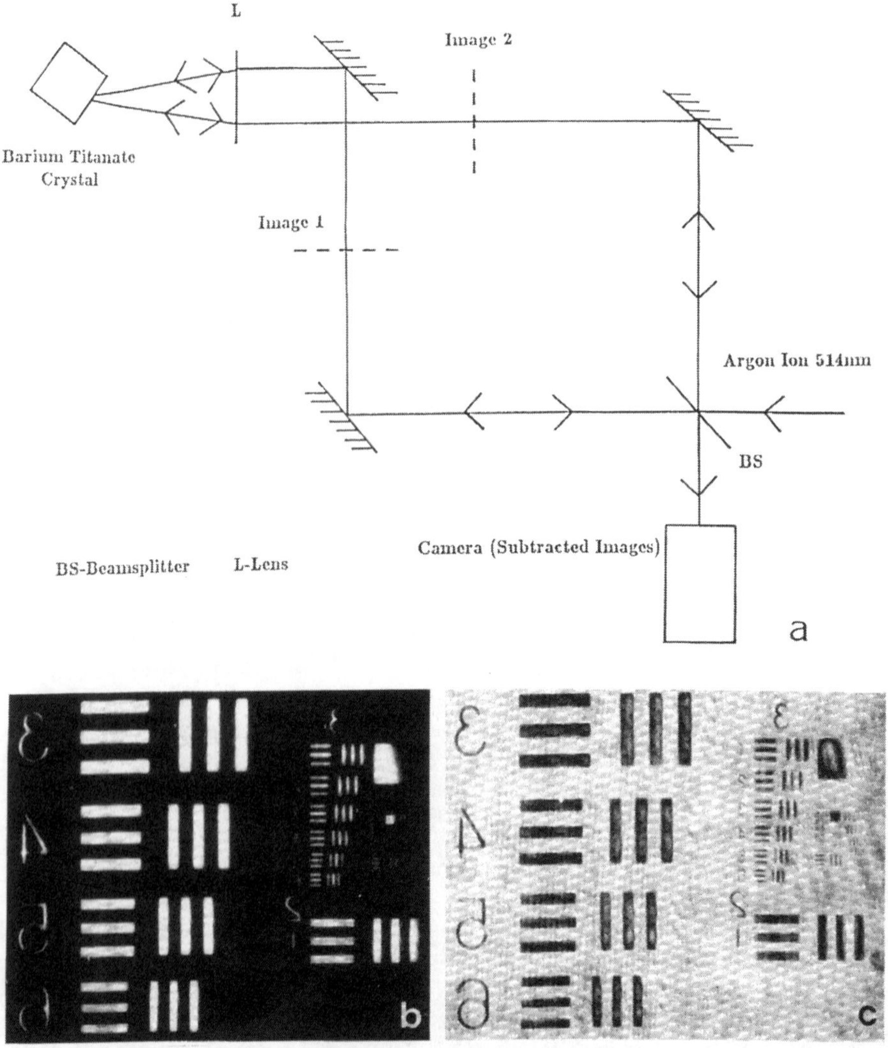

Figure 14. Optical image subtraction. (a) Experimental configuration. (b) Conjugate image with one arm of the interferometer blocked. (c) Plane beam subtracted from phase conjugate image.

The curious result that the two-wave mixing gain of $BaTiO_3$ improves when an incoherent erase beam is introduced (Fig. 16) is a direct consequence of the suppression of these competing processes. The combination of the distortion correction properties of phase conjugation and the gain provided by two-wave mixing should enable the stable iteration of images around an optical processor. Good results in this direction have been obtained by Klumb et al. using $BaTiO_3$ [24]. These schemes will become increasingly

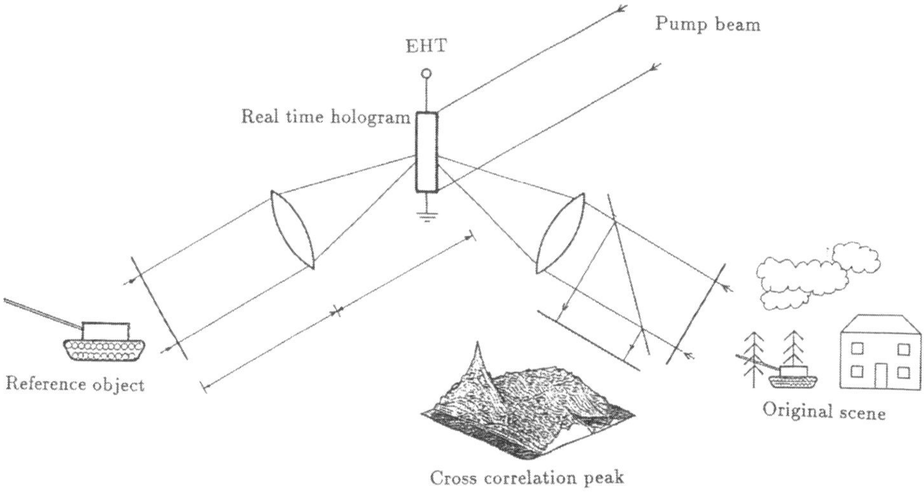

Figure 15. A real time two input optical correlator.

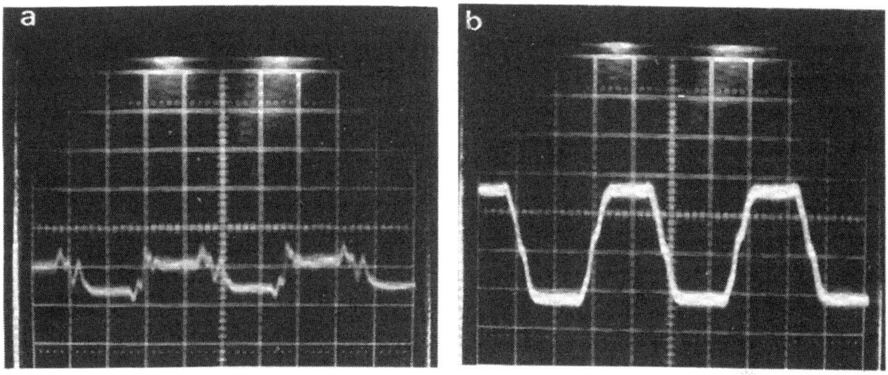

a) With no erase beam. b) With 6 mW erase beam.

Figure 16 Improved two-wave mixing with use of incoherent pump beam. Pump intensity = 1mW. Probe Intensity = $1\mu W$. a) Without erase beam. b) With erase beam.

more practical as the device and materials developments in phase conjugation described later progress.

The combination of phase conjugation with high gain can lead to another interesting phenomenon. If two $BaTiO_3$ crystals are separately pumped from the same source and placed with a line of sight between them it is found that a beam of light spontaneously oscillates between them (Fig. 17) [25]. If this path is now blocked but a second exists, provided, for example,

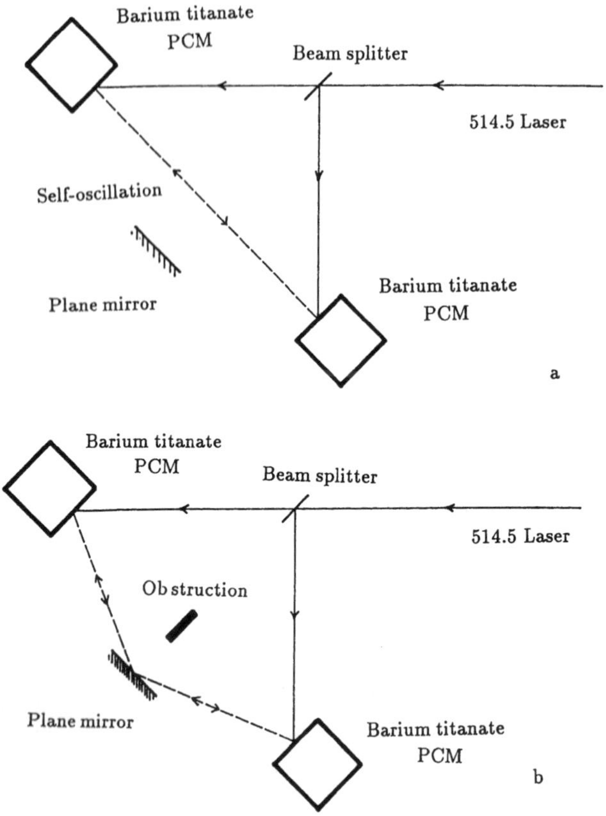

Figure 17 Self-oscillation between two phase conjugate mirrors. a) Direct path has lowest loss. b) Indirect path has lowest loss.

by a mirror, the beam will now take the second path. If both paths are possible the lowest loss path wins. This corresponds to the 'winner takes all' decision making process. If we now place a programmable correlator between our two phase conjugators the oscillation builds up along the path corresponding to the strongest correlation peak. This oscillation will therefore read out the hologram of the object most like the input. We therefore have an associative memory very similar to the Hopfield model of artificial neural networks (Fig. 18) [26]. Two-wave mixing may also be employed to construct 'intelligent' optical iterators [27, 28]. Schemes have also been proposed capable of learning optically [29].

Some of these applications of phase conjugation may seem remote but the same must have been thought by the first people introduced to the point contact transistor. As for the transistor successful applications will depend on improvements to materials, devices and systems integration. In order

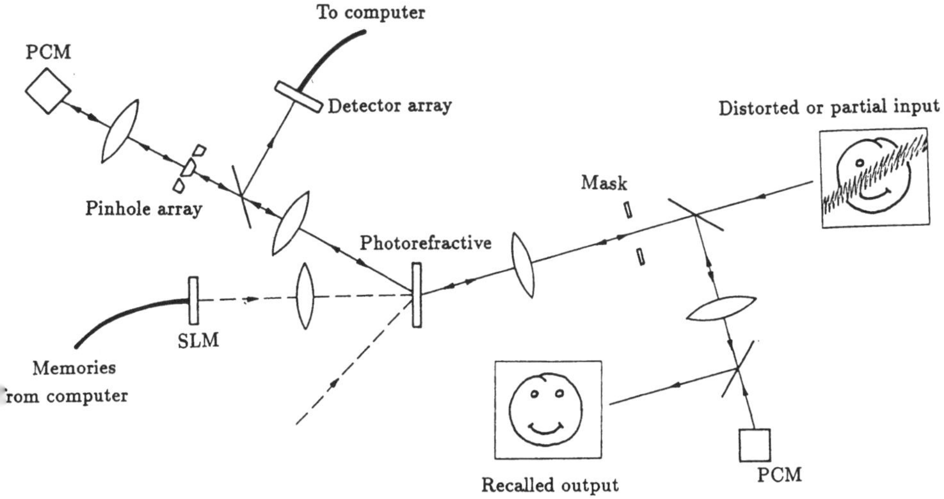

Figure 18. An analogue optical associative memory. After [26].

to justify materials programmes it is necessary to extract the maximum of performance from existing materials and to apply these optimised elements to applications which do not place too high demands on the element but offer huge improvements over other approaches. The application of phase conjugation to new semiconductor laser arrays and diode pumped solid state laser systems is promising in this respect. Strategies for extracting the maximum performance from existing materials are discussed here.

6. Material Requirements for Phase Conjugation

Self-pumped phase conjugation by both SBS and SPS have been described here. Both of these mechanisms have their advantages and disadvantages. SBS operates on fast timescales (ns) but requires high power pulsed lasers ($MWcm^{-2}$). SPS operates with moderate power lasers (Wcm^{-2}) but this is often slow (s). Both these mechanisms place some restrictions on the choice of operating wavelengths. The requirements for successful application of phase conjugation are freedom of wavelength choice, flexible speed intensity trade offs, zero power thresholds, high quality conjugation and a large aperture area acceptance angle product. The latter is particularly important in image processing applications where it is referred to as the space bandwidth product. Here the scope for improvement concentrating on photorefractive materials is discussed.

Photorefractive materials have tended to fall into two classes represented by the ferroelectrics on the one hand and the semiconductors and sillenites on the other. The ferroelectric materials have excellent electro-optic properties but poor photoconductive properties. The semiconductors by way of contrast have poor electro-optic properties but excellent photoconductive properties. The ideal photorefractive material should have excellent electro-optic and photoconductive properties.

The large electro-optic effect found with the ferroelectrics leads to non-linearities strong enough to demonstrate the applications. The poor photoconductive properties however result in excessive response times for practical applications. With these materials the non-linearity is, if anything, too large leading to complex and competing processes that defy quantitative theoretic description. Although this leads to many bizarre and interesting phenomena such as chaotic dynamical behaviour [30] the sheer complexity obscures the essential physics.

The large carrier mobilities found with the semiconductors leads to the best response times one can hope for as they are limited essentially only by the photon absorption rate. Their moderate electro-optic effects can however lead to inefficient light scattering, hindering successful application. The weakness of the non-linearity on the other hand allows a rather complete theoretical description.

In order to focus research on improved materials it is necessary to gain a thorough understanding of the physics responsible for the desired non-linear optical behaviour. Not only does this indicate what improvements can be made with the material of primary interest but also it allows one to determine whether the observed non-linear optical effect is unique to that material or should be observable in a wide variety of materials. It has been noted already that there is a striking similarity between stimulated photorefractive scattering and stimulated Brillouin scattering. This similarity stretches further into current research into self-pumped phase conjugation with resonators providing feedback and either photorefraction or Brillouin scattering providing the non-linearity. Although the material physics of the two processes are distinct there is a large amount of common physics in the non-linear optical part of the interaction. It is argued later that most phenomena currently observed with photorefraction should be observable in other materials with non-photorefractive and non-Brillouin third order optical non-linearities. The scope for materials improvement and flexibility in the choice of wavelength/power and speed is therefore enormous.

The group at King's College London have followed two strategies of investigation of phase conjugation in photorefractive materials. The first is to 'tame' $BaTiO_3$ by the use of erase beams and index matching. This provides a controllable non-linearity allowing the investigation of phase conjugating systems in the absence of competing processes and the determination of minimum levels of gain to observe a particular phenomenon. The elimination of competing processes not only leads to significantly improved two-wave mixing (Fig. 16) but also allows an accurate theoretical description.

This technique has been used in the investigation of phase conjugation systems using feedback. Feedback is provided by optical resonators and

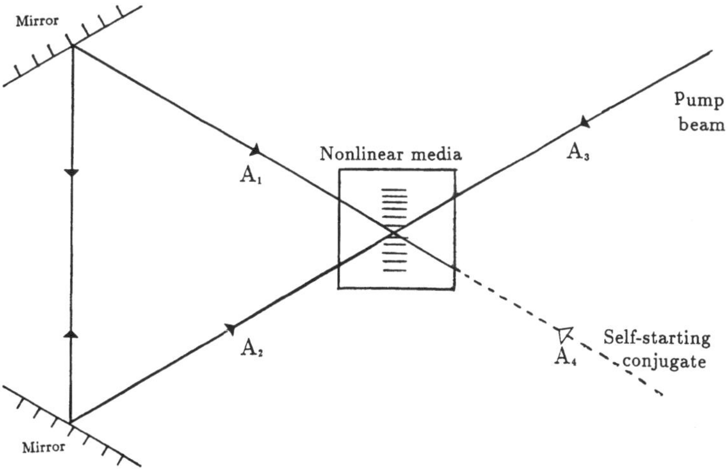

Figure 19 Ring passive phase conjugator

results in lower gain thresholds for self-pumped phase conjugation [31]. In the context of SBS this leads to reduced power requirements and in the context of photorefraction it allows self-pumping with materials in which SPS is absent. A particularly promising configuration is shown in Fig. 19. This resonator is essentially a Sagnac ring resonator with the beam splitter replaced with a photorefractive crystal [31]. Two-wave mixing gain is crucial to the operation of this phase conjugating system, although four-wave mixing also has a role. The incident wave A_3 passes through the crystal and is fed back via the mirrors as wave A_1. Any light A_4 propagating back along this path will see gain first in the presence of the pump A_3 and then, on reflection by the mirrors, sees gain in the presence of the pump A_1. The two gratings responsible for this gain only reinforce each other, increasing the gain further, when A_4 is the phase conjugate of A_1 and hence A_2 is the phase conjugate of A_3. Assuming the presence of a phase conjugate wave, the gratings then present will scatter the incident wave A_3 into the conjugate A_4. This four-wave mixing interaction maintains the seeding of the conjugate wave. Pump depletion is again responsible for ensuring that the phase conjugate wave is the dominant return. The conjugator is self starting and coherence length tolerant. It could therefore be a promising candidate for the end mirror of a solid state laser. It is possible to model the time development of the reflectivity (Fig. 20) and the effect of loss on the reflectivity (Fig. 21). Experiments with 'tamed' $BaTiO_3$ have confirmed these predictions (Fig. 22). It is clear that, although self-pumping can still be achieved, high reflectivities require careful attention to the minimisation of losses.

The second strategy the group at King's College London has followed is to increase the nonlinearity of photorefractive materials by employing

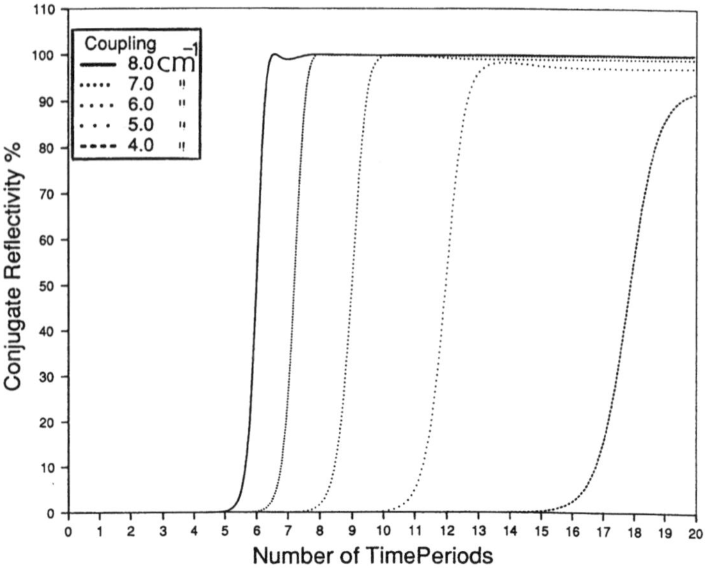

Figure 20 Theoretical dependence of reflectivity dynamics on coupling strength

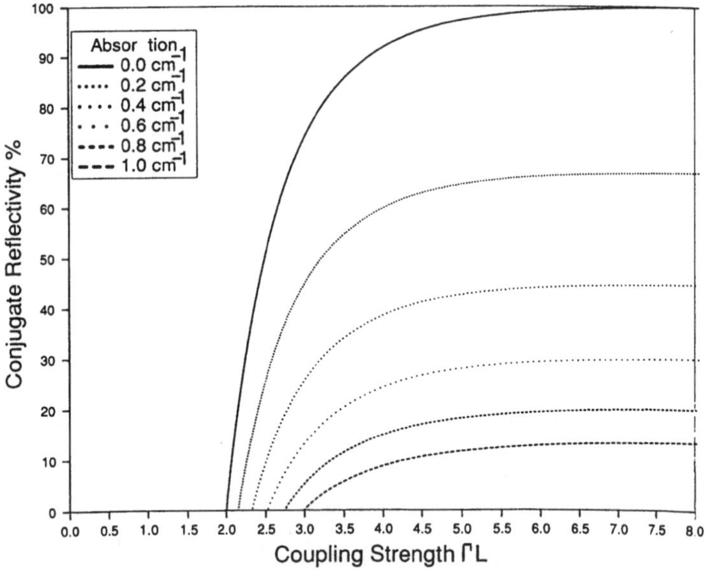

Figure 21 Theoretical dependence of reflectivity on coupling strength with absorptive losses.

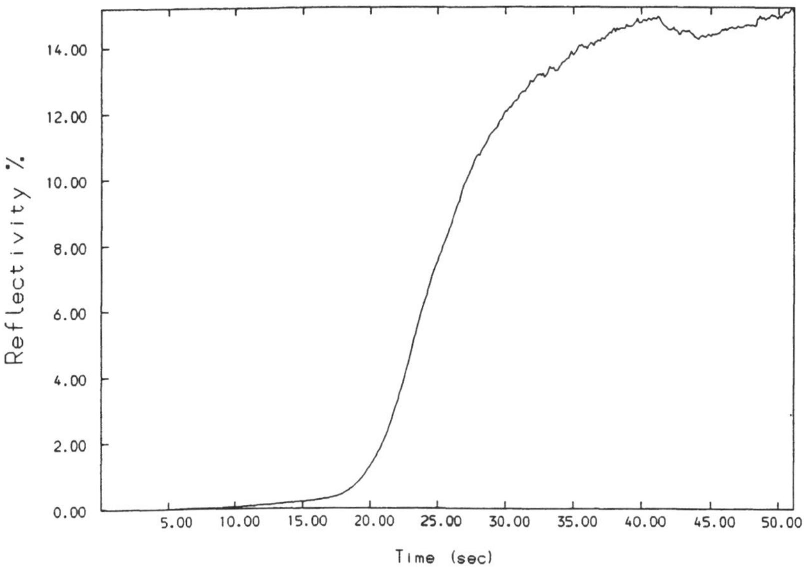

Figure 22 Experimental Time Development of Reflectivity in Barium Titanate

techniques to enhance the magnitude of the space charge electric field. The largest space charge electric field that can be supported in a photorefractive material is known as the saturation field E_S and is determined by the available charge density. The magnitude of the saturation field is inversely proportional to the spatial frequency of the grating. At low spatial frequencies (large grating periods) and in the absence of applied fields, the actual space charge electric field developed is determined by the diffusion field E_D. This is the field that balances the thermal diffusion of carriers and the drift of carriers in the electric field. It is directly proportional to the spatial frequency. In the case of no applied field there is therefore an optimum grating period typically around $1\mu m$ where the space charge electric field is maximised. If an electric field is applied to the crystal the magnitude of the space charge field is then determined by the applied field E_A, if this is less than the saturation field E_S. In the case of an applied electric field it is thus advantageous to operate at as large a grating period as possible and apply as large a field as possible without saturation occurring. Fields of 5-15 $kVcm^{-1}$ are commonly used. The rewards in terms of diffraction efficiency of the application of d.c. electric fields are clearly illustrated in Fig. 23.

When a d.c. electric field is applied the no-field simple exponential time response of the space charge electric field becomes oscillatory. It is possible to enhance the space charge electric field by frequency detuning one of the writing beams and tuning the resultant moving fringes into resonance

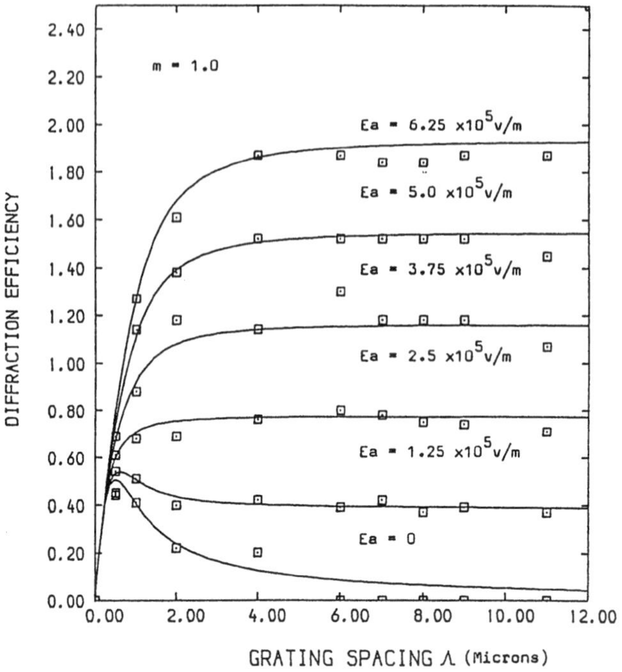

Figure 23 Enhancement of the diffraction efficiency in BSO with varied applied fields. Saturation is shown at high spatial frequencies and the solid lines are theoretical curves.

with this oscillation. The use of moving fringes allows the restoration of the phase shift between fringe pattern and grating necessary for two-wave mixing gain which is lost by the application of a d.c. electric field. The finite response time of the material results in the lagging of the space charge electric field behind the moving fringe pattern. The resulting phase-shift is near the optimum of $\pi/2$ at resonance. The enhanced diffraction efficiencies and two-wave mixing gains possible are illustrated in Figs. 24 and 25. This enhancement is only observed at small fringe visibilities (i.e. strong pump and weak signal) [32] and can never exceed the limits set by the saturation field [6]. Nevertheless, huge gains for weak signals are observed and several resonators have been demonstrated using this technique with both semiconductors and sillenites [33]. It is important to note that in self-pumping schemes it is the wave which sees the most gain that dominates. This wave will be the one that is detuned in frequency from the incident wave by exactly the amount needed to move the fringes at the optimum velocity. No special arrangements are necessary for producing frequency offset beams in this case.

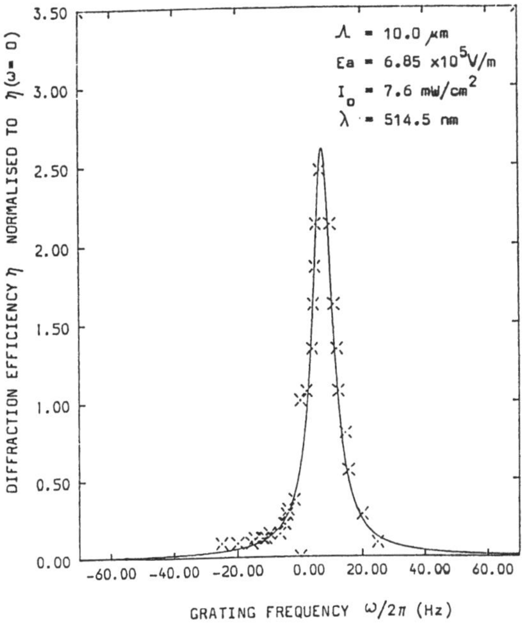

Figure 24
Diffraction efficiency η versus Grating frequency (Detuning of interfering beams Hz) with an applied d.c. field on BSO. Resonance occurs when the optimum phase shift between the interference pattern and the space charge field is achieved. The solid line is a theoretical curve.

The use of d.c. fields and moving fringes has a number of disadvantages. The crystal must be uniformly illuminated to ensure that the electric field is everywhere equally strong. More serious is the fact that the resonant frequency detuning depends on the applied electric field, the intensity and the grating period and the enhancement only occurs for a narrow band of spatial frequencies. Crucially it is impossible to achieve resonance simultaneously throughout the volume of a crystal with non-negligible absorption. It is possible to overcome these problems by the application of an a.c. electric field. The idea is that during a free carrier lifetime the carrier drifts in essentially a stationary field but over a grating build up time the field has changed sign so often that there is no preferred direction. The result is effectively an increased diffusion of carriers which is opposed by a necessarily increased space charge electric field. The lack of a preferred direction results in the optimum $\pi/2$ phase shifted grating and hence substantial gain (Fig. 26). Gain exceeding absorptive losses in GaAs has been demonstrated using this technique [34]. The technique has been used in conjunction with optical feedback to demonstrate self-pumping in BTO [35] and InP [36]. The main disadvantage of the technique is the accompanying modulation

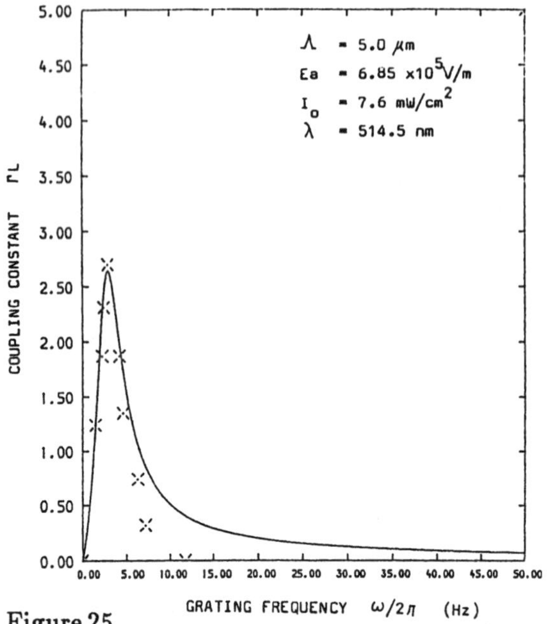

Figure 25
Two-wave mixing coupling strength versus Frequency detuning of the beams in BSO. Optimum coupling occurs when the moving fringe pattern restores the $\frac{\pi}{2}$ phase relationship with the space charge field, which is lost by the application of the d.c. field. The solid line is a theoretical curve.

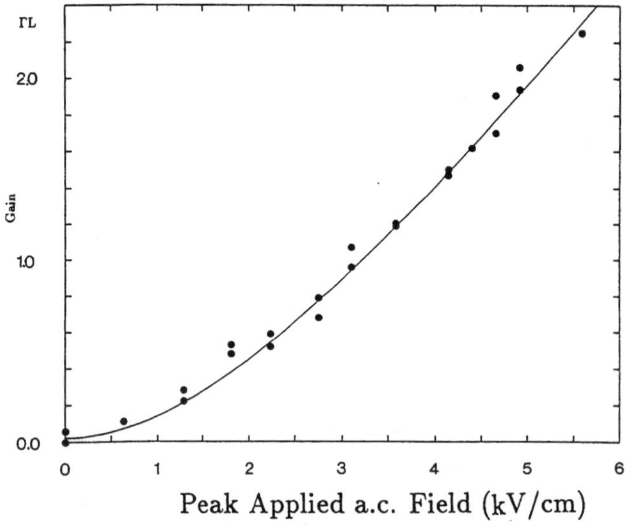

Figure 26. Photorefractive two-wave mixing gain in GaAs versus applied field. Absorption in the sample was $\alpha L \sim 2.0$ and hence net gain is achieved.

of the birefringence which always modulates the phase of the interacting waves but can also cause gain modulation. This is unavoidable in optically active materials [37]. A detailed comparison of enhancement techniques can be found in reference [38; see also 39].

The response time of the photorefractive effect is inversely proportional to intensity but every technique which enhances the magnitude of the steady state space charge electric field results in a slower response [6, 38]. The application of an electric field which does not increase the magnitude of an already saturated space charge electric field can, however, improve the response time [40]. What is required is that as large as possible refractive index change is induced by as few as possible absorbed photons. This requires that the ratio of electro-optic coefficient to d.c. dielectric constant be as large as possible. The large electro-optic coefficient provides a large refractive index change for a given space charge electric field. A small dielectric constant then allows a large space charge field for a given redistributed charge density. This consideration favours ferroelectric materials due to their enormous electro-optic coefficients and organic materials for their low dielectric constants. The material should clearly contain sufficient density of deep levels, ideally half occupied, to provide the required field. These levels are also responsible for the absorption. It is vital that this is the only significant source of absorption so that every absorbed photon is making a contribution to the non-linearity. Finally, the materials should have a mobility lifetime product sufficient for a photoexcited charge to move a grating period or more in one lifetime. Only one photon is then necessary to move the charge to the desired position. The time response of the material is then limited only by the amount of charge that needs to be relocated and the rate at which photons are absorbed. Good carrier transport properties favour the semiconductor materials most and the organics least. If the material does have good transport properties it is found that enhancement techniques increase the space-charge field by as much as they slow the response and so the energy required for a given refractive index change is unaffected. The steady state refractive index change on the other hand has been substantially improved.

The best strategy is therefore to select a material with a good electro-optic and carrier transport properties, and employ enhancement techniques until the material reaches saturation and hence becomes photon absorption rate limited. The speed of the photorefractive semiconductor materials is already close to this limit. The energy efficiency can only be improved further if the electro-optic effect is improved.

Substantial increases in the electro-optic effect can be obtained by working at optical frequencies nearly resonant with the band gap and exploiting the Franz-Keldysh effect. Gains as high as 20cm^{-1} have been achieved in GaAs in this way [41, 42]. This idea can be extended further by exploiting the Quantum Confined Stark Effect of multiple quantum well (MQW) materials [43]. Experiments at AT&T Bell [44] have demonstrated that these materials can be made sufficiently semi-insulating for photorefraction. The necessarily thin grating nature ($\simeq \mu$m) of these materials limits diffraction efficiency but fJ μm^{-2} writing energies appear to be achievable.

The thin material can be turned to advantage by applying a longitudinal electric field using transparent electrodes. Sufficient fields can then be generated with convenient voltages $\simeq$ 10V. This geometry has been exploited in the past in the Pockels Readout Optical Modulator (PROM) [45]. Its resolution is limited by the thickness of the material. In the case of MQW materials this is $\simeq 1\mu$m and is perfectly adequate for two and four-wave mixing applications. It might also be possible to reintroduce thick grating behaviour with concomitant large gains and diffraction efficiencies by cascading MQW layers with transparent spacers to form a stratified volume holographic element [46].

The introduction of band gap engineering techniques also allows the photoconductive properties of the material to be tailored. For example, asymmetric wells have been exploited to demonstrate a true photorefractive effect with a build up time of 5ps [47]. Now that it is slowly being realised that all optoelectronic devices that exploit photoconductive electro-optic effects (e.g. multiple quantum well self-electro-optic devices and optically addressed liquid crystal spatial light modulators) are in a sense photorefractive we can anticipate further rapid developments.

Although the major developments are occurring in the field of photorefraction other optical nonlinearities could become important. We have already become accustomed to the use of moving fringes to achieve two-wave mixing gain in photorefractive materials. Every optical non-linearity, however, has an associated response time. Moving fringe techniques or equivalently non-degenerate four-wave mixing should also display two-wave mixing gain. This is indeed the case and has been demonstrated using ruby [48]. Gain combined with feedback using a resonator should produce the conditions for self-pumping. The resonator could also automatically produce the correct detuning. This again has been demonstrated using sodium vapour [49]. This demonstration of self-pumped phase conjugation is significant since it is the first demonstration to our knowledge of self-pumped phase conjugation in using a non-photorefractive active and non-Brillouin active material.

These non-linearities are associated with considerable linear absorption which we have already noted has a deleterious effect on reflectivity. The significance of self-pumping is, however, that it does not require external pumps coherent with respect to the probe. Few applications would prevent the use of incoherent optical pumping to reduce the absorption of the material. Self-pumped phase conjugation in gain media as a consequence has exciting potential.

7. Conclusions

Self-pumped phase conjugating systems have shown themselves to be superior to systems based on four-wave mixing alone, by the generation of very high fidelity phase conjugated images and by their simplicity to incorporate into optical systems such as associative memories. The self-generation of pump beams in self-pumped systems like barium titanate always allows gain competition between nonlinear optical processes and consequently has a high selectivity for the generation of the true phase conjugate to the pump. In four-wave mixing, however, pump depletion and pump misalignment can never be corrected for in the wave mixing interaction, causing degradation of phase conjugate fidelity.

The possibility of using shifted gratings in a variety of materials permits the process of two-wave mixing. This can then be used for all optical beam amplification with fast response times possible with some materials. Using optical feedback in the form of a resonator, two-wave mixing may be used to achieve self-pumped phase conjugation, with all beams at the correct frequency for an optimum shifted grating being self generated.

Mirror losses and absorption have been shown to have severe effects upon the reflectivity of resonator based phase conjugating systems. Reflective losses may be overcome by the use of dielectric coatings, to either enhance or reduce reflections. Careful cavity design can overcome geometrical losses and reduce diffractive losses. In the case of absorption, it is not inconceivable to use an external source of incoherent light to excite a material so that its absorptive nature is lost. Extending this principle further, it may even be possible to combine self-pumped phase conjugation and laser action in the same material. Self-pumped phase conjugating systems, with gain, responding on nanosecond time scales are not only a distinct possibility for the near future but also have many exciting applications.

Acknowledgements

The authors are grateful to R.G. Hoptroff, A.G. Kirk and A.Bostel, all of King's College London, for assistance in the production of this chapter.

[1] Giuliano C.R., "Applications of optical phase conjugation", Physics Today, 27-35, (April 1981).
[2] Feinberg J., "Photorefractive nonlinear optics", Physics Today, 46-52, (Oct 1988).
[3] Zeldovich B.Ya., Pilipetsky N.F., Shkunov V.V., "Principles of Phase Conjugation", (Springer-Verlag, Berlin, 1985).
[4] Feinberg J., "Self-pumped, continuous wave phase conjugator using internal reflections", Opt. Lett., 7 (10), 486-488, (1982).

[5] Scott A.M., Waggott P., "Low-intensity phase conjugation by self-pumped Brillouin-induced four-wave mixing", J. Mod. Opt. 35 (3), 473-481, (1988).
[6] Hall T.J., Jaura R., Connors L.M., Foote P.D., "The photorefractive effect - a review", Prog. Quant. Electron. 10 (2), 77-146, (1985).
[7] Sutter K., Hulliger J., Günter P., "Photorefractive effects in the organic crystal 2 - cyclooctylamino - 5 - nitropyridine (COANP) doped with 7, 7, 8, 8 - tetracyanochinodimethane (TCNQ)", Société Francaise d'Optique technical digest, Topical meeting of Photorefractive Materials, Effects and Devices II, Jan 17-19 1990, Aussois (France), postdeadline paper PD1.
[8] Feinberg J., "Asymmetric self-defocusing in an optical beam from the photorefractive effect", J. Opt. Soc. Am., 72 (1), 46-51, (1982).
[9] Lam J.F., "Origin of phase conjugate waves in self-pumped photorefractive mirrors", Appl. Phys. Lett., 46 (10), 909-911, (May 1985).
[10] Foote P.D.,"Optically induced anisotropic light diffraction in photorefractive crystals", PhD Thesis, University of London (1987).
[11] Foote P.D., Hall T.J., Powell A.K., "Simulated photorefractive scattering in barium titanate', Int. Symp. on Technologies for Opto-Electronics, Cannes, 16-20 Nov 1987, Proc. SPIE, 864, 90-97.
[12] Mullen R.A., Vickers D.J., Pepper D.M., "Seeded stimulated photorefractive scattering", Société Francaise d'Optique technical digest, Topical meeting of Photorefractive Materials, Effects and Devices II, Jan 17-19 1990, Aussois (France), 204-207.
[13] McFarlane A., Steel D.G., "Laser oscillation using resonator with self-pumped phase conjugate mirror", Opt. Lett., 8 (4), 208-209, (1983).
[14] Chiou A.E., Yeh P., "Laser-beam clean-up using photorefractive two-wave mixing and optical phase conjugation', Opt. Lett., 11 (7), 461-463, (1986).
[15] MacCormack S., Eason R.W., "Sequential power transfer between stripes of a laser diode array via photorefractive two beam coupling in $BaTiO_3$.", Société Francaise d'Optique technical digest, Topical meeting of Photorefractive Materials, Effects and Devices II, Jan 17-19 1990, Aussois (France), 333-335.
[16] Fairchild P., Davis K., Valley M., "Coherent beam combination in barium titanate", J. Opt. Soc. Am. B, 5 (8), 1758-1762, (1988).
[17] Stephens R.R., Craig R.R., Narayanan A.A., Lind R.C., Giuliano C.R., "Single and multiple element 4-pass phase conjugate master oscillator power amplifier using diode laser", Optics News, 11-12, (Dec 1989).
[18] Rockwell D.A., Giuiano C.R., "Coherent coupling of laser gain media using phase conjugation", Opt. Lett., 11 (3), 147-149, (1986).
[19] Chiou A.E., Yeh P., Parallel image subtraction using a phase conjugate Michelson interferometer, Opt. Lett., 11 (5), 306-308, (1986).
[20] White J.O., Yariv A., "Real-time image processing via four wave mixing

in a photorefractive medium", Appl. Phys. Lett., 37 (1), 5-7, (1980).
[21] Pauliat G., Herriau J.P., Delboulbé A., Roosen G., Huignard J.P., "Dynamic beam deflection using photorefractive gratings in $Bi_{12}SiO_{20}$ crystals', J. Opt. Soc. Am. B, 3 (2), 306-313, (1986).
[22] Yeh P., Chiou A.E.T, Hong J., "Optical interconnection using photorefractive dynamic holograms", Appl. Opt., 27, 2093-2096, 1988.
[23] Foote P.D., Hall T.J., Connors L.M., 'High speed two input real time optical correlation using photorefractive BSO', Opt. and Laser Tech., 18 (1), 39-42, 1986.
[24] Klumb H., Herden A., Kobialka T., Laeri F., Tschudi T., "Active coherent optical feedback system with phase conjugating image amplifier", 5 (11), 2379-2385, (1988).
[25] Ewbank M.D., Yeh P., Khoshnevisan M., Feinberg J., "Time reversal by an interferometer with coupled phase-conjugate reflectors", Opt. Lett., 10 (6), 282-284, (Jun 1985).
[26] White H.J., Aldridge N. B., Lindsay I., "Digital and analogue holographic associative memories', Opt. Eng. 27 (1), 30-37, (1988).
[27] Anderson D.Z., Benkert C., Hermanns A., "Multistable and time sequencing photorefractive ring resonators", Société Francaise d'Optique technical digest, Topical meeting of Photorefractive Materials, Effects and Devices II, Jan 17-19 1990, Aussois (France), postdeadline paper PD4.
[28] Lininger D.M., Martin P.J., Anderson D.Z., 'Bistable ring resonator utilizing saturable photorefractive gain and loss", Opt. Lett., 14 (13), 697-699, (1989).
[29] Paek E.G., Wullert II J.R., Patel J.S., "Holographic implementation of a learning machine based on a multicategory perceptron algorithm", Opt. Lett., 14 (23), 1303-1305, (1989).
[30] Gauthier D.J., Narum P., Boyd R.W., "Observation of deterministic chaos in a phase conjugate mirror", Phys. Rev. Lett., 58 (16), 1640-1643, (April 1987).
[31] Cronin-Golomb M., Kwong S.K., Yariv A., "Optical oscillators with photorefractive gain", in Günter P. (Ed), Electro-optic and photorefractive materials, Part V, Springer Procs in Physics 18, (Springer-Verlag, Berlin, 1986), 291-307.
[32] Swinburne G.A., Hall T.J., Powell A.K., "Large modulation effects in photorefractive crystals", IERE Int. Conf. Holographic Systems, Components and Applications, Bath, UK, Sept 1989, 116-123.
[33] Rajbenbach H., Imbert B., Huignard J.P., Mallick S., "Near infrared four wave mixing with gain and self starting oscillators with photorefractive GaAs", Opt. Lett., 14 (1), 78-80, (1989).
[34] Walsh K., Hall T.J., "Gain Exceeding Absorptive Losses in Photorefractive GaAs"Appl. Opt., 28 (1), 16-17, (1989).
[35] Sochava S.L., Stepanov S.I., Petrov M.P., "Ring oscillator using a pho-

torefractive $Bi_{12}TiO_{20}$ crystal", Sov. Tech. Phys. Lett., 13 (6), 274-275, (1987).

[36] Glass A.M., Olson D.H., Cronin-Golomb M., 'Self pumped phase conjugation in InP:Fe", Appl. Phys. Lett., 54 (20), 1968-1970, (1989).

[37] Stace C., Powell A.K., Walsh K., HaLL T.J., "Coupling modulation in photorefractive materials by applying a.c. electric fields", Opt. Comm. 70 (6), 509-514, (April 1989).

[38] Walsh K., Powell A.K., Stace C., Hall T.J., "Techniques for the enhancement of space charge fields in photorefractive media", To appear J. Opt. Soc. Am. B, (Spring 1990).

[39] Stepanov S.I., Petrov M.P. "Efficient unstationary holographic recording in photorefractive crystals under an external alternating electric field", Opt. Comm., 53 (5), 292-295, (1985).

[40] Sayano K., Yariv A., Neurgaonkar R.R., "Order-of-magnitude reduction on the photorefractive response time in rhodium-doped $Sr_{0.6}Ba_{0.4}Nb_2O_6$ with a dc electric field", Opt. Lett., 15 (1), 9-11, (1990).

[41] Partovi A., Kost A., Garmire E. M., Valley G.C., Klein M.B., "Characterisation of band-edge photorefractive effect in compound semiconductors", Société Francaise d'Optique technical digest, Topical meeting of Photorefractive Materials, Effects and Devices II, Jan 17-19 1990, Aussois (France), p157.

[42] Garmire E., Jokerst N.M., Kost A., Danner A., Dapkus P.D., "Optical nonlinearities due to carrier transport in semiconductors', J. Opt. Soc. Am. B, 6 (4), 579-587, (1989).

[43] Miller D.A.B., "Electric field dependence of optical properties of quantum well structures", in Günter P. (Ed), Electro-optic and photorefractive materials, Part I, Springer Proceedings in Physics 18, 35-49, (Springer-Verlag, Berlin, 1986)

[44] Glass A.M., "Future trends in photorefractive and nonlinear optics", Société Francaise d'Optique technical digest, Topical meeting of Photorefractive Materials, Effects and Devices II, Jan 17-19 1990, Aussois (France), Oral Presentation.

[45] Horowitz B.A., Corbett F.J., "The PROM - Theory and applications for the Pockels readout optical modulator", Opt. Eng. 17 (4), 353-364, (1978).

[46] Johnson R.V., Tanguay A.R., "Stratified volume holographic optical elements", Opt. Lett. 13 (3), 189-191, (1988).

[47] Ralph S.E., Capasso F., Malik R.J., "Transient photorefractive effect in graded gap superlattices', Société Francaise d'Optique technical digest, Topical meeting of Photorefractive Materials, Effects and Devices II, Jan 17-19 1990, Aussois (France), 231-234.

[48] McMichael I., Yeh P., Beckwith P., "Nondegenerate two-wave mixing in ruby", Opt. lett. 13 (6), 500-502, (June 1988).

[49] Gaeta C.J., Lam J.F., Lind R.C., "Continuous-wave self pumped optical phase conjugation in atomic sodium vapor", Opt. Lett., 14 (4), 245-247, (1989).

Theory of Four-Wave Mixing in Photorefractive Oscillators

P.M. Petersen

Technical University of Denmark, Physics Laboratory III,
DK-2800 Lyngby, Denmark

Abstract. In this paper the usual numerical approach to four-wave mixing in photorefractive oscillators is compared with a new simple perturbational theory.

1. Introduction

The theory of four-wave mixing (FWM) in photorefractive oscillators /1-5/ is based on four coupled wave equations for the four optical waves. These equations cannot be solved analytically and therefore the usual approach to FWM in photorefractive oscillators is based on numerical solutions /1/ of the four coupled wave equations. In this paper the usual approach is compared with a new simple analytical theory which is valid for a restricted parameter range.

2. The Linear Passive Photorefractive Oscillator

The linear passive phase conjugate oscillator is shown in Fig. 1. It consists of a photorefractive medium between two mirrors M_1 and M_2. When the laser beam A_4 is incident on the photorefractive medium two pumping beams A_1 and A_2 are build up in the cavity formed by the mirrors M_1 and M_2. When the backward pump beam A_2 is diffracted in the photorefractive phase grating a fourth beam A_3 is generated. This beam A_3 is the phase conjugate of A_4. The four beams are coupled by the four coupled wave equations /1/

$$\frac{dA_1}{dz} = -\frac{\gamma}{I_o} gA_4 \tag{1A}$$

$$\frac{dA_2^*}{dz} = -\frac{\gamma}{I_o} gA_3^* \tag{1B}$$

$$\frac{dA_3}{dz} = \frac{\gamma}{I_o} gA_2 \tag{1C}$$

$$\frac{dA_4^*}{dz} = \frac{\gamma}{I_o} gA_1^* \tag{1D}$$

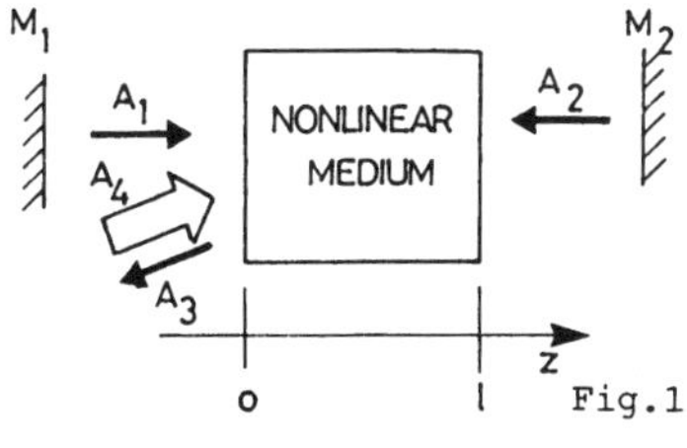

Fig.1 The linear passive phase conjugate oscillator

where γ is the complex coupling constant, $I_o = \Sigma|A_i|^2$, and $g = A_1A_4^*+A_2^*A_3$. The phase conjugate reflectivity $R = |A_3(0)|^2/|A_4(0)|^2$ can be found from these equations if we furthermore apply the two boundary conditions

$$M_1 = \frac{|A_1(0)|^2}{|A_2(0)|^2} \tag{2A}$$

and

$$M_2 = \frac{|A_2(\ell)|^2}{|A_1(\ell)|^2} \quad . \tag{2B}$$

The first solution to equations (1)-(2) was given by M. Croning-Colomb et al./1,3/. They found that the phase conjugate reflectivity was given by

$$R = \frac{(\Delta+1)^2|T|^2}{M_2|\Delta T + (\Delta^2+(\Delta+1)^2/M_2)^{\frac{1}{2}}|^2} \tag{3}$$

where

$$\Delta = I_2(\ell) - I_1(0) - I_4(0) \tag{4A}$$

and

$$T = \tanh\{(\gamma\ell/2)(\Delta^2 + (\Delta+1)^2/M_2)^{\frac{1}{2}}\} \ . \tag{4B}$$

In these equations Δ has to be found numerically as the zero values of the function

$$f(\Delta) = M_1M_2 - \left|\frac{T + (\Delta^2+(\Delta+1)^2/M_2)^{\frac{1}{2}}}{\Delta T + (\Delta^2+(\Delta+1)^2/M_2)^{\frac{1}{2}} + (\Delta+1)T/M_2}\right|^2 \ . \tag{5}$$

In equations (3)-(5) all intensities are normalized by the total intensity I_o. We discuss later on in this paper the behavior of $f(\Delta)$ and compare the solutions of the reflectivity R in Eq. (3) with a more simple theory which is outlined below.

If in Eq. (1) we assume that the input intensity $|A_4|^2$ is much larger than the pump beam intensities $|A_1|^2$ and $|A_2|^2$ we can neglect the term in the grating structure which does not involve A_4. In this case equations (1A-D) reduce to

$$\frac{dA_1}{dz} = -\frac{\gamma}{I_o} A_1|A_4|^2 \tag{6A}$$

$$\frac{dA_2^*}{dz} = -\frac{\gamma}{I_o} A_1A_4^*A_3^* \tag{6B}$$

$$\frac{dA_3}{dz} = \frac{\gamma}{I_o} A_1A_4^*A_2 \tag{6C}$$

$$\frac{dA_4^*}{dz} = \frac{\gamma}{I_o} A_4^*|A_1|^2 \tag{6D}$$

These equations can be solved analytically if we use the conservation law $A_1A_2+ A_3A_4 = C$. We obtain for the conjugate beam A_3 at $z = 0$

$$A_3(0) = \frac{A_4^*(0)}{A_1^*(0)} A_2(\ell) \frac{e^{-\beta\ell}(1-e^{\gamma\ell})}{\sqrt{(1+r^2)(1+r^2e^{2\alpha\ell})}} \exp(-i\frac{\beta}{2\alpha} \ln \frac{r^2+e^{-2\gamma\ell}}{1+r^2}) \tag{7}$$

where $\gamma = \alpha + i\beta$ and $r = |A_4(0)|/|A_1(0)|$. The parameters r, $|A_1(0)|$, and $|A_2(\ell)|$ can be obtained from the boundary conditions in equations (2A-B). We find

$$r^2 = \frac{B \pm \sqrt{D}}{N} \tag{8}$$

where $B = -2(M_1M_2e^{\alpha\ell}\cos\beta\ell - e^{2\alpha\ell})$, $D = B^2 - 4(M_1M_2-1)e^{2\alpha\ell}(M_1M_2-e^{2\alpha\ell})$, and $N = 2(M_1M_2 - e^{2\alpha\ell})e^{2\alpha\ell}$. The pumping beams are given by $|A_1(0)|^2 = I_f/(1+r^2)$ and $|A_2(\ell)|^2 = M_2I_f/(1+r^2e^{2\alpha\ell})$, where $I_f = |A_1|^2 + |A_4|^2$. The phase conjugate reflectivity can now be found from Eq.(7):

$$R = M_2 \frac{1+e^{2\alpha\ell} - 2e^{\alpha\ell}\cos\beta\ell}{(1+r^2e^{2\alpha\ell})^2} \ . \tag{9}$$

The phase conjugate reflectivity of the oscillator in Fig. 1 is within the approximations used in this paper given by Eq.(9) when we insert r^2 from Eq.(8).

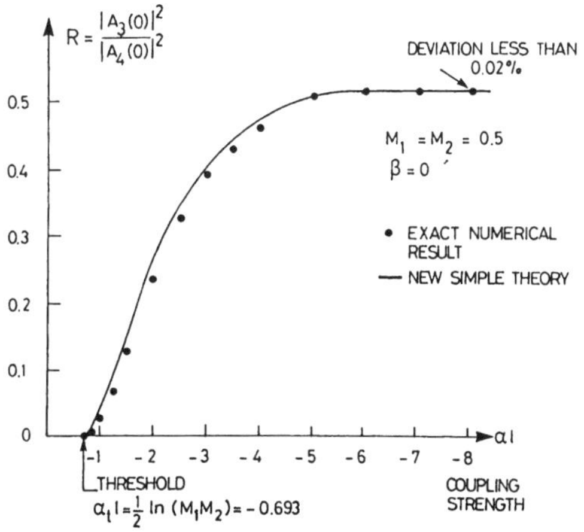

Fig. 2 The phase conjugate reflectivity versus coupling strength. The solid curve is calculated from Eq.(9) and the dots are calculated from Eq.(3). The parameters used are $M_1 = M_2 = 0.5$ and $\beta = 0$.

In Fig. 2 is shown the phase conjugate reflectivity versus coupling strength calculated from the simple theory in Eq.(9) when $M_1 = M_2 = 0.5$ and $\beta = 0$. This curve is shown as a solid line. Furthermore the exact values of the phase conjugate reflectivities calculated from Eq.(3) are shown in Fig. 1 as dots. The threshold value of the coupling strength for oscillations in the cavity is in our theory obtained when $r^2 \rightarrow \infty$. Using Eq.(8) we find $\alpha_t\ell = \frac{1}{2}\ln(M_1M_2) = -0.693$ which is in exact agreement with Eq.(3). Above threshold there is a minor deviation between the two theories. However, at high coupling strengths the simple theory again becomes in good agreement with the exact theory. At $\alpha\ell = -8$ the deviation is less than 0.02%.

In our perturbational approach we have assumed that the $A_1A_4^*$ interference grating is dominating and that the $A_2^*A_3$ interference grating can be neglected. Therefore, it is important that the reflectivity of mirror M_1 does not become too small. The behavior of the oscillator in Fig. 1 can be investigated by looking at $f(\Delta)$ from Eq.(5). In Fig. 3(a) we have shown $f(\Delta)$ when $M_2 = M_1 = 0.5$ and $\gamma\ell = -3$. In this case $f(\Delta) = 0$ only has one root $\Delta = -0.361$ corresponding to $R = 0.391$. The value of r^2 calculated from our theory in Eq.(8) is $r^2 = 23.3$ corresponding to $R = 0.405$ in fair agreement with the exact result. However, when M_1 becomes small a more complicated behavior of the phase conjugator takes place. In Fig. 3(b) we have shown $f(\Delta)$ for $M_1 = 0.01$, $M_2 = 1$, and $\gamma\ell = -3$. Now $f(\Delta) = 0$ has 3 roots and more than one state of operation for the phase conjugate oscillator is possible. The limit $M_1 \to 0$ gives the semilinear passive phase conjugate oscillator /1/. This device does not start by itself, but if a seed beam is present in start-up it is possible to build up oscillations which remain after the seed beam has been turned off. This type of operation cannot be explained by our simple theory in Eq.(9).

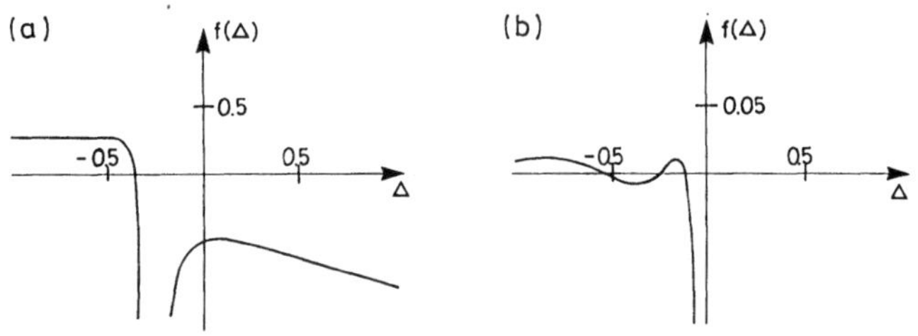

Fig. 3 $f(\Delta)$ calculated from Eq.(6) when: (a) $M_1 = 0.5$, $M_2 = 0.5$, $\gamma\ell = -3$
(b) $M_1 = 0.01$, $M_2 = 1$, $\gamma\ell = -3$.

3. Conclusion

In conclusion we have presented a simple theory for the linear passive photorefractive oscillator. The laser induced transmission grating in this device consists of two contributions, one for the forward-probe $A_1A_4^*$ interference and another from the backward-signal $A_3A_2^*$ interference. In our theory we have neglected the backward-signal interference term and we obtain values of the phase conjugate reflectivity which is in good agreement with the exact analysis. However, it should be pointed out that for certain values of the mirror reflectivities M_1, M_2 and the coupling strength $\gamma\ell$ there exist multistable solutions. This behavior is not included in our simple theory.

References

1. M. Croning-Golomb, B. Fisher, J. White, and A. Yariv "Theory and applications of four-wave mixing in photorefractive media", IEEE J. Quantum Electron, QE-20, 12 (1984).
2. J.O. White, M. Croning-Golomb, B. Fisher, and A. Yariv "Coherent oscillation by self-induced grating in the photorefractive crystal $BaTiO_3$", Appl. Phys. Lett. 40, 450 (1982).
3. M. Croning-Golomb, B. Fisher, J.O. White, and A. Yariv "Passive (self-pumped) phase conjugate mirror. Theoretical and experimental investigation", Appl. Phys. Lett. 41, 689 (1982).
4. B. Fisher, S. Sternklar, and S. Weiss "Photorefractive Oscillators", IEEE J. Quantum Electron QE-25, 550 (1989).
5. M. Croning-Golomb and A. Yariv "Plane-wave theory of nondegenerate oscillation in the linear photorefractive passive phase-conjugate mirror", Opt. Lett. 11, 242 (1986).

Optical Bistability

B.S. Wherrett and D.C. Hutchings

Department of Physics, Heriot-Watt University, Edinburgh, EH14 4AS, UK

Abstract. This article describes, at a tutorial level, those aspects of the linear Fabry-Perot cavity that are relevant to optical bistability, the sources for nonlinear refraction in such cavities, and the resulting cw nonlinear characteristics. Critical powers for switching and switching speeds are discussed and experimental results to date summarised. Alternative bistability and switching mechanisms are commented on.

1. Introduction

A bistable device is defined as one which, under steady-state conditions and given a unique set of inputs, is capable of existing in either of two internal states and of producing either of two different stable outputs. At any given time the internal state, and corresponding output, depend on the history of operation of the device; by temporary adjustment of the inputs it should be possible to set or reset the device to either state. Thus the latest temporary set/reset signal determines the device output long after that signal has been removed - the system in effect remembers the nature of the temporary signal. Clearly the output-versus-input response of such a device must be highly nonlinear. It is therefore not surprising that the concept of optical bistability was not discovered until the early 1960's when, following the advent of lasers, a vast range of nonlinear optical phenomena began to be investigated both experimentally and theoretically. Since that time optically bistable schemes have been studied for their intrinsic interest, for the role that bistability plays as a bridge between linear responses and the phenomena of instabilities and optical chaos, and for their device applications.

The use of bistables as optical memories or switches was suggested soon after 1960, as a basis for digital optical computing schemes. The recognition that similar, highly nonlinear but monostable responses, could be used for optical binary logic was implied by the earlier work but not discussed explicitly until the 70's and has led to interest in digital optic numeric and image processing. 2-D parallel arrays of such devices could be used as digital or analogue spatial light modulators, or as 2-D image displays. Other applications in fast switching for optical communications networks, signal amplifiers in communications, and conversely for noise reduction and power limiting have also been mooted.

In this article we will concentrate on one form of optically bistable system - the nonlinearly dispersive Fabry-Perot etalon; brief descriptions of alternative schemes are given in section (9). A Fabry-Perot etalon consists typically of a plane-parallel sided sample surrounded by partially reflecting mirrors. Following this Introduction the fundamentals of such a cavity are presented in Section (2) and the physics of nonlinearities of (i) electronic origin and (ii) thermal origin are covered in Section (3). This structure allows us to consider a nonlinear system in Section (4), having in mind precisely what is going on inside the active material of the Fabry-Perot and therefore enabling both the steady-state and dynamics of the bistable switching to be considered.

Springer Series in Wave Phenomena, Vol. 9 **Nonlinear Optics in Solids**
Editor: O. Keller

Device optimisation is covered in Section (5,6). A brief description of experimental results for dispersive Fabry-Perots is given in Section (8) and for alternative bistability schemes in Section (9), culminating in a discussion of achieved and predicted operating parameters.

Bistability occurs in many branches of physics: mechanics, magnetism, electronics, optics, etc. A simple mechanical scheme serves to describe some of the basic features of bistability. The easily constructed device depicted in Figure 1(a) consists of a pointer mounted on a spindle at position A. Two pieces of elastic are connected to the pointer, at position B. One is attached to a fixed point C; the other is free to move, it is held initially at some position on the x-y plane indicated. We are interested in the angle that the pointer makes to the vertical as the free end of the elastic is moved. Thus for example if the tension is weak, corresponding to position y_1 and the elastic is moved from side to side (varying x) the pointer will follow. One produces the essentially linear response indicated by the straight dashed line in Figure 1(b). However if the tension in the elastic is high, position y_2, then as the hold position is moved from the right to the left the pointer angle decreases slightly until at some critical point, with x to the left of centre (x_l), the pointer suddenly flips across from positive to negative angle. Conversely as the x position is retraced the pointer remains at negative angle until at an equivalent critical point, with x to the right of centre (x_r), the pointer jumps back to positive angle. The full line on Figure 1(b) shows the complete pointer response; a stable condition can be maintained at any place on this line, there being two possible stable states over the region x_l to x_r; this is the region of bistability. In the bistable region the actual state at a given time depends on whether one approaches from the left or the right. There is also, at the midpoint, a steady-state solution for which the pointer is vertical ($\theta = 0$). However this is unstable to a slight movement of x. In fact over the full bistable region there is a third unstable steady-state indicated by the

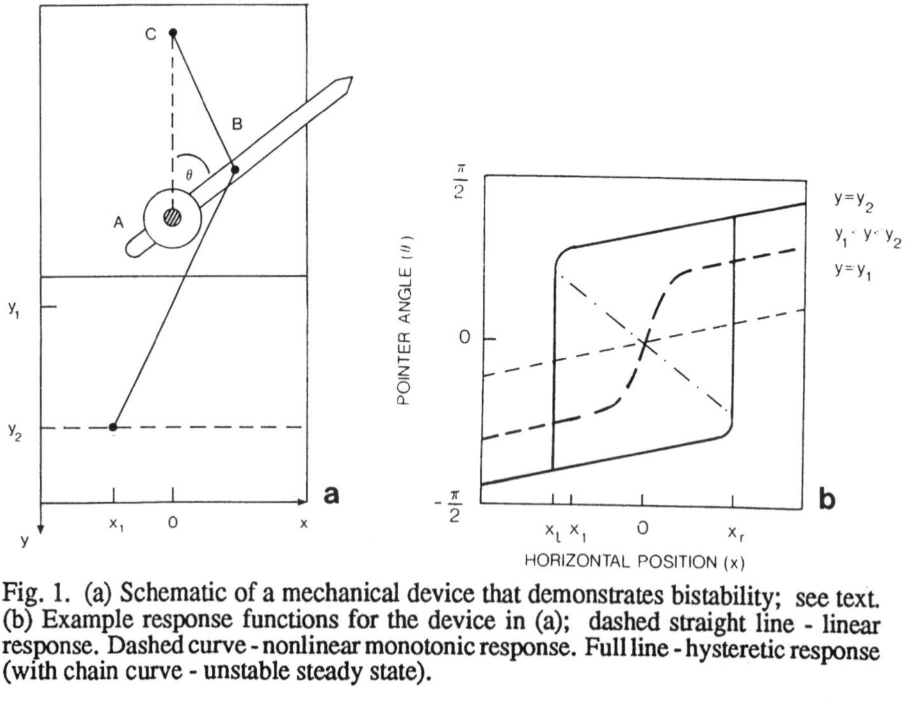

Fig. 1. (a) Schematic of a mechanical device that demonstrates bistability; see text. (b) Example response functions for the device in (a); dashed straight line - linear response. Dashed curve - nonlinear monotonic response. Full line - hysteretic response (with chain curve - unstable steady state).

chain-linked line in the figure. There are two 'control' parameters for this device - the positions x and y. Figure 1(b) gives an example of the family of response functions that is achieved for different y, as x is varied continuously. Note that if bistability is to be achieved then y (the tension) must exceed some critical value, and x must lie in a particular range of values. These same features will appear in the nonlinear Fabry-Perot case, where x is replaced by the optical irradiance or power level incident on the etalon and y represents the initial detuning of the etalon with respect to a Fabry-Perot transmission peak. There will be three steady-state solutions for a range of irradiances, given a large enough initial detuning, and the central solution will be shown to be unstable. In the optical case it is also possible to achieve bistability at fixed irradiance (x) as the detuning parameter (y) is varied.

The first mention of optical versions of bistability in the scientific literature occurred in 1964, when schemes involving lasers with intracavity saturable absorbers were described conceptually [1]. If we consider such a scheme under lasing conditions the internal irradiance level is high and the absorber will be saturated; this is consistent with the achievement of the lasing threshold and therefore with the high internal irradiance. It is then supposed that a temporary external source can be used to quench the laser medium; this causes the system to drop below threshold, the internal irradiance falls and the absorber becomes operable. Hence when the external source is removed the presence of the absorption prevents laser acting being re-achieved. Hence for the same pumping conditions one either has lasing with the absorber saturated or no lasing with a large absorption - two, self-consistent steady-states. To switch back from below to above threshold one would introduce a second external source this time onto the absorber, in order to re-initiate the saturation. The first theoretical description of optical bistability also involved the use of nonlinear absorption. In 1969 Szöke [2] formulated the response of a Fabry-Perot containing a saturable absorber to a single laser input of variable irradiance. Again, two self-consistent cw solutions can be expected qualitatively for the same input level; one with internal irradiance high and the absorber inactive, and vice versa. To switch from one level to the other the input is either reduced or increased temporarily. Once again the system should 'remember' the last temporary signal. It was in an attempt to perform such an experiment, using a cell containing sodium vapour, that in 1976 Gibbs and co-workers [3] first observed cw dispersive optical bistability. Their results are reproduced in Figure 2; a family of nonlinear responses was observed, for different cavity lengths (equivalent to varying y above) as a function of the incident irradiance level (x). Following this experiment it was recognised that the high sensitivity of a Fabry-Perot to small changes of the optical path length should be more effective in producing bistability than the weaker sensitivity to absorption changes. This has led to submilliwatt bistability in cavities containing thin (1-10 μm) layers of semiconductors or liquid crystals, as described in later sections of this article.

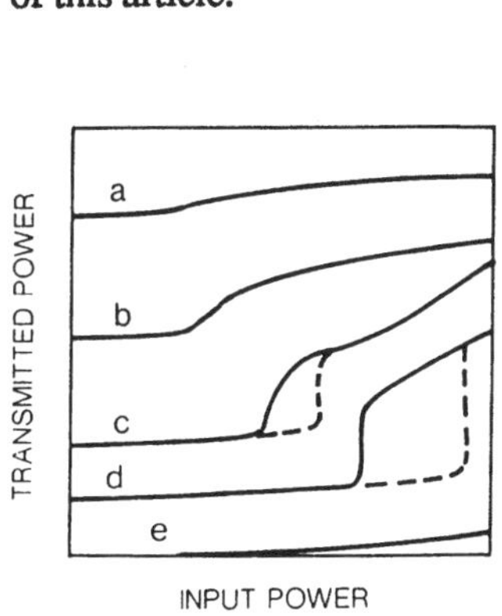

Fig. 2. Dispersive optical bistability in Na-vapour. After Gibbs et al. [3]. The letters a - e signify different cavity detunings.

2. The Linear Fabry-Perot Etalon

The simplest Fabry-Perot cavity is a plane-parallel sided, linear medium, in air, with no absorption; Figure 3. The electric fields (E_1' to E_2''') are evaluated in terms of the incident field E_I by solving Maxwell's wave equation for the linear propagation inside the medium, with continuity of the tangential components of the E and H fields at the two semiconductor-air interfaces. Thus

$$E_2 = E_1 \exp(ik_0 D) \quad , \quad E_1'' = E_2''' \exp(ik_0 D) \ . \tag{1}$$

$$E_I + E_1''' = E_1' + E_1'' \quad , \quad E_2 + E_2''' = E_2' \ . \tag{2}$$

$$E_I - E_1''' = nE_1' - nE_1'' \quad , \quad nE_2 - nE_2''' = E_2' \ . \tag{3}$$

With the field definition $E(t) = \{E(\omega) \exp(-i\omega t) + \text{c.c.}\}$ the irradiance in a medium of index n is

$$I = 2nc\varepsilon_0 \, |E(\omega)|^2 \ , \tag{4}$$

and the solutions for the transmitted and average internal irradiances are

$$I_t = \frac{I_0}{1 + F \sin^2 k_0 D} \quad ; \quad I = \left(\frac{1+R}{1-R}\right) I_t \ , \tag{5}$$

with surface reflectivity $R = \{(n-1)/(n+1)\}^2$ and $F = 4R(1-R)^2$. Thus for example in InSb, with an index $n = 4$, the natural surface reflectivity is 36% and the internal irradiance 'on-resonance' is twice the incident level. Figure 4 shows, for InSb, the

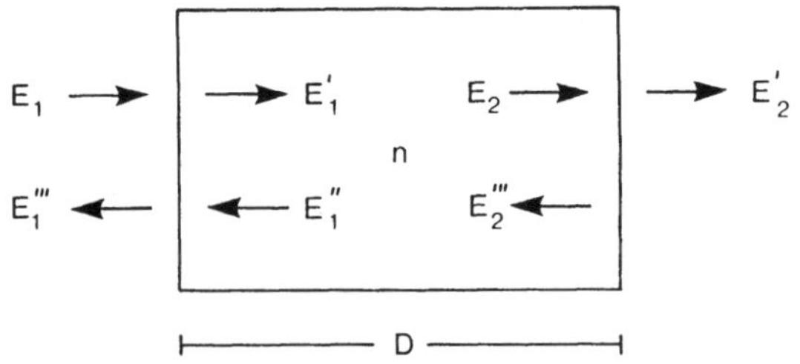

Fig. 3. Boundary fields defined in a Fabry-Perot cavity.

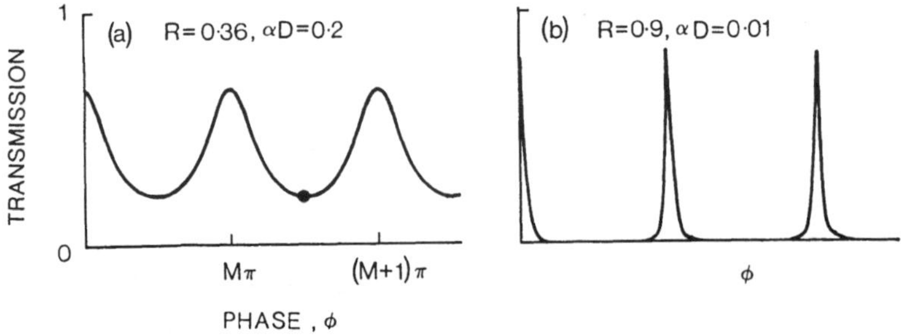

Fig. 4. Fabry-Perot transmission for (a) InSb and (b) an $R = 0.9$ cavity.

classic Fabry-Perot transmission as a function of the single-pass cavity phase, $\phi = k_0 D$. Resonance occurs, as usual in bounded systems, when the cavity length is a half integer number of optical wavelengths, $D = (M/2)\ \lambda_v/n$. Here we make use of the relation $k_0 = 2\pi n/\lambda_v$.

For more complicated cavities, described by front- and back-face reflectivities R_f, R_b and including absorption, one obtains

$$I_t = \frac{(1-R_f)(1-R_b)a}{(1-R_\alpha)^2}\frac{I_0}{1+F_\alpha \sin^2\phi} \quad , \quad I = \frac{(1+R_b a)(1-a)}{(1-R_b)\,\alpha D a} I_t \quad , \tag{6}$$

with $R_\alpha = (R_b R_f)^{1/2} a$, $F_\alpha = 4R_\alpha/(1-R_\alpha)^2$, $a = \exp(-\alpha D)$, and where ϕ includes phase changes at the surfaces (ϕ_f, ϕ_b),

$$2\phi = 2k_0 D + \phi_f + \phi_b = 4\pi\, nD/\lambda_v + \phi_f + \phi_b \quad . \tag{7}$$

A significant point to note is that the internal irradiance may be far higher than the incident for suitable cavity parameters. Thus with $\alpha D = .01, R_b = 1, R_f = a^2$, the internal level is 100 times the incident and any nonlinearities should be considerably enhanced by the cavity feedback.

The cavity finesse, defined as the resonance full width at half height divided by the peak separation, is $\mathcal{F} = \pi F^{1/2}/2$. From Figure 4(a), for example, it is immediately obvious why highly nonlinear responses might be expected from a Fabry-Perot etalon. Suppose that the cavity phase, i.e. the optical path length, depends on the internal irradiance and that at low irradiance one is at an anti-resonance position - marked by the full circle on the figure. If the incident irradiance, and hence the internal irradiance, is increased then the phase will alter and the cavity transmission will increase. But the internal irradiance is directly proportional to that transmitted (equations 5,6). Hence the internal irradiance increases further. This process can avalanche so that from an initial low incident irradiance and low transmission one can in principle switch to a very much higher transmission (close to the resonance peak) for only a small additional signal. The high finesse response, Figure 4(b), should give similar avalanching for very small phase changes, given an initial phase detuning only slightly away from a peak.

The varying phase required here is achieved if either the physical thickness of the cavity, D, or the refractive index n, is altered by the internal irradiance. In practice both occur, however in almost all practical cases the index change dominates. We shall be concerned primarily with two physical origins for such a change: (i) the index change associated with carriers generated by the radiation in a semiconductor, and (ii) the thermo-optic index change due to absorption-induced temperature changes. Thus in general

$$\frac{\Delta(nD)}{nD} = \frac{1}{n}\frac{\partial n}{\partial N}\Delta N + \frac{1}{n}\frac{\partial n}{\partial T}\Delta T + \frac{1}{D}\frac{\partial D}{\partial T}\Delta T \quad , \tag{8}$$

where $\partial n/\partial N$ is the refractive cross-section (the index change per photogenerated carrier pair σ_n), $\partial n/\partial T$ is the thermo-optic coefficient and $D^{-1}\ \partial D/\partial T$ is the thermal expansion coefficient. The first two terms are addressed in the following section.

3. Nonlinear Refraction

In nonlinear optics it has been conventional to describe nonlinearities by appropriate susceptibilities, through the polarisation in the medium that is established by the incident field(s). Thus at the frequency ω we are interested in a polarisation $P(t) = \{P(\omega) \exp(-i\omega t) + c.c.\}$ where,

$$P(\omega) = \varepsilon_0 [\chi^{(1)} E(\omega) + \chi^{(3)} |E(\omega)|^2 E(\omega) + \ldots] \quad . \tag{9}$$

This contains those components of all powers of $E(t)$ that have $\exp(-i\omega t)$ time-dependence. $\chi^{(n)}$ is known as an n-th order nonlinear susceptibility, it depends on the material and the radiation frequency, but not on the field strength. One can define a generalised dielectric constant, which has been used to compare $\chi^{(3)}$ with a nonlinear refractive index coefficient n_2:

$$\varepsilon(\omega, E) = i + P(\omega)/\varepsilon_0 E(\omega) \quad . \tag{10}$$

By expressing ε in terms of irradiance-dependent refraction and absorption coefficients,

$$\varepsilon^{1/2} = (n_0 + n_2 I + \ldots) + i(\alpha_0 + \alpha_2 I + \ldots)c/2\omega \quad , \tag{11}$$

one obtains relations between the susceptibilities and these coefficients. In particular

$$n_2 = \frac{1}{4 n_0^2 c \varepsilon_0} \mathcal{R}e\, \chi^{(3)} \quad . \tag{12}$$

Irradiance, or intensity, dependent refraction has been recognised since 1963, soon after the first applications of lasers to nonlinear optical phenomena [4]. Early measured nonlinear refractive susceptibilities were of order 10^{-16} esu (H_2O) up to 10^{-12} esu (CS_2), corresponding to n_2 values between 10^{-19} per W cm^{-2} and 10^{-15} per W cm^{-2} [5,6]. Closely associated values measured in semiconductors until the mid 70's were all less than 10^{-9} esu. In the late 70's however it was realised that by operating in frequency regions close to absorption resonances (e.g. In the band-tail of the semiconductor fundamental absorption edge) considerably larger coefficients were achievable, large enough to lead to optical bistability at milliwatt power levels rather than at the kilowatt power levels typically employed in nonlinear optics at that time. The reason for these contrasting measurements lies not in the material but in the choice of frequency. Working in the transparency regime, nonlinear processes involve virtual transitions to intermediate states in which the material excitation can exist only on timescales determined by the uncertainty principle; $\Delta t \approx 1/\Delta\omega$, where $\Delta\omega$ is the mismatch between the transition frequency and the radiation frequency. Typically $\Delta t \approx 10$ fs. For resonant conditions however the excitation remains over the material recombination timescale, which may be as long as microseconds for electronic excitations and effectively milliseconds for thermal excitation. As a result, a given excitation level can be built up over a period of time, using far lower irradiances than in the virtual transition limit. It is the achieved excitation level that determines the magnitude of the change in the material optical properties, e.g. $\Delta n = \sigma_n \Delta N$.

3.1 Electronic Nonlinearities

If the excitation recombination time is a constant, τ, then the excited state population is obtained from the rate equation:

$$\frac{d}{dt}\Delta N = \frac{\alpha_I I}{\hbar\omega} - \frac{\Delta N}{\tau} \quad . \tag{13}$$

In steady-state:

$$\Delta N = \alpha_I I \tau / \hbar\omega \quad , \tag{14}$$

where α_I is the interband absorption coefficient. Hence there is an effective nonlinear refractive index:

$$n_2 = \Delta n / I = \sigma_n \, \alpha_I \, \tau / \hbar\omega \quad . \tag{15}$$

For a two-level transition, time-dependent perturbation theory can be used to give the refractive cross-section:

$$\sigma_n = -\frac{4\pi}{n\hbar} | e r_{10} |^2 \frac{1}{\omega_{10} - \omega} \quad , \tag{16}$$

where ω_{10} is the transition frequency and er_{10} is the transition electric-dipole-moment. In the semiconductor case the dipole moment is conventionally replaced by the Kane momentum matrix element, $P = |r_{cv}| E_g$, and one obtains [7]:

$$\sigma_n \approx -K \frac{e^2 P^2}{n E_g^3} \left(\frac{E_g}{E_g - \hbar\omega} \right) \quad . \tag{17}$$

The constant K is determined by summing over bands and averaging over directions in k-space. The significant factor in σ_n is the inverse cubic dependence on the bandgap E_g. The coefficient P and the linear index n vary little from one material to another by comparison to E_g. As with all nonlinear phenomena, small-gap semiconductors give larger effects [8]. The measurement of n_2 in InSb, close to the 5 μm bandgap, with values as large as -10^{-3} per W cm^{-2} [9] is a consequence of this gap dependence and of the long carrier lifetime (200 ns - 1 μs) of InSb; the corresponding susceptibility value is 1 esu, fifteen orders of magnitude greater than the original H_2O nonlinearity.

In general the carrier lifetime depends on the carrier concentration itself, particularly at high densities, and equation (14) must be modified. Hence the cross-section result is a more general formulation than one using n_2. Alternatively one can express the index change in terms of the change in absorption coefficient, appealing to the Kramers-Krönig (K-K) relation, which expresses causality:

$$\Delta n(\omega) = \frac{c}{\pi} \int_0^\infty \frac{\Delta\alpha(\omega')}{(\omega'^2 - \omega^2)} d\omega' \quad . \tag{18}$$

Physically the internal irradiance, at frequency ω, generates carriers which redistribute in the valence and conduction bands (e.g. forming a Boltzmann distribution in the non-degenerate limit). As a result the linear absorption is altered at all frequencies. The K-K relation describes the index change at ω in terms of this change in the entire spectrum; it allows one to determine Δn purely from experimental absorption measurements. Both equation (18) and more especially equation (16) show that for frequencies below a saturating transition the refractive index decreases. Above the transition frequency an increasing refraction is observed.

3.2 Thermal Nonlinearities

The temperature rise (ΔT) of a sample of density ρ, specific heat C_p, and thermal conductivity κ, obeys the dynamic equation:

$$\frac{d}{dt} \Delta T = \frac{Q}{\rho C_p} + \frac{\kappa}{\rho C_p} \nabla^2 (\Delta T) \quad . \tag{19}$$

For plane-wave irradiation the rate of heat input power unit volume is $Q = \alpha I$. To a first approximation, in the thermal diffusion term one may replace $\nabla^2(\Delta T)$ by $(-\Delta T/A)$, where the effective area A depends on the sample heat sinking geometry. This results in a recombination time, $\tau^T = \rho C_p A/\kappa$. Under steady-state conditions the resulting temperature change is

$$\Delta T = \alpha I\, \tau^T / \rho C_p \quad . \tag{20}$$

The equivalence of equations (19, 20) to equations (13, 14) is immediate, and one has an effective thermal index coefficient:

$$n_2^T = \frac{\partial n}{\partial T}\frac{\alpha \tau^T}{\rho C_p} = \frac{\partial n}{\partial T}\frac{\alpha A}{\kappa} \quad . \tag{21}$$

4. Dynamics and Steady-State Solutions of the Nonlinear Etalon

There are a number of timescales that are relevant to bistability. In practical devices the cavity round-trip and build-up times are usually very short compared to the material response times. As a consequence it is assumed that the internal irradiation level adjusts instantaneously to the relatively slow excitation dynamics (i.e. for a given index change the irradiance is always taken to be that for a cavity of that index). This is known as adiabatic elimination of the field time-dependence. Either of equations (13) or (19) then form ideal starting points to describe the two stable steady-states that are achievable for a nonlinearly dispersive etalon. We will select the thermal case for illustration, substituting the Fabry-Perot expression for the internal irradiance I, using equation (6).

$$\frac{d}{dt}\Delta T = \frac{B\, I_0}{1 + F_\alpha \sin^2(\phi_0 + b\Delta T)} - \frac{\Delta T}{\tau^T} \quad . \tag{22}$$

The coefficient B contains cavity reflectivities and αD [10]. ϕ_0 is the initial detuning from the nearest cavity resonance at the ambient temperature, and b expresses the nonlinear refraction,

$$b = \frac{2\pi D}{\lambda_v}\frac{\partial n}{\partial T} \quad . \tag{23}$$

In the dynamic equation the first term on the right-hand-side can be thought of as the driving force for temperature changes, the second term is the thermal sink. We are concerned with the optical response of the etalon, in particular the transmitted irradiance (equation 6):

$$I_t = \frac{C\, I_0}{1 + F_\alpha \sin^2(\phi_0 + b\Delta T)} \quad . \tag{24}$$

It is important to note that I_t is directly proportional to the thermal driving force. Equations (22, 24) are key to understanding the steady-state, stability and dynamic switching of nonlinear Fabry-Perot etalons. They apply equally to the electronic case - simply substituting ΔN for ΔT - or to other excitation schemes. The equations are also expressable very simply in terms of the cavity phase change $\Delta\phi = b\Delta T$.

Fig. 5(a) demonstrates the method of solution. The two contributions to $d(\Delta T)/dt$ are plotted against ΔT itself. The driving force is the Fabry-Perot Airy function, of peak height equal to BI_0 and of value $BI_0/(1 + F_\alpha \sin^2\phi_0)$ at $\Delta T = 0$. The heat sink is simply a straight line of gradient $1/\tau^T$. At low incident irradiance (curve (1)) there is

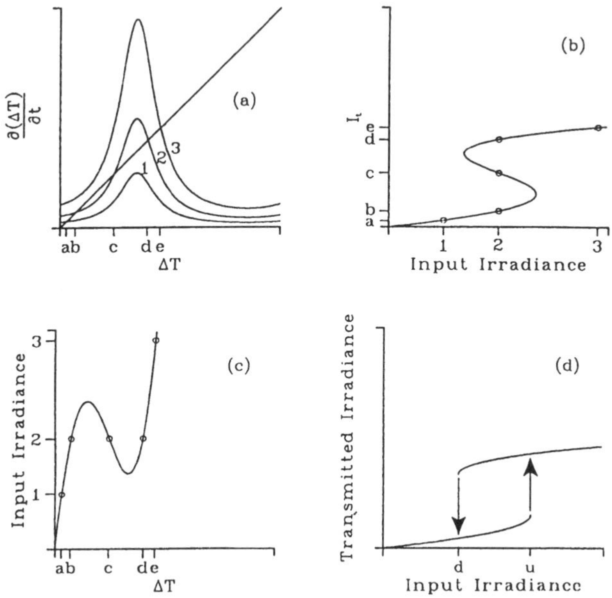

Fig. 5. Illustration of graphical solutions for optical bistablity, from the steady-states of the dynamical equation (22). (a) Thermal driving term in $d(\Delta T)/dt$ for three irradiance levels, and the heat sink; plotted versus the temperature change. (b) Intercepts obtained from (a), giving the transmitted versus incident (scaled) irradiances by comparison of equations (24, 22); showing non-monotonic solutions. (c) Temperature change versus incident irradiance, from the intercepts of (a); showing non-monotonic solutions. (d) Stable steady-states in the transmitted irradiance response, showing switching and hysteresis (bistability).

one intersection of the curve and the line. This is the steady-state solution. The horizontal position of the intersect is the steady-state temperature rise; the vertical position is, through equation (24), simply B/C times the transmitted irradiance.

At high incident irradiance (curve (3)) there is again a single steady-state, at higher ΔT and I_t. The temperature rises in this case is sufficient that the Fabry-Perot resonance has been swept through the operating wavelength ($\{\phi_0 + b\Delta T\} > 0$, whereas initially $\phi_0 < 0$).

Curve (2) shows the intermediate condition. For a range of I_0 values there are not one but three intersections of the Airy curve with the load-line. There are in principle three self-consistent solutions for which the internal irradiance achieved for the given cavity phase is consistent with the phase that such an irradiance produces. One achieves bistability however because one of the steady-state solutions is unstable.

A qualitative test of stability is obtained by assuming one particular steady state condition and perturbing the system slightly. Thus starting at position b, consider a small decrease in ΔT at fixed I_0. The driving force curve is then above the heat sink line and hence $d(\Delta T)/dt$ is positive, so that the temperature is restored towards that at position b. For a small increase in ΔT the curve falls below the line, $d(\Delta T)/dt$ is negative and again the system stabilises onto b. The same argument applies to position d and to the monostable solutions a and e that occur for the difference I_0 values.

For position c however a small decrease in ΔT takes the curve below the line and $d(\Delta T)/dt$ is negative, the temperature will continue to fall until position b is reached. A small increase in ΔT from c leads to heating until position d is reached. Hence those intermediate steady-state solutions, where the gradient of the line is less than that of the driving curve, are **unstable**; all other steady-state solutions are stable.

In Figures 5(b,c) the steady-state transmission and temperature are plotted as functions of I_0, showing the particular solutions discussed in Figure 5(a) and the regions of stability (positive gradient) and instability (negative gradient).

The above technique, based on the dynamic equations; and the graphical method introduced by Marburger and Felber [11] that is based on the steady-state equations alone, are particularly useful for an understanding of the forms of the nonlinear responses. In practice the simplest technique for plotting the precise response is to treat $(b\Delta T)$ as a dummy variable in equations (22, 24). Thus for the steady-states only:

$$I_0 = bB\tau^T\,[1 + F_\alpha \sin^2(\phi_0 + x)] \quad , \tag{25a}$$

$$I_t = \frac{cx}{bB\tau^T} \quad . \tag{25b}$$

It is a trivial computer exercise to step the x variable and plot the resulting values of I_t versus those of I_0.

Figure 5(d) shows what would be observed in practice as I_0 is gradually increased and then reduced. The transmission increases smoothly until at the switch-up irradiance I_u (for which the line is at the tangent to the Airy function) there is a sudden jump in the transmission, followed by a further smooth increase. On reducing I_0 the sudden jump down occurs when the top of the Airy curve is tangential to the line, at a lower, switch-down irradiance I_d. There is thus a hysteresis loop around the region of bi-stability, $I_d < I_0 < I_u$.

Figure 6 shows the range of possible transmission responses achievable for different initial detunings ϕ_0. If ϕ_0 is small one obtains nonlinear but monostable solutions, unless the irradiance is so high that the next order of Fabry-Perot fringe is swept through the operating wavelength. Monostability will occur if the gradient of the thermal driving term does not exceed $1/\tau^T$ at any point of intersection. There is thus a critical condition:- an initial detuning ϕ_c and a specific incident irradiance I_c, for which the intersection occurs at the point of inflexion of the Airy function. Because the gradient of the curve

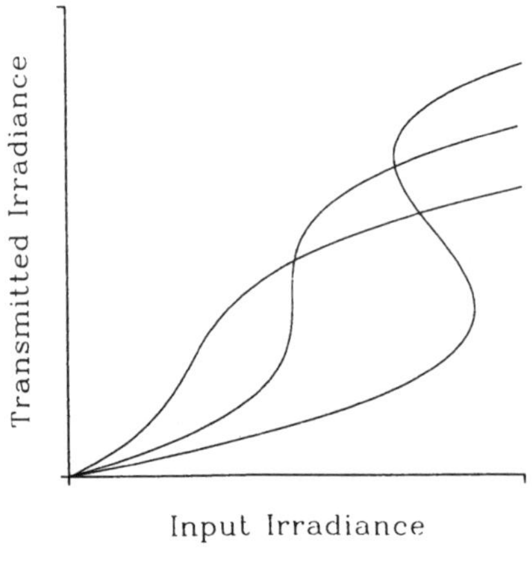

Fig. 6. Family of steady-state solutions to equation (22), obtained for three different initial detuning (ϕ_0) values.

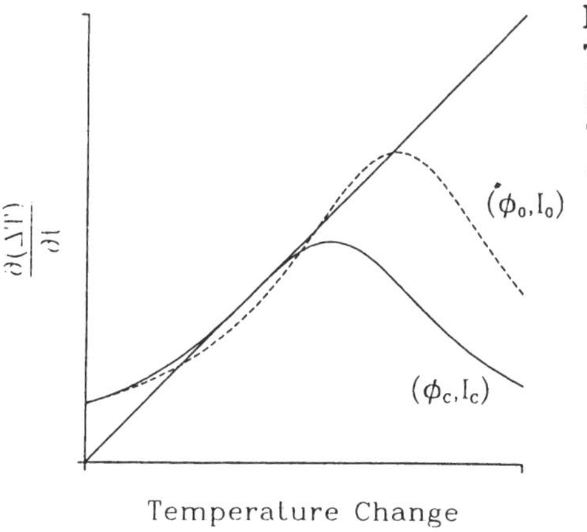

Fig. 7. Critical conditions on I_0 and ϕ_0. The line passes through the point of inflexion of the full curve, at a tangent. The dashed curve indicates bistable conditions; $\phi_0 > \phi_c$, $I_0 > I_c$.

is proportional to I_0, there is the possibility of bistability for $I_0 > I_c$, provided also that ϕ_0 is greater than ϕ_c, as shown in Figure 7. It is possible to obtain analytic solutions for the critical conditions, in the n_2-limit. These have the form [12]

$$\phi_c = f_\phi(F_\alpha) \quad , \tag{26}$$

$$I_c = \frac{\lambda_v \alpha}{2\pi |n_2|} f(R_f, R_b, \alpha D) \quad . \tag{27}$$

Note that the parameters ϕ_0 and I_0 correspond to the control parameters described in the introduction for mechanical bistability and that similar sets of responses are achieved. The unstable intermediary solutions were also described for the mechanical example. The general form of the solutions as functions of ϕ_0 and I_0 form a "cusp catastrophy" with the cusp at the point ϕ_c, I_c.

For practical purposes one is usually interested in obtaining a low value of I_c, this being the incident irradiance that has to be exceeded in order to achieve bistability. Before considering the optimisation of I_c it is, however, worth looking at the cavity internal irradiance value (I) that must be achieved for switching. In general the n_2-limit is achievable only if this internal irradiance is maintained below a certain level - either because of saturation of the fundamental nonlinearity (σ_n) or because the carrier recombination time decreases at higher densities effectively reducing n_2. Equally, higher order thermo-optic terms could prevail in the thermal example. At the critical condition, I is given by the functional form

$$I_{ic} = \frac{\lambda_v \alpha}{2\pi |n_2|} \frac{g(F_\alpha)}{\alpha D} \quad . \tag{28}$$

The critical excitation levels are

$$\Delta N_c = \frac{\alpha \lambda_v}{2\pi |\sigma_n|} \frac{g(F_\alpha)}{\alpha D} \quad ; \quad \Delta T_c = \frac{\alpha \lambda_v'}{2\pi |\partial n/\partial T|} \frac{g(F_\alpha)}{\alpha D} \quad . \tag{29}$$

5. Optimisation

Incident Irradiances. To reduce the switching irradiances for a given material one must first design an optimised cavity. The optimum conditions will depend on whether or not n_2 is itself a function of the cavity length; this results in different conditions for the electronic and thermal cases.

Cavity optimisation relies on achieving sufficient absorption to generate the relevant excitation but not so much that the finesse is destroyed. Figure 8(a) shows the cavity factor f of equation (27), plotted for fixed reflectivities and for varying αD. Pronounced minima exist at the conditions,

$$\alpha D = (2 - R_f - R_b)/4 \quad , \tag{30}$$

and the value of f at these minima decreases with increasing finesse. The implication is that, starting with an electronic nonlinearity with D-independent n_2, one would design a high finesse, small αD cavity. From equations (15,17),

$$I_c = \frac{\hbar c}{\tau} \frac{1}{|\sigma_n|} f \rightarrow \frac{\hbar c 3\sqrt{3}}{\tau |\sigma_n|} \quad \alpha D \quad . \tag{31}$$

The material factor peaks in the vicinity of the semiconductor band edge [13], thereby determining α at the optimum operating wavelength. One therefore knows the required D for chosen reflectivity conditions. The limit to the reduction of I_c is set by: (i) a minimum value of D of approximately $\lambda_v/2$ below which there are no cavity resonances, (ii) fabrication restrictions either for low D or for high reflectivity coatings or (iii) the fact that unless carrier confinement is achieved (for example in quantum well material) the effective carrier lifetime will be reduced by surface recombination for small D [14].

It τ is dominated by surface recombination it becomes effectively proportional to D. This same dependence is achieved for thermal heat sinking longitudinally through the interface into a substrate, as is appropriate in the interference filters described in Section (8). In this case

$$I_c \propto \frac{f}{\alpha D} \quad . \tag{32}$$

The αD-dependence is similar to that of f for given reflectivities (Figure 8) but now the minimum values do not decrease at high finesse, instead $f/\alpha D$ tends to a fixed value of order 2.6, at the optimum conditions [10]:

$$\alpha D = (2 - R_f - R_b) \quad . \tag{33}$$

Internal irradiances and excitation levels. At the minima of I_c (conditions (30, 33)) the function $g/\alpha D$ in equations (28, 29) is essentially constant; in particular for high finesse cavities $g/\alpha D = \sqrt{3}$ under condition (30) and $g/\alpha D = \sqrt{3}/2$ under condition (33). This result manifests the fact that in the electronic case the reduction in switching conditions achievable in high reflectivity cavities is almost entirely due to the enhancement of the internal fields. That is, to achieve the required internal irradiance I_{ic} one needs far lower incident irradiances if the cavity finesse is high. The required refractive index **change** is thus also independent on the cavity - under the optimum condition (30),

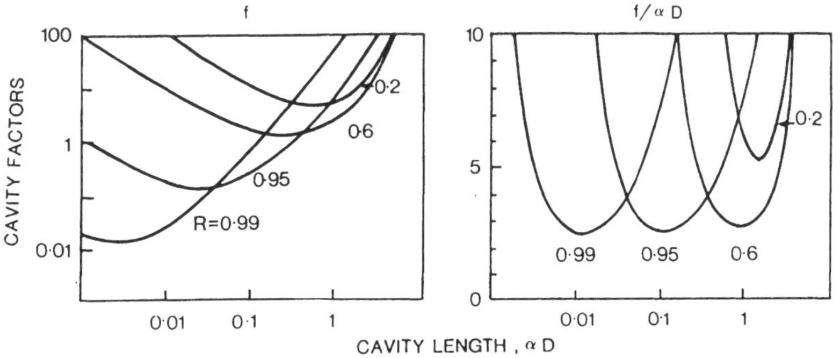

Fig. 8. (a) The cavity factor f that determines the relative switching irradiances for difference cavity reflectivities (R_f=R_b=R), plotted against the cavity length (αD); see equation (27) for optoelectronic nonlinearities. (b) The corresponding factor $f/\alpha D$ that pertains for optothermal nonlinearities [10].

$$\Delta n_c \approx \frac{\sqrt{3}}{2\pi}\lambda_v \alpha \quad . \tag{34}$$

Typically we are interested in α values of $2 \rightarrow 200$ cm^{-1} so that for visible radiation refractive index changes must be in the range $10^{-4} \rightarrow 10^{-2}$. This explains why materials such as InSb with n_2 values as high as 10^{-3} W^{-1} cm^2 are necessary if cw low power bistability is to be observed. For low α values the nonlinearities are usually considerably smaller and require pulsed laser intensities in order to meet condition (34).

Result (34) also demonstrates clearly that if a given excitation (band or excitonic) is to be used to achieve bistability then the total index change associated with the complete saturation of the near-resonance absorption must certainly be comparable to $\alpha\lambda_v$. As α here refers to the total absorption coefficient (that associated with the excitation plus parasitic effects) the importance of removing background absorption is emphasized.

Under condition (33) the effective coefficient $|\, n_2^T \,|$ decreases at high finesse. Thus in the thermal case, in contrast with the above considerations, it is beneficial to use a relatively low finesse cavity if the internal irradiance is to be minimised.

6. Switching Powers and Power-Time Products

6.1 Powers and Energies

In practice one must deal with finite beam sizes (e.g. Gaussian beams) rather than the plane-waves discussed above. At first sight one would not expect bistability for such beams because, whilst the beam centre transmission will increase suddenly when the peak irradiance reaches the switch-up value, this will only effect an infinitesimally small spatial region. As the incident power is increased a gradually larger portion of the spatial beam profile will exceed the switching irradiance and a smoothly increasing power transmission would be expected. Similarly on decreasing the incident power, as each part of the beam reduces below the switch down irradiance value one expects the portion of the beam that is in the upper bistable branch to diminish smoothly. This is not the case, either in practice or if the full theoretical treatment is made. The reason is that transverse spatial effects, diffraction, carrier diffusion or thermal diffusion, couple different parts of the beam cross-section. Hence if the beam centre switches up then transverse coupling leads to higher excitation levels in the vicinity of the centre, and can cause a spread of the switching. This will occur essentially out to irradiance values at the switch-down level. The phenomenon is known as whole-beam-switching [15] and leads to hysteretic responses very similar to those predicted from the plane-wave treatment. In addition the trends (optimisation) given by the plane-wave model are obeyed qualitatively, though not in quantitative detail, for finite beams.

The time taken to switch is discussed in more detail in Section (7). For practical devices the time is controlled by the material time constant, τ or τ^T. The product of the switch-power (P) and switch-time has been called the switching energy ($\mathcal{E}$). Whilst this does not necessarily represent the energy that must be dissipated at each switch (because only a fraction of the power is absorbed), it is a useful measure for the device. Perhaps more important, in view of the demand for high parallelism, is the switching power itself and hence the number of devices that could be operated simultaneously for a given available power.

The corresponding critical values, for Gaussian beams of $1/e^2$ irradiance radii equal to r_0, are

$$P_c=\frac{\pi}{2}r_0^2 I_c \quad , \quad \mathcal{E}_c=\frac{\pi}{2}r_0^2\tau I_c \quad . \tag{35}$$

It is interesting to look at the limits to these parameters for (i) an idealised two-level system, (ii) electronic nonlinearities associated with carrier generation in semiconductors and (iii) thermal nonlinearities.

6.2 The Ideal Two-Level System

The optimum value of f tends to $\alpha D 3\sqrt{3}$ at high finesse, with $D \geq \lambda_v/2n$. For a concentration of N completely isolated two-level systems in a cavity, from equation (31):

$$I_{c(min)} \rightarrow \frac{\hbar c}{\tau}\frac{1}{|\sigma_n|}\frac{\alpha\lambda_v}{n}=\frac{\hbar c 4N}{\tau}\frac{\gamma}{|\omega_{10}-\omega|} \quad , \tag{36}$$

where γ is the absorption linewidth.

The point is that there is no realistic theoretical limit; as one works further from the line centre and for very dilute systems then I_c, P_c and E_c can all decrease toward single photon values. In practice parasitic absorption and the necessity for high reflectivity, very low loss mirrors reduce the device realisation of such a system.

6.3 Semiconductor, Electronic Nonlinearities

The enhancement of the carrier refractive cross-section described by equation (12) is modified in the immediate vicinity of resonance and at finite temperature. Thus for example in the Boltzmann limit [16]

$$I_c \propto \frac{\hbar c}{e^2}\frac{n}{P^2}J^{-1}\left(\frac{E_g-\hbar\omega}{kT}\right)kT\,E_g^2\frac{f}{\tau} \quad . \tag{37}$$

The factor J^1 has a minimum of order unity at $\hbar\omega=E_g$. Hence the minimum power for bistability, which will always occur for small spot-sizes, is of order,

$$P_{c(min)} \approx \lambda_v^2 I_c \approx (\hbar c/E_g)^2 I_c \quad . \tag{38}$$

Because smaller spot sizes are achievable for shorter wavelengths the advantage of large σ_n or n_2 in small gap materials can in principle be lost; the minimum critical power level is essentially independent of material gap but is proportional to τ^{-1}, favouring long recombination times. Even this factor drops out in the characteristic switching energy:

$$\mathcal{E}_{c(min)} \approx \frac{\hbar c}{e^2}\left(\frac{\hbar c}{P}\right)^2 kT\, f(R_f, R_b, \alpha D) \quad . \tag{39}$$

The importance of cavity optimisation is emphasised here again, $\mathcal{E}_c/f$ has a value of order picoJoules at room temperature. As the temperature is reduced the factor kT falls but must eventually be replaced by the dephasing lifetime of the excitation, nevertheless femtoJoule limits are conceivable.

6.4 Thermal Nonlinearities

In the thermal case consider a sample that is "pixellated" so that individual elements are isolated from each other and have transverse dimension of the order of the radiation spot size (r_0). In this configuration the effective thermal sink area A is of order $r_0 D\, \kappa_s/\kappa$, where κ_s is the substrate conductivity. In the diffraction limit therefore

$$P_{c(min)} \approx \frac{\alpha\lambda_v^2\kappa_s}{|\partial n/\partial T|}\frac{1}{\beta}\frac{f}{\alpha D} \quad , \tag{40}$$

and the thermal conductivity time is of form

$$\tau^T \approx \frac{\rho C_p}{\kappa_s}\lambda_v^2\beta' \quad . \tag{41}$$

β and β' are thermal geometry factors of similar magnitudes. There is therefore considerable advantage in this example, in going to short wavelength operation and to small active cavities,

$$\mathcal{E}_{c(min)} \approx \frac{\alpha\lambda_v^4\rho C_p}{|\partial n/\partial T|} \quad . \tag{42}$$

To a first approximation $\partial n/\partial T$ is inversely proportional to the semiconductor bandgap and the interband absorption coefficient is directly proportional. Hence the switching energy scales roughly as E_g^{-2}. $|\partial n/\partial T|$ is of order 10^{-4} K^{-1} for most semiconductors, particularly those with visible bandgaps, and ρC_p is of order 10^{-3} K kg^{-1} J^{-1}. A typical value for $\mathcal{E}_c/\alpha$ is therefore picoJoules, with α in cm^{-1}. It is important to operate with the minimum absorption values for which the cavity conditions on αD can be satisfied. Again in principle subpicoJoule operation is conceivable, although typical observations to date (in unoptimised devices) have been nanoJoule or above.

7. Switching Dynamics

7.1 Switching Times

The graphical method of Section (4) can be used to give a qualitative description of the time-dependence of either the transmission or the internal excitation level of a nonlinear Fabry-Perot, and in particular a description of the phenomenon called critical slowing down.

A typical experimental situation is the so-called hold-and-switch configuration in which a cw (hold) beam is used to bias a device into the bistable region, as shown in Figure 9 by the input level I_i. An additional switching signal is such that it takes the total input irradiance beyond I_u (the switch-up level). After sufficient time the transmission will reach a level such that on removal of the switching signal the transmission will settle in the upper bistable branch. For theoretical analysis it is simpler to consider a step increase in the incident irradiance from I_i to I_f and discuss the rate of excitation change. Again the thermal case will be used for illustration, assuming a constant thermal recovery time. Figure 10(a) shows the dynamics. The initial steady-state is position (i). If the incident level is instantaneously raised to I_f then the system must eventually reach the steady-state point (f), at a new temperature ΔT_f above ambient. The height of the I_f Airy curve above the sink line indicates the rate at which ΔT is changing at any given ΔT value. In the example shown $d(\Delta T)/dt$ is therefore moderately fast initially, then slows down in the region of proximity of the curve and

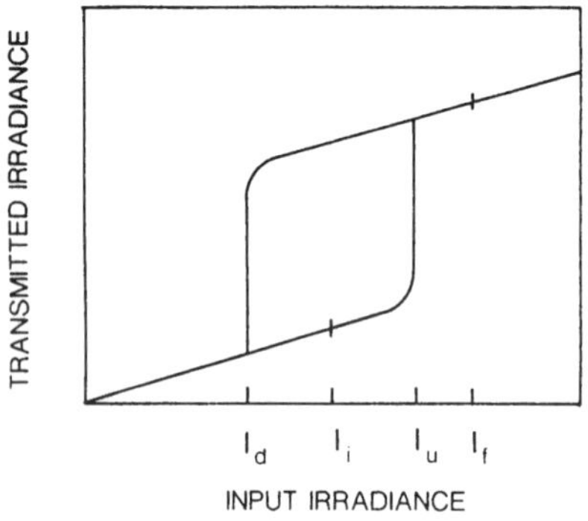

Fig. 9. Schematic of the hold-and-switch conditions employed in bistable devices.

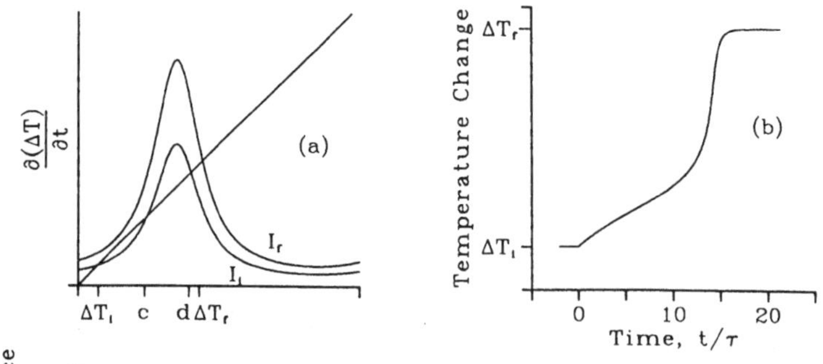

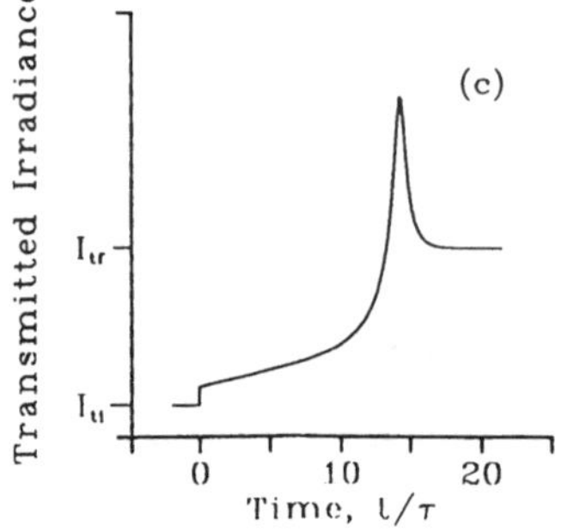

Fig. 10. Analysis of the dynamics of switching. (a) Thermal driving force as a function of temperature for initial I_i and final I_f irradiances, and thermal sink. (b) Resulting time-dependence of the temperature. (c) Corresponding time-dependence of the transmitted irradiance.

the line it speeds up dramatically as ΔT passes through the resonance condition at ΔT_p and finally slows down so that ΔT asymptotically approaches ΔT_f. This behaviour is shown in Figure 10(b). The corresponding transmission time-dependence, taken from the height of the I_f Airy function, is shown in Figure 10(c). The transmission peak occurs at the point of inflexion of Figure 10(b), where $\Delta T = \Delta T_p$.

There is therefore an observable characteristic switching time t_p at which the transmission peaks. It is guaranteed that the device will switch if the signal is removed after this time, and the system will settle to position d on Figure 10(a). In principle the switch-on will occur provided that the temperature change exceeds ΔT_c, so that if the input is reduced back to I_i the thermal driving term will take the system from c to d. Thus in general

$$t_{\text{switch-on}} = \int_{\Delta T_i}^{\Delta T_c} d(\Delta T)\left[\frac{BI_f}{1+F_\alpha \sin^2(\phi_0+b\Delta T)} - \frac{\Delta T}{\tau}\right]^{-1} = t(\Delta T_i\,,\Delta T_c) \quad . \tag{43}$$

A characteristic time for switching from $I_i = 0$, is

$$t_{on} = t(0, \Delta T_p) \quad , \tag{44}$$

where ΔT_p is such that $(\phi_0 + b\Delta T_p) = 0$.

The switching time is dramatically dependent on the final irradiance, I_f, and on the detuning, ϕ_0. Thus for the detuning used in Figure 10, if I_f is such that the line is almost tangential to the curve, as shown in Figure 11(a), then $d(\Delta T)/dt$ is very small, and it takes a long time for the temperature change to move through the region indicated. The switching time diverges to infinity as I_f is reduced to the value of I_u of Figure 9, at which the switch is expected. This effect, known as critical slowing down, is particularly significant experimentally as it means that whilst a small incremental signal can in principle be used to cause switching it will be at a cost of device speed. Further, if one works close to the critical condition the curve and line are almost tangential over a longer ΔT-range and the effect is accentuated. Note however that the initial proximity

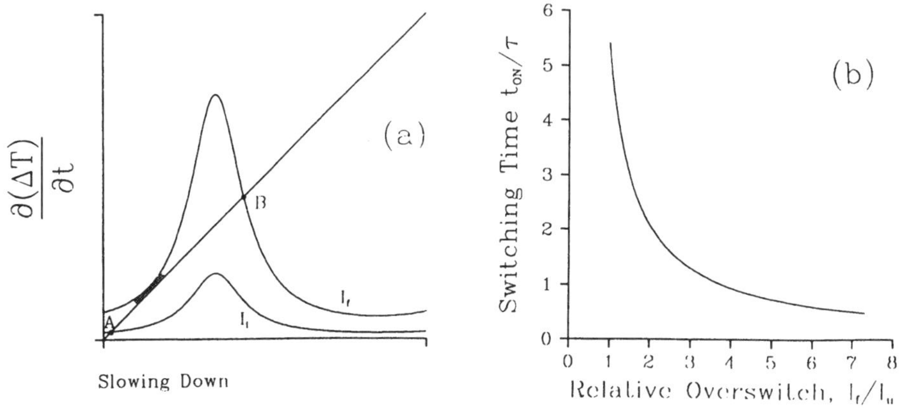

Fig. 11. (a) Illustration of critical slowing down. In the region indicated $d(\Delta T)/dt \approx 0$ for $I = I_f$. The eventual steady-state solution is at B however. (b) Switch-ON time versus over-switch conditions, showing critical slowing down divergence for $(I_f/I_u) \rightarrow 1$.

to the switch-up point for input I_i, does not significantly alter the switch speed. Figure 11(b) shows the switching time as a function of the relative over-switch (I_f/I_u). For heavily overdrive cases, $I_f >> I_u$, the limit to the switch time is set only by the time taken to achieve the desired excitation in the absence of recovery. Thus an extremely brief, high energy pulse may be used to switch the transmission of a cw beam of level I_i, from low to high transmission, on extremely short time scales.

Note that similar effects to the above occur for switch-down from the upper bistable branch if the input level is reduced below I_d.

7.2 Off-Axis Address

Up to now all of the optical beams have been assumed to be collinear on the sample, also of the same wavelength and polarisation, and therefore experiencing the same feedback properties. Nonlinearity and bistability can also be achieved for a fixed irradiance beam experiencing feedback in the presence of a second, signal beam that may have different cavity resonances or indeed experience no feedback at all (e.g. one that is always entirely absorbed). The signal may even have a different physical form - for example e-beam or direct thermal heating of a thermo-optic cavity.

This configuration is known generically as "off-axis address" - the axis referring to the propagation direction of the holding beam. In the extreme case that the additional signal experiences no feedback, in the thermal model

$$\frac{d\Delta T}{dt} = \frac{BI_0}{1+F_\alpha \sin^2(\phi_0 + b\Delta T)} + \sigma_E I_E - \frac{\Delta T}{\tau} \quad , \tag{45}$$

where I_E is the rate at which energy is supplied to the sample by the signal, per unit area, and σ_E is the temperature change per unit fluence for the thermo-optic material in the cavity.

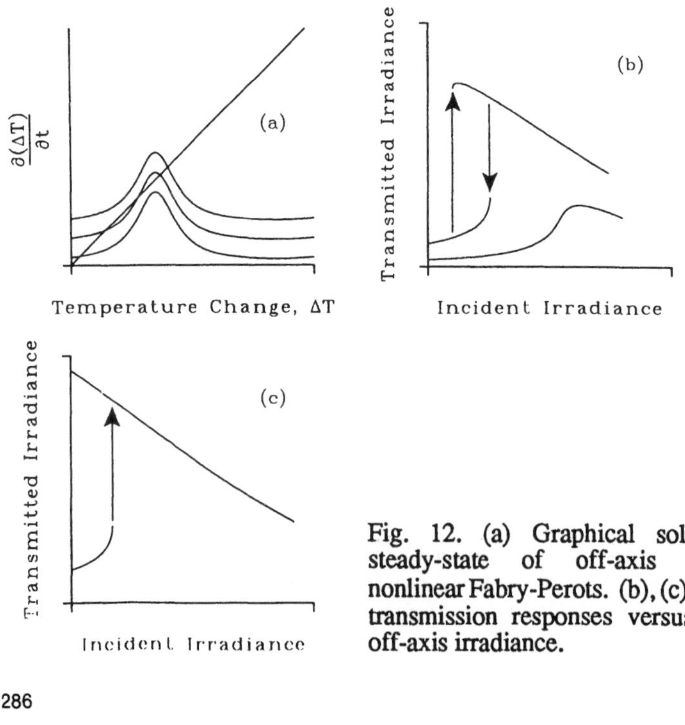

Fig. 12. (a) Graphical solution for steady-state of off-axis addressed nonlinear Fabry-Perots. (b), (c) Resultant transmission responses versus incident off-axis irradiance.

Similar graphical analysis is possible as for the on-axis case, but now I_0 is fixed and the Airy curve is vertically displaced by the additional input, rather than scaling with I_0, Figure 12(a). The transmitted irradiance varies between CI_0 and $CI_0/(1+F_\alpha)$ as the address is varied; example nonlinear responses are shown in Figure 12(b). As the I_0, ϕ_0 combination could be set in a region of bistability it is also possible to achieve the response shown in 12(c).

7.3 Transphasor Operation

For detunings less than ϕ_c, the response is nonlinear but non-hysteretic. In this region the device may be used for signal gain or limiting, as indicated in Figure 13. If the system is biased to level I_0, then a small additional modulated signal can result in a large modulation of the total transmission. The dynamics of this operation obey classic gain-bandwidth relations. An off-axis modulated address suffices to indicate the form of this relation:

$$\frac{d(\Delta T)}{dt}=\frac{BI_0}{1+F_\alpha \sin^2(\phi_0+b\Delta T)}+\sigma_E I_E\,(1+\cos\Omega t)-\frac{\Delta T}{\tau} \quad . \tag{46}$$

If the modulated signal is weak then the solution will be of form

$$\Delta T(t)=\Delta T_0+\sigma_E I_E\,(\beta+g\cos\,\{\Omega t+\theta\}) \quad , \tag{47}$$

where ΔT_0 is the temperature rise in the absence of the modulated signal. A perturbation treatment gives [17]

$$\beta=\left[1+\frac{BI_0F_\alpha \sin^2\{\phi_0\, b\Delta T_0\}}{1+F_\alpha \sin^2(\phi_0+b\Delta T_0)}\right]^{-1} \quad , \tag{48}$$

$$g=\frac{\beta}{(1+\Omega^2\tau^2/\beta^2)^{1/2}} \quad , \tag{49}$$

$$\theta=-\tan^{-1}(\Omega\tau/\beta) \quad . \tag{50}$$

The device gain at frequency Ω, defined as the amplitude of modulation of the transmitted hold beam divided by the amplitude of the modulated signal, I_B, is directly

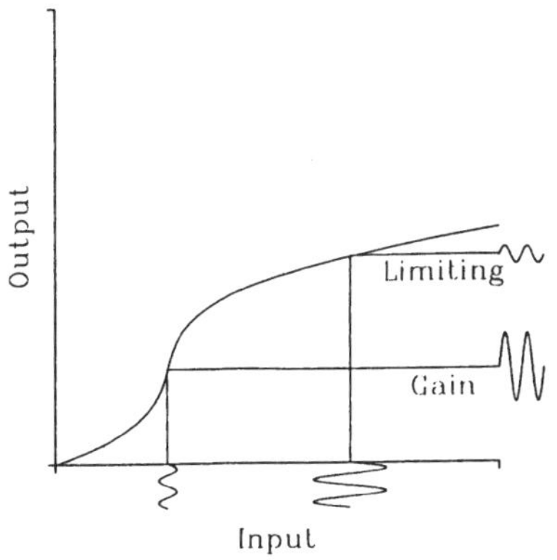

Fig. 13. Non-hysteretic response characteristic with schematic of analogue signal gain and signal limiting.

proportional to g. The point is that the gain depends on Ω only through the denominator in equation (50). In particular this means that the product of the zero-Ω gain and the frequency at which the gain drops to one half of this value - the gain-bandwidth product - is independent of the initial cavity detuning, it is a direct measure of the recombination time. This is another manifestation of the slowing down phenomenon. For ϕ_0 close to the critical value the cw gain is at a maximum but the response rate is at its slowest.

By contrast to the gain experienced for a small modulated signal added to the bias at I_0, a limiting or noise-reduction analogue-mode is achieved for a bias I_1. Again noise or signal modulation in excess of a characteristic frequency will not experience the device nonlinearity. This results in the possibility of using bistable devices for high-bandwidth (rapid modulation) transparent transmission, switched by relatively long pulses.

8. Experimental Results

As a result of the discovery of giant refractive nonlinearities in InSb, experiments were undertaken with plane-parallel sided samples, using the natural material reflectivity (36%) to produce a cavity with the passive response shown in Figure 4(a). The feedback from such a cavity was sufficient to give the nonlinear response of Figure 14, the first observation of all optical bistability in a semiconductor using continuous-wave lasers [18]. Non-hysteretic power nonlinearities are detected at below 100 mW of incident power and two bistable regions exist near 130 and 240 mW. The latter are associated with the movement of two adjacent Fabry-Perot peaks through the operating wavelength. This result was obtained in 1979 at the same time as bistability was reported for pulsed laser operation in GaAs [19].

Under pulsed conditions one measures the transmitted versus incident power as a function of time. If switching occurs there should be a sharp step in the transmission at some power level P_u; when the input power decreases after the peak of the pulse a corresponding step down is expected but at a lower power level P_d (Figure 15). The step may of course be smeared out by time-averaged detection but one is left with different transmissions for the range of power levels between P_u and P_d on the rise and decay sides of the pulses. Plotted in x-y form (P_t versus P_i) bistability is observed. It is necessary however to know that the excitation recovery time of the sample is very short compared to the pulse duration. If this is not the case then x-y hysteresis is always

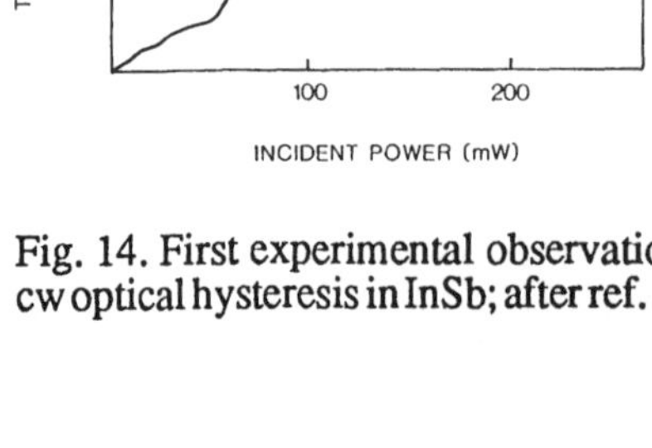

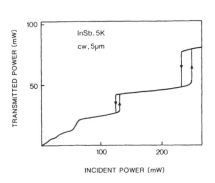

Fig. 14. First experimental observation of cw optical hysteresis in InSb; after ref. [18].

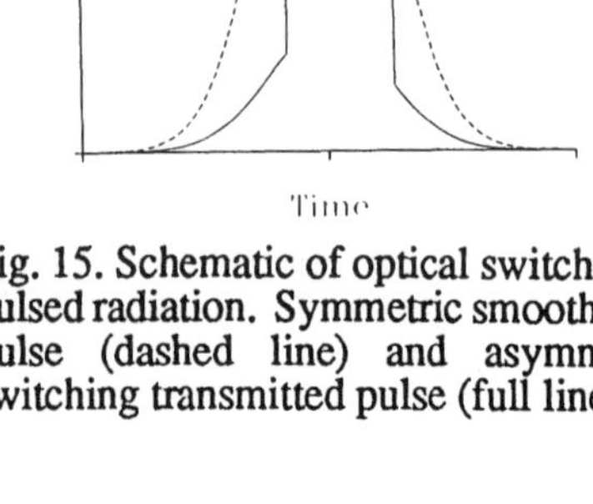

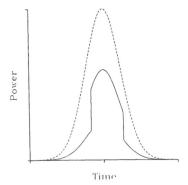

Fig. 15. Schematic of optical switching of pulsed radiation. Symmetric smooth input pulse (dashed line) and asymmetric, switching transmitted pulse (full line).

expected, because the latter part of the pulse experiences the increased excitation level generated by the front part. For true bistability the transmission should be 'stable' at each instantaneous irradiation level.

The "Optical Bistability" conference series [20-22] provides useful references for systems in which bistability has now been observed, also the text "Optical Bistability" by H.M. Gibbs [23], the review article "Optical Bistability and Related Devices" by E. Abraham and S.D. Smith [24] and the project report "From Optical Bistability Towards Optical Computing" [25].

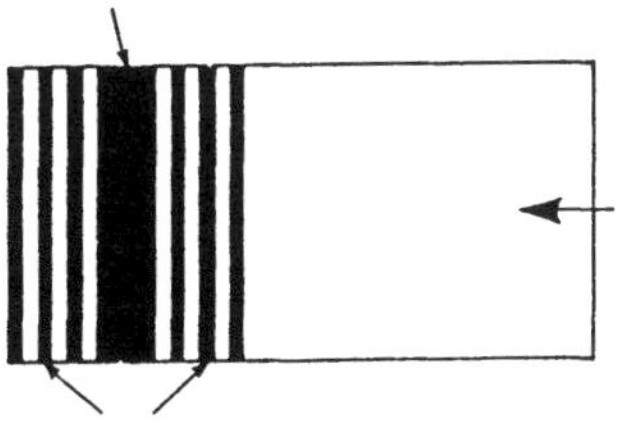

Fig. 16. Structure of a narrow-band pass dielectric multilayer interference filter.

The majority of systems in which hysteresis associated with electronic excitation has been observed have employed pulsed radiation. This is either because the nonlinearity is too weak to allow cw operation, or because of the effect of unwanted laser heating occurring either as the electronic excitations decay or because of parasitic absorption. The thermal systems make use of the otherwise detrimental heating effect. The major sample configuration used for cw opto-thermal bistability is the nonlinear interference filter. Figure 16 shows the structure of a narrow-band-pass filter in which a central spacer region, typically two optical wavelengths thick, is surrounded by high reflectivity stacks made up of alternate low and high index dielectric materials of $\lambda/4$ optical thicknesses. This forms a high-finesse Fabry-Perot cavity. Karpushko and Sinitsyn [26] first reported cw bistability for such a structure, using ZnSe as the spacer material. This work was taken up by a number of other groups, in particular by Smith et al., and forms the basis for prototype circuitry. The 0.5 μm band-gap of ZnSe means that operation is possible in the visible regime (argon-ion lasers have a powerful cw line at 514 nm) and at room temperature (compared to the 5 K or 77 K operation of InSb). In the larger gap ZnSe the electronic nonlinearity is extremely small and the thermal effect dominates, allowing bistability in the 1 - 100 mW regime to date, Figure 17 [27]. The filter system has been used to confirm the optimisation principles set out in Sections (5,6). For example Figure 18(a) shows the spot-size scaling of the switching powers, in agreement with Section 6.4, and Figure 18(b) demonstrates both the reduction of switching times with spot-size and the critical slowing down that occurs for marginal switching [28,29].

Bistable studies have not been restricted to semiconductors, for example a variety of mechanisms lead to nonlinearity in liquids. A recent example of extremely low power bistability ($P_c = 14$ μW for a 25 μm spot size) in the nematic liquid crystal cyanobiphenyl is shown in Figure 19 [30]. This material has a thermo-optic coefficient well above the semiconductor values, rising to around 0.1 K^{-1} at temperatures very close to the nematic-to-isotropic phase transition. Because the material is essentially transparent in the 633 nm wavelength range used for this result the cavity construction was slightly different from the conventional filters, with the rear reflector consisting of an absorbing, metallic mirror. Such a structure removes the wavelength selectivity

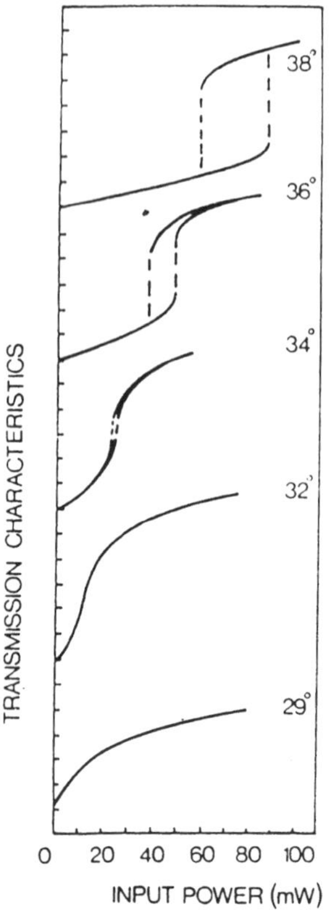

Fig. 17. Nonlinear and bistable transmission response curves achievable using ZnSe interference filters; after [27].

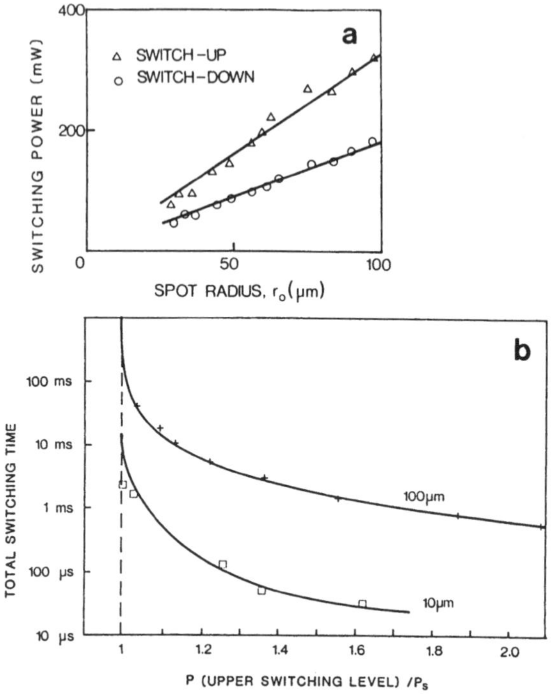

Fig. 18. (a) Linear scaling of switch-powers with spot-radius; after [28]. (b) Spot-size and over-switch dependence of filter switching times; after [29].

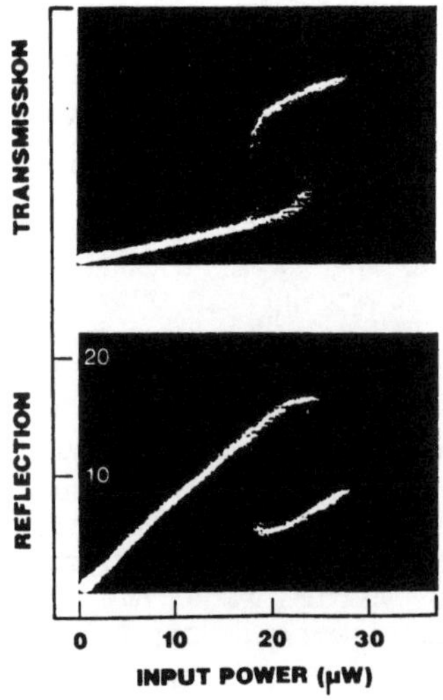

Fig. 19. Very low power cw optothermal Fig. 20. Dynamics of switching times for a bistability in a Fabry-Perot cell containing liquid crystal cell; after [30]. cyanobiphenyl; after [30].

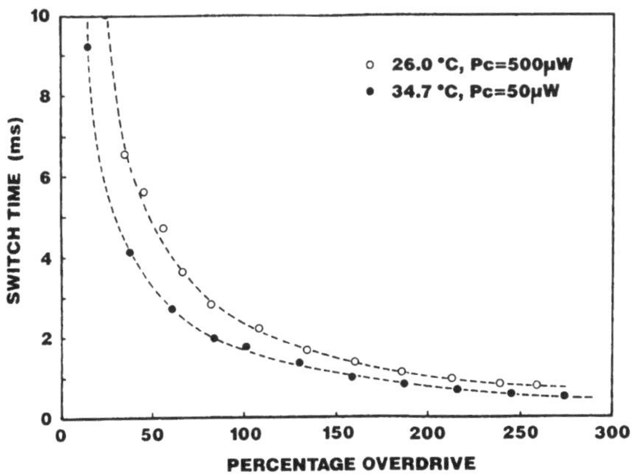

Fig. 20. Dynamics of switching times for a liquid crystal cell; after [30].

of the material absorption and can therefore be used over a range of wavelengths determined only by the cavity Fabry-Perot peak positions and by the metal properties [31]. Figure 20 demonstrates, for such cavities, the dynamics of switching, as described in Section 7.1.

An example of fast, low energy pulsed operation is the low temperature CdS work of Dagenais [32], Figure 21. Here the appropriate material excitation is a discrete, bound-exciton peak. Similar excitonic nonlinearities have been studied in detail for GaAs and for the multiple quantum well GaAs/AlGaAs structures [33,34]. Whilst the value of σ_n is high for such systems, because of the screening effect of generated carriers, the requirement on the total index change (equation (34)) means that the full excitonic oscillator strength is barely sufficient to produce bistability without the additional contribution from interband carrier blocking, unless background absorption can be reduced.

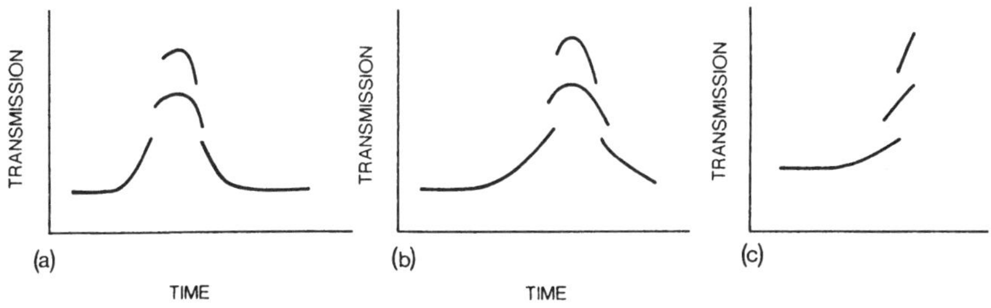

Fig. 21. Nonlinear transmission in CdS with optical pulses of (a) 100 ns, (b) 500 ns and (c) corresponding hysteresis in input/output characteristics; after [32].

9. Alternative Bistability Mechanisms

We have concentrated in this article on refractive bistability in Fabry-Perot cavities with reflective feedback. For completeness brief comments on a number of the alternative schemes that have been studied are included here; these involve either refractive or absorptive nonlinearities with feedback provided either by cavity reflection, or intrinsically in the nonlinear material itself, or by an external non-optical mechanism.

9.1 Bistability by Increasing Absorption

Consider a semiconductor fundamental absorption edge. For most semiconductors the bandgap energy decreases as the sample temperature is raised. Hence starting at a moderate absorption level at low incident power in a poorly heat sunk sample, as the power is gradually increased the sample temperature will increase and hence so will the absorption coefficient. There is therefore the possibility of a thermal runaway, with avalanching from low absorption and low temperature to a high temperature and high absorption. Under appropriate initial conditions there is a hysteresis associated with the effect and bistability of the form shown in Figure 22 occurs [35]. The control parameters in this example are the incident power level and the laser frequency (and hence the initial absorption coefficient). This is just one example of how a coupling between the intrinsic excitation level and the optical power levels can lead to bistability in the absence of cavity feedback [36,37].

9.2 Laser Diode Amplifier Bistability

Given that an absorbing nonlinear Fabry-Perot results in hysteresis it is not surprising that a cavity with material gain is a more effective device in terms of optical power alone. Consider a laser diode held just below the lasing threshold for some electric current input. An incident radiation beam near the notional oscillation frequency experiences amplification inside the cavity, but insufficient amplification to give net round-trip gain. The amplification is caused by the stimulated emission associated with conduction-to-valence transitions of the injected carriers. Hence as the incident power level is increased the population inversion alters very slightly. In turn there is an associated refractive index change and for the current initial condition the cavity can be pulled onto resonance, which is essentially a very high finesse Fabry-Perot resonance. The bias power (essentially of order 1 mW) is now provided by the electric current, but the optical power for such a device can easily be reduced to the microwatt level [38,39].

9.3 Laser Diode Bistability

In contrast to the previous example one now considers a diode laser biased above the threshold current and therefore lasing. The first proposals for bistable systems involved the use of diode lasers with intra-cavity saturable absorbers [1.40]. In the bistable regime the two self-consistent states are (i) low light level in the cavity and hence a consistent absorption and (ii) a high light level and saturation of the absorption. Switching from one level to the other is achieved by temporary input signals either onto the absorber alone (to temporarily saturate the absorption until the laser level has built up sufficiently to sustain the saturation) or onto the emitter region (such as to quench the emission, reduce the cavity irradiance and allow the absorption saturation to recover). A number of multi-cavity and cleaved cavity structures have recently been fabricated to enable such bistability. For example optical switching in 5 ns using 10 μW pulses has been achieved in a two contact, 1.3 μm structure [41]. Refractively induced laser bistability is also possible, for example by use of an external grating for feedback [42].

9.4 Modal Switching

In parallel stripe waveguides radiation coupling exists if the guides are physically close. The energy of a beam injected into one guide will oscillate between the two as it propagates, being entirely in the second guide after a specific distance. The strength of the coupling depends critically on the guide refractive indices. Hence a beam that has crossed over the guide at low light levels, will in principle switch and be propagated entirely within one guide if the level is increased and if the guide material is nonlinear [43].

A similar effect occurs for twin-stripe lasers, where the spatial mode of the system depends strongly on the details of the index profile in the stripe region. The competition between the gain guiding within each stripe and the index guiding has been used to demonstrate laser modal switching in GaAs/AlGaAs structures for, typically, milliwatt power levels [44].

9.5 Polarisation Bistability

Fabry-Perot cavities containing birefringent spacer material exhibit different resonant wavelengths for the ordinary and extraordinary rays; indeed TE and TM beams at non-normal incident on isotropic, dielectric-coated cavities also show birefringence due to the different phase changes on reflection for the two beams. By arranging for one beam to be initially slightly off-resonance, the other on-resonance, it is in principle possible to switch between polarisation, by selective transmission or reflection [45]. Similar effects have been used to achieve polarisation-switching in double-heterostructure lasers [46].

9.6 Switching at Nonlinear Interfaces

Optical waveguiding and input coupling into waveguides is sensitive to small refractive index differences between the guiding and cladding materials or between for example input prism coupler indices and the guide indices. It is therefore expected that optical nonlinearities in any of the relevant media will lead to highly nonlinear responses. A simple case is frustrated internal reflection: here a beam inside one medium is incident on the second at an angle just above the critical total internal reflection angle and so is almost totally reflected. If the second medium has a positive nonlinear refractive index then as the beam power is increased the evanescent wave causes the index to increase and the critical angle increases. Hence the beam becomes refracted rather than reflected - the irradiance in the second medium becomes even higher and one has a self-consistent condition with no internal reflection [47].

9.7 Hybrid Bistable Schemes and SEED Devices

Early schemes used feedback via electronics to demonstrate the principles of optical bistability. Figure 23 is a schematic of a dispersive bistable cavity with an electro-optic spacer material. Initially it is off-resonance and the transmission low. Part of the transmitted power reaches a photodetector and the electronic signal is used to produce a voltage, proportional to the transmitted power, across the electro-optic material. The refractive index is thereby altered, the cavity is brought towards resonance and switching, with bistability, can be achieved in precisely the manner of a nonlinearly-refractive cavity [48].

A similar voltage feedback can be used to produce bistability by increasing absorption, provided that the absorption coefficient is a function of electric field. Excitonic absorption in multiple-quantum-well structures has attracted particular attention in this respect. Figure 24(a) shows the effect of applied field on the excitonic level and 24(b) the resulting bistability - of similar form to Figure 22 [49]. because the thickness of quantum well material can be extremely small field strengths of 10^4 - 10^5 V cm^{-1} are obtained for only a few volts; this is achievable directly from the electrical response of the structure to the radiation field. As a result it is possible to

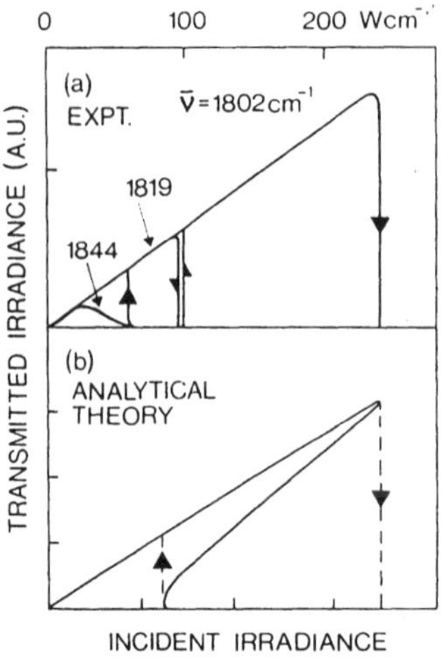

Fig. 22. Experimental and theoretical plots of bistability by increasing absorption; after [35].

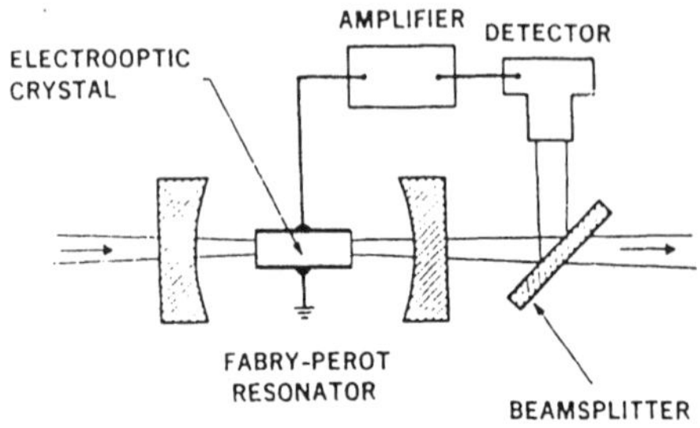

Fig. 23. Schematic of bistability induced by hybrid feedback; after [48].

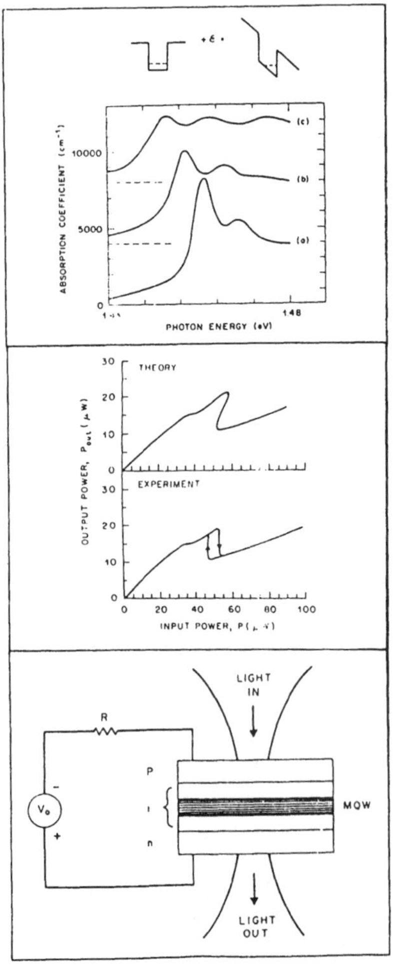

Fig. 24. (a) Effect of applied electric field in the excitonic absorption of GaAs/AlGaAs MQW. (b) Bistability in GaAs/AlGaAs MQW. (c) Schematic of SEED device; after [49].

fabricate a self-electroabsorptive device with no electrical amplification, Figure 24(c) [50]. The constant voltage supply produces a field across the MQW device. As the light level increases the device produces a photocurrent so that the voltage drop across the MQW is reduced. The initial conditions are set so that the absorption increases as a consequence (position A on figure 24(a)). Hence the photocurrent increases and one has the avalanching that forms the basis for switching and bistability. Named the SEED (self electro-optic effect device), switching energies of nanoJoules have been observed. A number of closely similar devices based on the self electro-optic effect have been designed. Most recently a symmetric-SEED structure in which two devices are coupled together has been investigated. Bistability occurs in the ratio of the two input levels and hence can be driven to very low powers. Switching times down to 40 ns (8 mW) and switching powers as low as 10 nW have been observed [51].

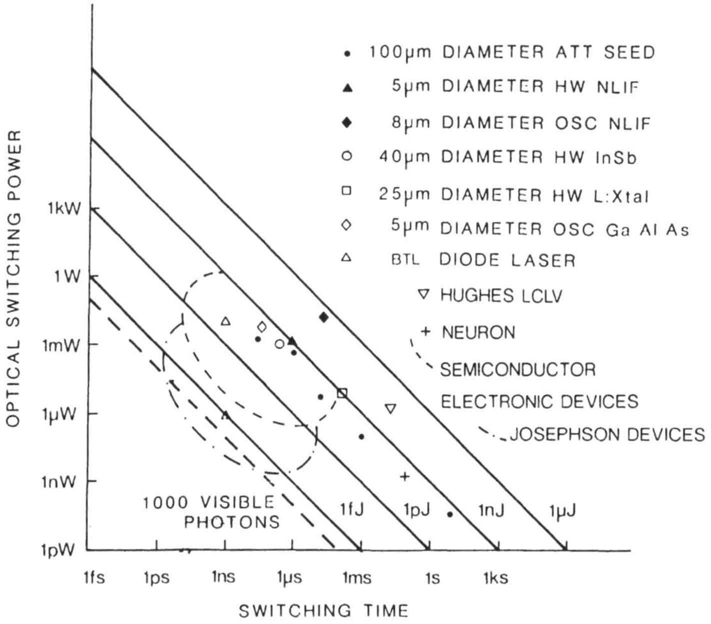

Fig. 25. Power-time results for various optically bistable devices. The operating regions for conventional and Josephson semiconductor devices and neurons are indicated for comparison.

10. Summary

A vast number of materials and techniques have now been used to achieve optical bistability. No attempt has been made to give a comprehensive list in this article. A selection for the power-time results is shown in Figure 25, which follows the style originally presented by P.W. Smith [52]. For comparison the operating regions of conventional and Josephson junction electronic switches are indicated, and of neurons. Whilst the achievement of low power-time products is clearly important it should be noted that low power and, it follows, relatively slow devices have an important role to play in highly parallel switching schemes such as are mooted for image processing and neural networks. Equally there is still a possibility that ultrafast, and by implication high-power, optical switches will have a role in high bandwidth data analysis where the duration and repetition rate of optical pulses prevents the use of direct electronic processing.

References

[1] W.F. Kosomosky, Proc. Symp. Optical and Electro-Optical Information Processing, Boston, ed. J. Tippett et al., p. 269 (1964).
[2] A. Szöke, V. Daneu, J. Goldhar and N.A. Kumit, Appl. Phys. Lett., **15**, 376 (1969).
[3] H.M. Gibbs, S.L. McCall and T.N.C. Venkatesan, Phys. Rev. Lett., **36**, 1135 (1976).
[4] P.A. Franken and J.F. Ward, Rev. Mod. Phys., **35**, 23 (1963).
[5] G. Mayer and F. Gires, C.r. Hebd. Seanc. Acad. Sci., Paris, **258**, 2039 (1964).
[6] P.D. Maker, R.W. Terhune and C.M. Savage, Phys. Rev. Lett., **12**, 507 (1964).
[7] B.S. Wherrett. A.C. Walker and F.A.P. Tooley, in "Optical Nonlinearities and Instabilities in Semiconductors", ed. H. Huag, Academic Press (1988).
[8] B.S. Wherrett, J. Opt. Soc. Am., **B1**, 67 (1984).
[9] D.A.B. Miller, C.T. Seaton and S.D. Smith, Phys. Rev. Lett., **47**, 197 (1981).
[10] B.S. Wherrett, D. Hutchings and D. Russell, J. Opt. Soc. Am., **B3**, 35 (1986).
[11] J.H. Marburger and F.S. Felber, Phys. Rev. **A17**, 335 (1978).
[12] D.A.B. Miller, IEEE J. Quantum Electron., **QE-17**, 306 (1981).
[13] H.A. Al-Attar, D. Hutchings, D. Russell, A.C. Walker, H.A. MacKenzie and B.S. Wherrett, Opt. Commun., **58**, 433 (1986).
[14] D. Frank and B.S. Wherrett, J. Opt. Soc. Am., **B4**, 25 (1987).
[15] W.J. Firth, I. Galbraith and E.M. Wright, J. Opt. Soc. AM., **B2**, 1005 (1985).
[16] B.S. Wherrett, Phil. Trans. Roy. Soc. Lond., **A313**, 213 (1984).
[17] F.A.P. Tooley, W.J. Firth, A.C. Walker, H.A. MacKenzie, J.J.E. Reid and S.D. Smith, IEEE J. Quantum Electron., **QE-21**, 1356 (1985).
[18] D.A.B. Miller, S.D. Smith and A.M. Johnson, Appl. Phys. Lett., **35**, 658 (1979).
[19] H.M. Gibbs, S.L. McCall, T.N.C. Venkatesan, A.C. Gossard, A. Passner and W. Wiegmann, Appl. Phys. Lett., **35**, 6 (1979).
[20] Ed. C.M. Bowden, M. Ciftan and H.R. Robl, "Optical Bistability", Plenum Press, New York (1981).
[21] Ed. C.M. Bowden, H.M. Gibbs and S.L. McCall, "Optical Bistability II", Plenum Press, New York (1984).
[22] Ed. H.M. Gibbs, P. Mandel, N. Peyghambarian and S.D. Smith, "Optical Bistability III", Springer Proc. Phys., **8**, p. 364 (1986).
[23] H.M. Gibbs, "Optical Bistability", Academic Press, p. 471 (1985).
[24] E. Abraham and S.D. Smith, Rep. Prog. Phys., **4**, 815 (1982).
[25] Ed. P. Mandel, S.D. Smith and B.S. Wherrett, "From Optical Bistability Towards Optical Computing", Elsevier Pub. Co., Amsterdam, p. 362 (1987).
[26] F.V. Karpushko and G.V. Sinitsyn, Zh. Prikl. Spektros., **29**, 820 (1978).
[27] S.D. Smith, J.G.H. Mathew, M.R. Taghizadeh, A.C. Walker, B.S. Wherrett and A. Hendry, Opt. Commun., **51**, 357 (1984).
[28] S.D. Smith, A.C. Walker, B.S. Wherrett, F.A.P. Tooley, J.G.H. Mathew, M.R. Taghizadeh and I. Janossy, Appl. Opt., **25**, 1586 (1986).
[29] J.G.H. Mathew, M.R. Taghizadeh, E. Abraham, I. Janossy and S.D. Smith, Springer Proc. Phys., **8**, 57 (1986).
[30] A.D. Lloyd and B.S. Wherrett, "Optical bistability IV", Ed. Firth, Peyghambarian and Tallet, J. de Physique, **C2**, 6, 141 (1988)
[31] D. Hutchings, A.D. Lloyd, I. Janossy and B.S. Wherrett, Opt. Commun., **61**, 345 (1987).
[32] M. Dagenais and W.F. Sharfin, Appl. Phys. Lett., **46**, 230 (1985).

[33] N. Peyghambarian and H.M. Gibbs, J. Opt. Soc. A., **B2**, 1215 (1985).
[34] H.M. Gibbs, S.S. Tarng, J.L. Jewell, D.A. Weinberger, Appl. Phys. Lett., **41**, 222 (1982).
[35] B.S. Wherrett, F.A.P. Tooley and S.D. Smith, Opt. Commun., **52**, 301 (1984).
[36] J. Hajto and I. Janossy, Phil. Mag., **B47**, 347 (1983).
[37] D.A.B. Miller, J. Opt. Soc. Am., **B1**, 957 (1984).
[38] M.J. Adams, Solid State Electron., **7**, 707 (1964).
[39] W.F. Sharfin and M. Dagenais, Appl. Phys. Lett., **48**, 321 (1986).
[40] G.J. Lasker, Solid State Electron., **7**, 707 (1964).
[41] Y. Odagiri, K. Komatsu and S. Suzuki, CLEO Conf. Digest, THJ3 (1984).
[42] H. Kawaguchi, Opt. & Quantum Electron., **19**, S1 (1987).
[43] S.M. Jensen, IEEE J. Quantum Electron., **QE-18**, 1580 (1982).
[44] I.H. White and J.E. Carroll, Electron. Lett., **19**, 337 (1983).
[45] A.D. Lloyd, Opt. Commun., **64, 302 (1987)**.
[46] Y.C. Chen and J.M. Liu, Appl. Phys. Lett., **46**, 16 (1985).
[47] I.C. Khou and J.Y. Hou, J. Opt. Soc. Am., **72**, 1761 (1982).
[48] P.W. Smith and E.H. Turner, Appl. Phys. Lett., **30**, 280 (1977).
[49] D.S. Chemla, D.A.B. Miller and P.W. Smith, Opt. Eng., **24**, 556 (1985).
[50] D.A.B. Miller, D.S. Chemla, T.C. Damen, A.C. Gossard, W. Wiegmann, T.H. Wood and C.A. Burrus, Appl. Phys. Lett., **45**, 13 (1984).
[51] A.L. Lentine, H.S. Hinton, D.A.B. Miller, J.E. Henry, J.E. Cunningham and L.M.F. Chirovsky, Appl. Phys. Lett., **52**, 1419 (1988).
[52] P.W. Smith, Bell Syst. tech. J., **61**, 197 (1982).

Nonlinear Optical Properties of ZnO at Room Temperature Investigated by Laser-Induced Self-Diffraction

J.N. Ravn

Physics Laboratory III, Technical University of Denmark,
DK-2800 Lyngby, Denmark

ABSTRACT

The nonlinear optical properties of ZnO have been investigated with an excimer pumped ultraviolet dye-laser. The wavelength dependence of the first order diffracted intensity in a self-diffraction experiment has been measured for photon energies below but near the fundamental absorption edge. The measurements show a resonant behaviour of the third order nonlinear susceptibility. Further the temporal behaviour of the diffracted light has been measured.

1. INTRODUCTION

The laser-induced diffraction technique is a powerful tool for investigation of optical and other properties of materials. The technique can be used to determine fundamental parameters such as the third order nonlinear susceptibility, carrier lifetime, and diffusion constants. The method was first used by Woerdman et al.[1,2] to investigate grating origin and electronic processes in Si. Eichler et al. [3] first used the technique to determine heat diffusivities in ruby and glycol. Dean et al. [4] studied transient phase grating in ZnO induced by two-photon absorption. More recently ZnO has been investigated near the fundamental absorption edge at low temperatures by Kalt et al.[5]. In this note we report preliminary investigations of the third order optical susceptibility in ZnO just below the bandgap. Earlier Petersen[6] has investigated the nonlinear susceptibility in CdS.

2. EXPERIMENTAL SETUP

The experimental setup is shown in Fig. 1. A Lambda Physik EMG 50 excimer laser (XeCl, 308 nm) pumps a Lambda Physik FL 3002 E dye laser (without etalon). The dye laser uses BBQ diluted in ethanol and toluene[7]. The laser wavelengths range from 380 nm to 398 nm (1/e power limits), the laser linewidth is 0.03Å, and the pulse width τ_L = 10 ns (FWHM). The laser light is guided by two mirrors through a 2 mm aperture, blocking fluorescence and amplified spontaneous emission (ASE) from the laser system. After the Glan-Thompson prism polarizer (GTP), the linearly polarized light is split into two orthogonally polarized beams by a Rochon prism (RP) placed in the front focal point of the first quartz lens f_1 = 0.22 m. The Nicol prism polarizer (NP) is oriented 45 degrees away from the polarization directions of the laser beams making light interference between the two beams possible in the ZnO-crystal placed at the back focal point of the lens f_2 = 80 mm. Both beamsplitters (BS) are simple glass plates. The first BS splits off light to one (PD_T) of three fast silicon planar PIN photodiodes, 4220 Hewlett-Packard. The photodiode active area is $2\cdot10^{-3}$ mm^2 and the risetime is less than 1 ns. Photodiode PD_T generates a trigger signal to the Hewlett-Packard oscillo-

Springer Series in Wave Phenomena, Vol. 9 **Nonlinear Optics in Solids**
Editor: O. Keller

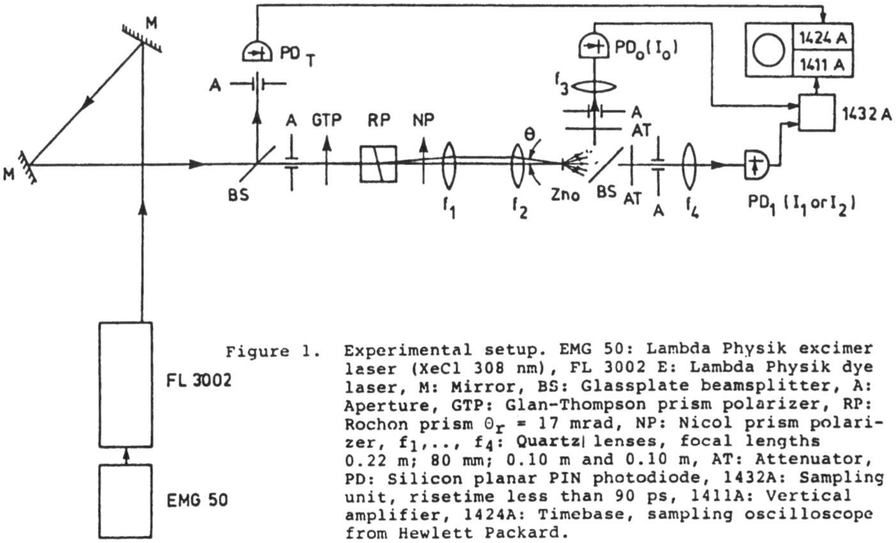

Figure 1. Experimental setup. EMG 50: Lambda Physik excimer laser (XeCl 308 nm), FL 3002 E: Lambda Physik dye laser, M: Mirror, BS: Glassplate beamsplitter, A: Aperture, GTP: Glan-Thompson prism polarizer, RP: Rochon prism Θ_r = 17 mrad, NP: Nicol prism polarizer, $f_1, \ldots, f_4$: Quartz lenses, focal lengths 0.22 m; 80 mm; 0.10 m and 0.10 m, AT: Attenuator, PD: Silicon planar PIN photodiode, 1432A: Sampling unit, risetime less than 90 ps, 1411A: Vertical amplifier, 1424A: Timebase, sampling oscilloscope from Hewlett Packard.

scope made of time base 1424 A, vertical amplifier 1411 A and sampler 1432 A with less than 90 ps response time. Care must be taken in order to trigger on dye pulse and not on fluorescence or ASE. The zero order diffracted beam I_o is measured by PD_o. The first I_1 and second I_2 order diffracted beams are measured by PD_1. The attenuators (AT), 2 mm apertures (A) and lenses f_3 = f_4 = 0.10 m adjust the light power levels for the diodes and remove unwanted light.

The ZnO crystal is grown at our laboratory by the vapour phase reaction method[8] and polished to a thickness of 26 µm in the direction of the c-axis as indicated in Fig. 2. This configuration gives an optical field perpendicular to the crystal c-axis.

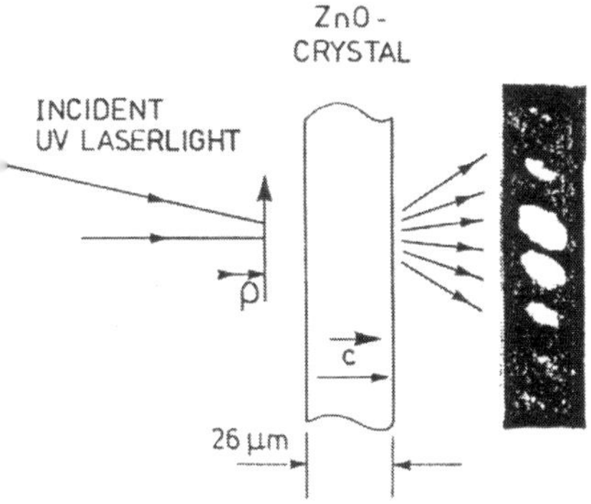

Figure 2. Photograph of diffracted beams showing diffraction up to second order. $\vec{\rho}$: Field polarization, $\vec{c}$: ZnO-crystal c-axis.

3. EXPERIMENTAL RESULTS AND DISCUSSION

In Fig. 2 is shown a photograph of the diffracted beams from the ZnO crystal. At 388 nm we observed diffraction up to three orders at high excitation levels indicating a high third order nonlinearity in ZnO. The grating period is 8.3 µm. The wavelength dependence of the excitation light I, zero, and

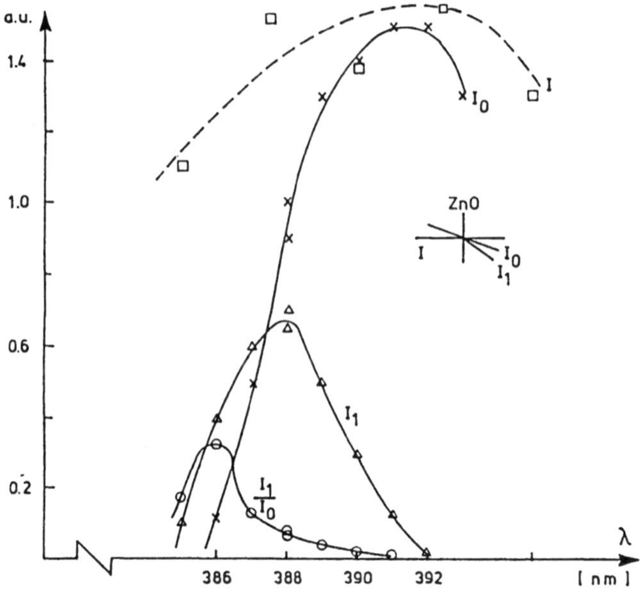

Figure 3. Spectral dependence of diffraction. I: Excitation intensity, I_o: Zero order diffraction intensity and I_1: First order diffraction intensity.

first order diffraction measured at 23 degrees centigrades is shown in Fig. 3. I_o is maximum at 392 nm. At shorter wavelength I_o decreases because of light absorption. This is indicated by smaller slope of I than I_o. The first order diffracted beam shows a maximum at 388 nm. The decrease at shorter wavelengths is again due to absorption whereas the decrease of I_1 at longer wavelengths must be due to a decrease in crystal nonlinearity when the photon energy decreases, i.e. moves away from the bandgap, and the A,B exciton energies. A crude estimate of the nonlinearity can be calculated as I_1 divided by I_o. As shown in Fig. 3 this figures decreases beyond 388 nm underlining the fall off in nonlinearity when the photon energy decreases. The maximum in I_1/I_o at 386 nm is not a maximum in nonlinearity but is due to absorption and therefore dependent on crystal thickness. In Fig. 4 is shown the low power transmission spectrum of the 26 µm thick ZnO-crystal. The transmission increases from 8% at 385 nm to 20% at 391 nm. Thomas[9] has measured the A and B exciton resonance energy at 300 degrees Kelvin to 3.31 eV corresponding to 376 nm photon wavelength. The exciton binding energy at 300 degrees Kelvin in ZnO is 53 meV which shows that the bandgap at 370 nm is 6 nm above the exciton resonance. Measurements closer to the bandgap demand much thinner crystals.

In Fig. 5a and 5b are shown the time dependence of crystal diffractions. It is seen that I_1 is delayed about 1.3 ns compared to I_o whereas I_1 and I_2 are emitted simultaneously from the crystal. This illustrates that light induced diffraction only occurs at high light intensities. I_o generates the grating and when the grating is established diffractions in 1st and 2nd orders occur. The fluctuating tails of I_1 and I_2 in Fig. 5b have no physical origin but reflect trigger problems in the 22 year-old sampling oscilloscope. These two pulses are also broader than they ought to be compared to Fig. 5a.

The diffraction efficiency also depends on the grating decay time constant τ_D due to carrier diffusion[2]

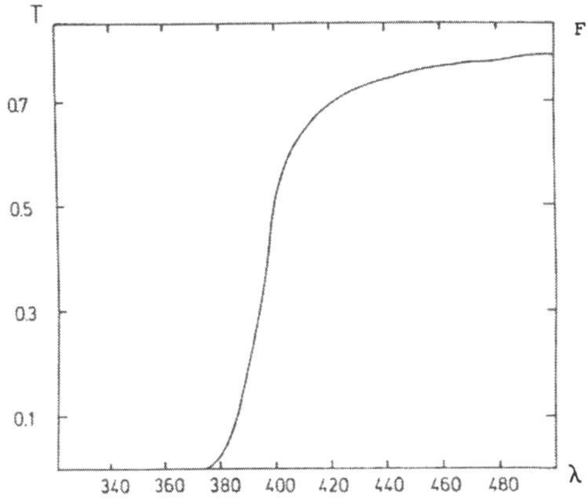

Figure 4: ZnO-crystal transmission. Crystal thickness 26 µm.

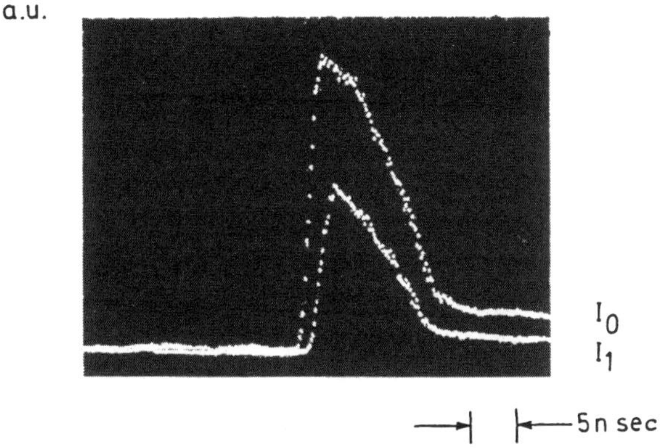

Figure 5a.

Zero and first order diffracted pulses.

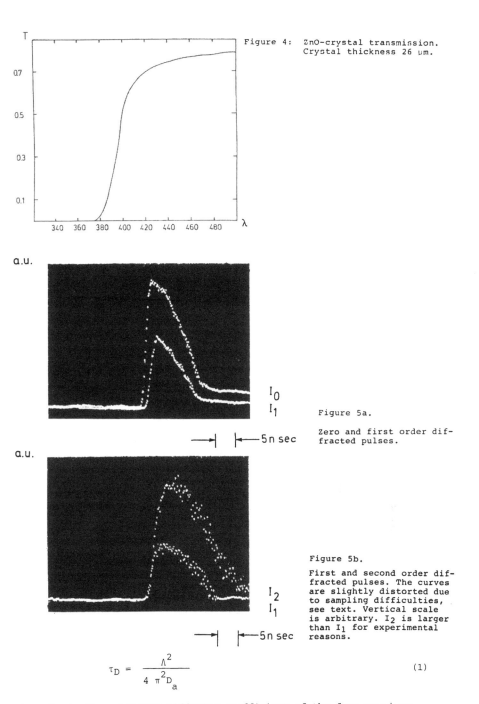

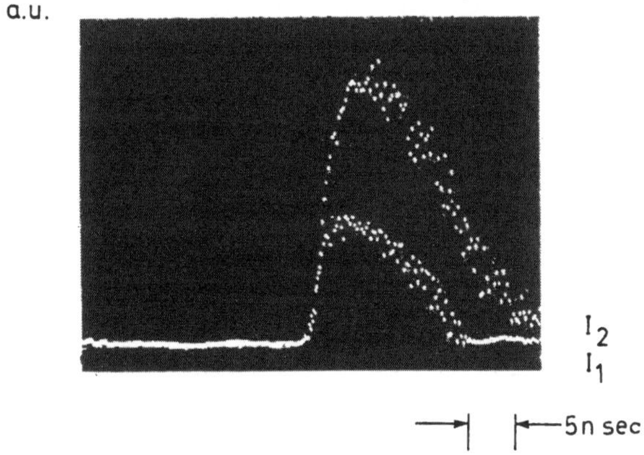

Figure 5b.

First and second order diffracted pulses. The curves are slightly distorted due to sampling difficulties, see text. Vertical scale is arbitrary. I_2 is larger than I_1 for experimental reasons.

$$\tau_D = \frac{\Lambda^2}{4\pi^2 D_a} \tag{1}$$

where D_a is the ambipolar diffusion coefficient of the free carriers

$$D_a = \frac{2D_n D_h}{D_n + D_h} \quad . \tag{2}$$

D_n and D_h are the diffusion coefficients of electrons and holes respectively. Inserting $D_h \cong D_n$ 3.82 cm^2/s, using data from [10], we get τ_D = 4.6 ns which is of the same order as τ_L. Diffusion is therefore decreasing the diffraction efficiency.

4. CONCLUSION

I have investigated the third order nonlinear susceptibility qualitatively by means of self-diffraction at wavelengths from 385 nm to 392 nm. ZnO is highly nonlinear at 388 nm and up to 3rd order diffraction can be observed. The measurements show a resonant behaviour of the nonlinear susceptibility below the bandgap. The nonlinearity is caused either by electronic excitation or thermal heating in the crystal. Investigation of ZnO nearer the A and B exciton resonances at 376 nm at room temperature demands very thin (<< 26 μm) crystals.

This preliminary note will be followed by a more extensive paper when I have finished the investigations.

5. ACKNOWLEDGEMENTS

A lot of people have been involved in this work. I would like to thank T. Skettrup, I. Filinski, Line G. Andersen, T. Guldbrandsen, M.H. Jørgensen, S.V. Skaarup, A. Lauritsen and the staff in the mechanical workshop, E. Søndberg, G. Højland, U. Christensen, I. Nyberg, M. Wisborg, P.V. Petersen, J. Schjær-Jacobsen and the staff in the electronics workshop and K.F. Nielsen. The name order contains no priority. I am very pleased for the help I have received during the work. Thanks a lot.

6. REFERENCES

1. J.P. Woerdman and B. Boelger, Phys. Lett. 30A, 164 (1969).
2. J.P. Woerdman, "Some optical and electrical properties of a laser-generated free-carrier plasma in Si", Phillips Res. Repts. Suppl., No. 7 (1971).
3. H.J. Eichler, G. Salje and H. Stahl, J. Appl. Phys. 44, 5383 (1973).
4. D.R. Dean and R.J. Collins, "Transient phase gratings in ZnO induced by two-photon absorption, "J. Appl. Phys. 44, 5455 (1973).
5. H. Kalt, R. Renner and C. Klingshirn, "Resonant self-diffraction from dynamic laser-induced gratings in II-VI compounds", IEEE J. QE-22, 1312 (1986).
6. P.M. Petersen, "Self-induced changes in the refractive index in CdS at λ = 532 nm", Springer Proceedings in Phys. 18, 357 (1987).
7. F. Bos, "Optimization of spectral coverage in an eight-cell oscillator-amplifier dye laser pumped at 308 nm", Appl. Optics 20, 3553 (1981).
8. K.F. Nielsen, "Growth of ZnO single crystals by the vapor phase reaction method", J. of crystal growth 3, II-7 (1968).
9. D.G. Thomas, "The exciton spectrum of zinc oxide", J. Phys. Chem. Solids 15, 86 (1960).
10. N.I. Meyer and M.H. Jørgensen, "Acoustoelectric effects in piezoelectric semiconductors with main emphasis on CdS and Zn"", Festkörperprobleme X, Advances in Solid State Physics, 29 (1970).

Part VI

Organic Materials

Nonlinear Optical Effects in Organic Materials

P.N. Prasad

Photonics Research Laboratory, Department of Chemistry,
State University of New York at Buffalo, Buffalo, NY 14214, USA

Abstract. The physics of nonlinear optical processes in organic materials relates to the molecular nature of these systems and their unique chemical bonding. In this article, these concepts are discussed at both the microscopic and bulk levels. Experimental methods of studies of optical nonlinearities are described. Ultrafast techniques to measure dynamics of resonant third-order nonlinear optical response are discussed along with contributions from various photoexcitation processes in organic materials. Organic systems also exhibit a large variety of carriers which can be either photoexcited or produced by doping. The role of these carriers in relation to optical nonlinearity is examined. In regard to device processes, nonlinear optical processes in an optical waveguide and optical bistability in a Fabry-Perot etalon are discussed. To conclude, a discussion of current status, future directions of research and exciting research opportunities in this new multidisciplinary field of nonlinear optical effects in organic materials is presented.

1. Introduction

The physics of nonlinear optical processes has received considerable attention during recent years. At a fundamental level the issue is how the electrons in a material respond to an applied intense electromagnetic field. Another issue is the dynamic response of optical nonlinearity of a material to an applied ultrashort laser pulse. Finally, how the nonlinear optical processes are influenced by the presence of a specific excitation such as electron-hole pairs, excitons, etc. also yields challenging opportunities for fundamental research.

The interest in this area is also technology driven. Nonlinear optical processes provide mechanisms for optical switching and frequency conversion which are essential for the development of the newly emerging technology of photonics in which photons instead of electrons will be used to acquire, store, transmit and process information [1,2].

Springer Series in Wave Phenomena, Vol. 9 **Nonlinear Optics in Solids**
Editor: O. Keller

Organic materials have recently emerged as an important class of nonlinear optical materials which offer unique opportunities for both fundamental research as well as technological applications [3]. Organic materials are molecular materials which represent an ensemble of chemically bonded molecular units only weakly interacting with each other in the bulk through Van der Waals interactions. In such a case the nonlinear optical response of organic systems can be described primarily as derived from deformation of electron clouds within each molecule arising due to the presence of the intense electric field of an applied optical pulse. In other words, the optical nonlinearity is primarily molecular in nature. This behavior is very different from inorganic semiconductors or ionic crystals in which no single molecular unit in the bulk can be identified; consequently, the nonlinearity in these cases is a bulk effect [4]. Organic structures also exhibit two different kinds of bondings: σ and π-types. A σ bond is formed by an overlap of atomic orbitals of two chemically bonded atoms, along the internuclear axes. A single bond formed by a carbon atom is always of σ-type. The π-bonds are formed by a lateral overlap of the transverse 2p-orbitals on two chemically bonded atoms. Because each carbon atom can form a maximum of four bonds, involving one 2s and three 2p orbitals, it allows them to form multiple bonds in which one bond is of σ type and the remaining bonds of π-type. For example in a bond -C≡C-, the two carbon atoms are bonded by a σ bond which is formed by two σ-electrons and two π-bonds which involve four π-electrons.

In conjugated organic structures consisting of alternate single and multiple bonds, the π-electrons are delocalized over an effective conjugation length. The conjugated structures exhibit large optical nonlinearity even under non-resonant condition, the nonlinearity being primarily derived from the delocalized π-electrons [3].

Anisotropic molecules, especially in a fluid phase, can also exhibit orientational optical nonlinearity derived from the molecular alignment either in a low frequency electric field (e.g. electric field induced second harmonic generation or EFFISH) or in an optical field. For small molecules, the alignment by optical field can be very fast, in subpicoseconds.

From the technological point of view, organic polymers offer flexibility of fabrication into various device structures. Polymeric materials can be cast into films and fibers. Lithographic techniques and other processing methods can be used to make channeled waveguides for photonic circuits. Various processing techniques are available for casting films and fibers of polymers. Also, polymers can be stretch oriented to make devices which would utilize large anisotropy of nonlinear optical properties.

2. Microscopic Theory of Optical Nonlinearity

As discussed above the nonlinearity of organic systems can be discussed at the molecular level by considering the interaction of a molecule with the radiation field. Under dipolar approximation this interaction distorts the electron cloud of the molecule creating an induced dipole moment which is given by the following expression [3,5]:

$$\mu_{ind} = (\mu-\mu_o) = \alpha \cdot E + \beta : E + \gamma \vdots EEE + . \quad (1)$$

In the above equation μ_{ind} is the induced dipole moment; μ and μ_o are the total and permanent dipole moments. The term α is the linear polarizability which describes the linear response to the applied field. The coefficients β and γ describe, respectively, the microscopic second and third order nonlinear optical response of a molecule and are called first and second hyperpolarizabilities. They are third and fourth rank tensors respectively. The nonlinear optical resonse can also be described in terms of Stark energy shifts of a molecule created by dipolar interaction with the electric field. The dipolar interaction is given as

$$\mu \cdot E = -\mu_o \cdot E - \frac{1}{2}\alpha : EE - \frac{1}{3}\beta \vdots EEE - \frac{1}{4}\gamma \vdots EEEE . \quad (2)$$

According to equations (1) and (2), the microscopic nonlinear optical coefficients β and γ can be theoretically computed from a quantum mechanical calculation of the induced dipole moment or the energy in the presence of an applied electric field by obtaining the various derivatives with respect to the applied field. For example, β is obtained by taking the second derivative of the induced dipole moment or the third-derivative of the Stark energy with respect to the electric field. This is the derivative method [6,7]. The calculation may be performed at the ab-initio level in which all electrons are considered. This approach is practical only for small molecules. In this approach the molecular orbitals are constructed as a linear combination of atomic orbital basis functions [6,7]. The geometry of the molecule is optimized by using the self-consistent field method. Then the electric field interaction term is included to calculate the energy (or induced dipole moment). The derivatives can be obtained by using either the numerical method which is called the finite field method. A more precise method which avoids any problem due to numerical instability is the analytical method of obtaining the various derivatives of energy using the coupled perturbed Hartree-Fock method [8].

In the quantum mechanical calculations of microscopic nonlinearities we find that the choice of basis functions plays an important role [6]. To properly account for optical nonlinearities the inclusion of diffuse and polarization functions is necessary. This way one can properly describe the tail portion of the wavefunctions which relate to the anharmonic behavior of electrons, the latter being responsible for the optical nonlinearities. As one generally uses a static electric field in the hamiltonian, this method of computation yields only the static hyperpolarizabilities β and γ which represent optical nonlinearities in the zero frequency limit. In order to calculate optical nonlinearities at optical frequencies, one has to use a time-dependent Schrödinger equation, making the computation much more complex [9].

For large molecules or polymeric structures, ab-initio calculations are not practical. In these cases approximate quantum chemical methods are used which utilize approximate hamiltonians as opposed to all electron hamiltonians. Various approximate methods have been used. For conjugate organic structures, some approximate methods such as Hückel's theory or Pople-Pariser-Parr (PPP) method consider only the π-electrons [3].

An alternate approach for calculating the microscopic optical nonlinearities utilizes the sum-over-states (SOS) method [10,11]. This method is based on perturbation expansion of the various Stark energy terms. The sum-over-states approach describes the nonlinearities as derived from dipolar mixing between various states. The expressions for α, β and γ are as follows [10]:

$$\alpha(\omega) = \sum_n \left[\frac{\langle g|\mu|n\rangle\langle n|\mu|g\rangle}{h(\omega_{ng}-\omega+i\Gamma_{ng})} + \frac{\langle g|\mu|n\rangle\langle n|\mu|g\rangle}{h(\omega_{ng}+\omega-i\Gamma_{ng})}\right] \quad (3)$$

$$\beta(-\omega_3;\omega_1,\omega_2) = P\sum_{n,m} \frac{\langle g|\mu|n\rangle\langle n|\mu|m\rangle\langle m|\mu|g\rangle}{2h^2(\omega_{ng}-\omega_1+i\Gamma_{ng})(\omega_{ng}-\omega_3-i\Gamma_{mg})} \quad (4)$$

$$\gamma(-\omega_4;\omega_1,\omega_2,\omega_3) = P\sum_{n,m,l}\left(\frac{1}{4h^3}\right)\frac{\langle g|\mu|n\rangle\langle n|\mu|m\rangle\langle m|\mu|l\rangle\langle l|\mu|g\rangle}{(\omega_{ng}-\omega_1+i\Gamma_{ng})(\omega_{mg}-\omega_1-\omega_2+i\Gamma_{mg})(\omega_{lg}-\omega_4+i\Gamma_{lg})} . \quad (5)$$

In the above equations g represents the ground electronic states; n, m and l are various excited states, ω_{ng} are the transition frequencies $(\omega_n-\omega_g)$; Γ_{ng} are the corresponding damping terms. P stands for various permutation terms. In the above description, the representations used for β and γ are as follows. The second order process is represented as a three wave mixing in which two waves at frequencies ω_1 and ω_2 mix to produce an output at a frequency $\omega_3(=\omega_1\pm\omega_2)$. For second harmonic generation $\omega_1=\omega_2=\omega$ and $\omega_3=2\omega$. Similarly, a third order nonlinear optical process involves four wave mixing in which three waves at ω_1, ω_2 and ω_3 mix to produce a fourth wave at frequency ω_4. For third-harmonic generation $\omega_1=\omega_2=\omega_3=\omega$ and $\omega_4=3\omega$. Another manifestation of third-order optical nonlinearity is the intensity dependence of refractive index in which case the appropriate γ is $\gamma(-\omega; \omega,-\omega,\omega)$. The sum-over-states description can also be viewed in the case of non-resonant processes as virtual excitations to levels n, m and l giving rise to nonlinear optical response.

In the sum-over-states method, one computes the various dipole matrix elements $\langle g|\mu|n\rangle$ and transition frequencies ω_{ng} by quantum mechanical methods, generally using a semi-empirical hamiltonian [11]. For exact calculations, the sum should be performed over all excited states n, m and l. However, in practice the sum is limited to only some low lying excited states, generally involving π-electron states (π-π^* transitions). This artificial truncation of the sum is the limitation of this method but may provide insight into resonance enhancement effects. Equation (3) shows that the polarizability will be greatly enhanced at a frequency $\omega=\omega_{ng}$ ie. at one photon resonance. Similarly, equation (4) shows that for second harmonic generation, ie. when $\omega_1=\omega_2=\omega$ and $\omega_3=2\omega$), resonance enhancement occurs when $\omega_{ng}=\omega$ or when $\omega_{mg}=2\omega$. In other words, both the one photon and two-photon resonances are important for second harmonic generation. For third harmonic generation, for which $\omega_1=\omega_2=\omega_3=\omega$ and $\omega_4=3\omega$, equation (5) shows that one photon, two-photon and three photon resonances enhance γ. For intensity dependence of the refractive index described by $\gamma(-\omega; \omega,-\omega,\omega)$, only one and two-photon resonances are important.

To get some useful insight into molecular structural requirements for enhanced optical nonlinearity, a useful model for the second order effect has been the two level model. In this case one treats a molecule as if having only two levels, a ground state, g, and an excited state, i. In such a case, far from resonance β is given as [12]

$$\beta(-2\omega; \omega,\omega) = \left(\frac{3}{2h^2}\right) \frac{\omega_{ig}^2}{(\omega_{ig}^2-\omega^2)(\omega_{ig}^2-4\omega^2)} f\Delta\mu \quad . \tag{6}$$

In the above equation f is the oscillator strength of the transition $g\rightarrow i$ and $\Delta\mu$ is the difference of dipole moment between the excited state and the ground state. Therefore, a molecular structure which possesses a low lying excited state with a large oscillator strength and more ionic character (to give large $\Delta\mu$) will possess large β. A suitable structure for this purpose is given below [3,12]:

In the above structure, a conjugated unit (benzene ring in the above example) separates an electron donor group (D) and an electron acceptor group, A. The lowest lying excited state in such structures involves a charge transfer from group D to group A which gives rise to a large change of dipole moment.

Our theoretical understanding of optical nonlinearity, however, is still very limited. We have found that there is a significant number of molecular compounds which do not conform to the above structures but exhibit large values of β. For third-order optical nonlinearity, our understanding is even more limited. Two level models can not be seriously taken for third-order effect. However, all the existing theoretical descriptions predict a strong dependence of γ on the effective conjugation length (delocalization length for the π-electrons) [3]. Therefore, conjugated polymeric structures such as poly-paraphenylene vinylene (shown below) have emerged as an important class of third-order nonlinear optical materials [3,13].

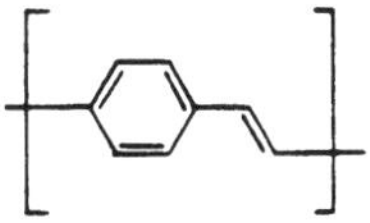

So far we have discussed the nature of electronic nonlinearities of molecules described by simple field expansion of the induced dipole moment (or energy). There are other contributions to optical nonlinearities which under certain conditions may play important roles. Two specific effects discussed here are (i) cascading effect and (ii) molecular reorientation effect. The cascading effect describes a process in which nonlinear polarizations produced in a lower-order nonlinear optical process generates an intermediate macroscopic electric field which can interact with the applied field to contribute to higher order optical nonlinearity. An example will be a molecule with a large β, for which the polarization produced by β generates a field which can interact with the applied field through another β to produce an overall effect similar to the one exhibited by a pure γ term. In other words, a cascaded second-order effect gives rise to a third-order effect. Meredith has discussed this effect in detail [14]. For example, for the third-order process describing intensity dependence of refractive index, the total γ can be written as

$$\gamma(-\omega;\, \omega,-\omega,\omega) = \gamma + \frac{2}{3}\left[\langle \beta^{\lambda}(-\omega;\, 2\omega,-\omega)\; c_1^{\lambda}\; \beta^{\lambda}(-2\omega;\, \omega,\omega)\rangle + 2\langle \beta^{\lambda}(-\omega;\, \omega,o)\; c_2^{\lambda}\; \beta^{\lambda}(-o;\, \omega,-\omega)\rangle\right]. \qquad (7)$$

In the above equation, the terms c describe various intermediate fields which couple the two β terms. For p-nitroaniline, Meredith has shown that 20% of the third harmonic generation value is due to the cascading effect [14]. Perrin and Prasad [15] find that for degenerate four wave mixing studies which use intensity dependence of refractive index, the cascading effect is ~28% for the same molecule.

The molecular reorientation effect, as briefly mentioned above, describes the reorientation of an anisotropic molecule in an applied electric or optical field [16]. In an optical field, this alignment occurs because of difference in the α-tensor components which produce different induced dipole moments. As a result, the molecule reorients to align along the polarization direction of the optical field. This makes the medium anisotropic and contributes only to the third-order optical nonlinearity which describes the intensity dependence of refractive index (and not to third harmonic generation). The orientational γ contribution is given as [17]

$$\gamma_{orientation}(-\omega; \omega,-\omega,\omega) = \frac{1}{135kT}[(\alpha_{11}-\alpha_{22})^2 + (\alpha_{22}-\alpha_{33})^2 + (\alpha_{11}-\alpha_{33})^2] \quad (8)$$

where α_{ii} are the components of the linear polarizability tensor. The above equation describes the orientational contribution in equilibrium situation i.e. for optical fields present long enough to allow the molecules to completely reorient. In short pulse experiments, one may not have the equilibrium situation and the orientational contribution may be less than the equilibrium value given by equation (8).

3. Bulk Optical Nonlinearity

To describe bulk optical nonlinear response one can use the Bloembergen expansion for the bulk polarization P, which is given as follows [5]:

$$P = \chi^{(1)} \cdot E + \chi^{(2)}:EE + \chi^{(3)} \; EEE + . \quad (9)$$

This equation is the bulk analog of equation (1). The terms $\chi^{(n)}$ describe the nth order bulk susceptibilities. They can be related to the microscopic coefficients α, β and γ of equation (1) if one uses the weak intermolecular coupling limit of an oriented gas model [3,10,12]. Under this model, the bulk susceptibilities $\chi^{(n)}$ are derived from the corresponding microscopic coefficients by using simple orientationally-averaged site sums with appropriate local field correction factors which relate the applied field to the local field at a molecular site. Under this approximation [3]

$$\chi^{(2)}(-\omega_3; \omega_1,\omega_2) = F(\omega_1)F(\omega_2)F(\omega_3) \sum_n \langle \beta^n(\theta,\phi)\rangle \quad (10)$$

$$\chi^{(3)}(-\omega_4; \omega_1,\omega_2,\omega_3) = F(\omega_1)F(\omega_2)F(\omega_3)F(\omega_4) \sum_n \langle \gamma^n(\theta,\phi)\rangle \; . \quad (11)$$

In the above equation, β^n and γ^n represent the microscopic coefficients at site n which are averaged over molecular orientations θ and φ and summed over all sites n. The terms $F(\omega_i)$ are the local field corrections for a wave of frequency ω_i. Generally, one utilizes the Lorentz-Lorenz approximation for the local field, in which case [3,5]

$$F(\omega_i) = \frac{n_o^2(\omega_i) + 2}{3} \ . \qquad (12)$$

In this equation, $n_o(\omega_i)$ is the linear refractive index of the medium at frequency ω_i.

From equation (10), it is clear that even for molecular systems with $\beta \neq 0$, the bulk second order nonlinearity determined by the second order nonlinear susceptibility $\chi^{(2)}$ will be absent if the bulk structure is centrosymmetric, in which case $\sum_n \langle \beta^n(\theta,\phi) \rangle = 0$. Therefore, for a molecular system to give rise to second-order effect the conditions are that (i) $\beta \neq 0$ and (ii) the bulk structure is non-centrosymmetric.

Since γ is a fourth rank tensor, its average does not vanish even in a centrosymmetric structure. Therefore, even an isotropic medium such as a glass or liquid or gas will exhibit third-order nonlinear optical response. However, a system may still show large differences in the $\chi^{(3)}$ value if it is oriented. An example is a conjugated polymeric structure which has the largest component of the γ tensor along the polymer chain [3]. If only this component contributes, then the largest value of $\chi^{(3)}$ will correspond to a bulk in which all the polymeric chains are oriented in the same direction. The largest component of $\chi^{(3)}$ in this case will be along the orientation direction. In contrast, the $\chi^{(3)}$ value in a truly amorphous phase of the same polymer will be reduced by a factor of five. These orientation effects have been observed in stretch-oriented polymers [3,13].

4. Experimental Studies of Optical Nonlinearities

For the study of $\chi^{(2)}$ processes, the two experimental techniques widely used are second harmonic generation and electro-optic effects [12]. For the study of third-order optical nonlinearities, the methods used are degenerate four wave mixing, third-harmonic generation, transient absorption, and optical Kerr gate techniques [3].

For second harmonic generation (SHG) studies one can use both nanosecond and picosecond laser sources. In our Photonics Research Laboratory the nanosecond pulsed laser system is based on a Quantel Nd-Yag laser which is a 10 Hz rep. rate high power pulse laser. With an appropriate cavity modification it produces a smooth temporal pulse shape with the pulse duration of 10 ns. Another nanosecond laser system is a commercial excimer laser (Lambda Physik) coupled to a dye laser. Pulse duration for the dye laser is about 30 nanosecond. This system is tunable from uv to near IR using appropriate dyes and a Raman shifter. Such a wide tuning range permits us to compare resonant vs. nonresonant nonlinearities.

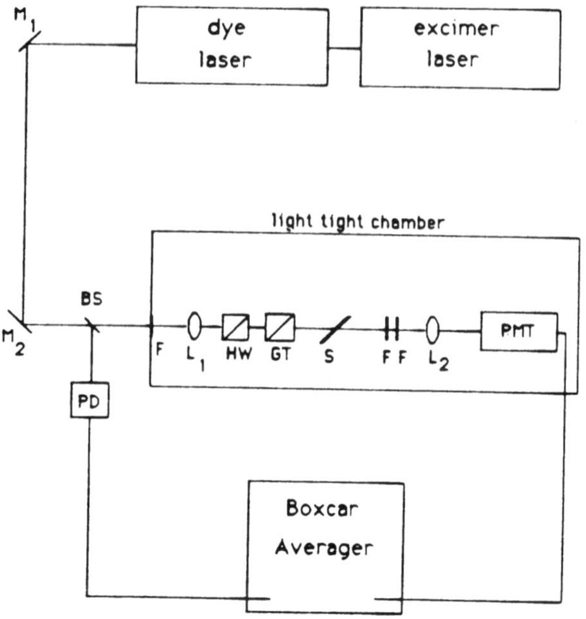

Figure 1. Apparatus used for SHG experiments: BS, beam splitter; L_1, lens; HW, half-wave plate; GT, Glan-Taylor prism; F, color filters; L_2, lens; PD, photodioide used as trigger and PMT, photomultiplier used for detection of SHG.

The picosecond laser source used for the SHG studies utilizes a cw mode-locked and Q-switched Nd-Yag. It is frequency doubled in a CD^*A crystal, and used to pump a cavity-dumped dye laser. It provides stable, relatively high repetition rate (500 Hz) pulses (Q-switch frequency) of 50 picosecond pulse duration, and, most importantly, with tunability within a broad wavelength range. Furthermore, by selecting a pulse of the Nd-Yag fundamental at 1.064 nm one can use it in a well defined TE_{00} mode for SHG experiments. A typical experimental arrangement used for second harmonic generation is shown in Figure 1. Here, polarization of the fundamental light is set by a half-wave retarder and a Glan-Thompson polarizer combination. Samples can be rotated about an axis perpendicular to the incidence plane; the second harmonic signal is separated by a set of filters or a monochromator and detected with a photomultiplier. An appropriate data collection system is able to correlate the movement of the sample with the SHG signal.

For experiments with crystalline powders an integration arrangement (a parabolic mirror) is mounted around the sample, and measurements are conducted relative to known standards like crystalline quartz. The powder

SHG method is used for a qualitative and quick characterization of the $\chi^{(2)}$ behavior of a new material [12].

For electro-optic measurements a highly sensitive method which can be used for ultrathin (mono and multilayer) Langmuir-Blodgett (L-B) films involves surface plasmons. Surface plasmons are electromagnetic waves which propagate along the interface between a metal and a dielectric material. Since the surface plasmons propagate in the frequency and wave vector ranges for which no propagation is allowed in either of the two media, no direct excitation of surface plasmons is possible. The most commonly used method to generate a surface plasmon wave is known as attenuated total reflection [18].

In our laboratory, the Kretschmann configuration of attenuated total reflection has been used to couple surface plasmons. A microscopic slide is coated with a thin film of metal (usually 400-500 Å thick silver film by vacuum deposition). Then the L-B film is coated on the metal surface. The microscopic slide now is coupled to a prism through an index matching fluid. A laser beam (HeNe or argon ion laser beam) is incident at the prism at an angle larger than the critical angle. The total attenuated reflection of the laser beam is monitored. At a certain angle $\theta_{s.p.}$, the electromagnetic wave couples to the interface as a surface plasmon. At this angle the ATR signal drops. The angle is determined by the relationship [18]

$$k_{s.p.} = kn_p \sin \theta_{sp} \ . \tag{13}$$

In this equation $k_{s.p.}$ is the wave vector of the surface plasmon, k is the wave vector of the bulk electromagnetic wave and n_p is the refractive index of the prism. The surface plasmon wave vector $k_{s.p.}$ is given by [18]

$$k_{s.p.} = \frac{\omega}{c} \left(\frac{\varepsilon_m \varepsilon_d}{\varepsilon_m + \varepsilon_d}\right)^{1/2} , \tag{14}$$

where ω is the optical frequency, c, the speed of light, ε_m and ε_d are the dielectric constants of the metal and the dielectric respectively. In the case of a bare silver film, ε_d is the dielectric constant of air and the dip in reflectivity occurs at one angle. In the case of silver coated with a L-B film this angle shifts. A significant shift is observed in the resonance dip caused by the L-B film deposition which clearly demonstrates monolayer sensitivity of this method [19,20]. In this experiment one measures the angle for the reflectivity minimum, the minimum value of reflectivity, and the width of the resonance curves. We use these observables for a computer fit of the resonance curve using a least squares fitting procedure with the Fresnel reflection formula which yields three parameters: real and imaginary part of the refractive index and the film thickness. For electro-optic studies, the experimental arrangement is slightly modified. A mylar spacer is placed between the L-B film and a transparent electrode. A voltage is applied between the silver film and the transparent electrode. If the film has a $\chi^{(2)}$ component along the direction of the field, the refractive index of the L-B film for the TM polarization will change upon the application of the field which will, therefore, cause a shift of the surface plasmon resonance angle. This approach has been used by Cross et al. [21].

For the studies of third-order nonlinearity, our preferred method is degenerate four-wave mixing (DFWM) which provides us with information on intensity dependent refractive index and its time response. The four-wave mixing process can be best described by using the general picture of a laser-induced dynamic grating [22]. Two coherent beams $I_1(\omega_1)$ and $I_2(\omega_2)$ crossing at an angle θ set up an intensity grating in the material which produces a modulation of the refractive index. A third beam $I_3(\omega_3)$ is Bragg diffracted from this grating to generate the signal $I_4(\omega_4)$. For degenerate four-wave mixing $\omega_1=\omega_2=\omega_3=\omega_4=\omega$. This process is described by the grating efficiency $\eta = I_4/I_3$ which is given as [22]

$$\eta^{(\lambda)} \alpha \left(\frac{\partial n}{\partial x}\right)^2 g(\lambda)^2 \, I^{2\beta}(\lambda) f^2(t,\theta) \ . \qquad (15)$$

For a resonant process producing material excitation (excited states, charge carriers etc.), x relates to the excitation density and $g(\lambda)$ is the quantum yield (photogeneration efficiency) at wavelength λ. The term f is a function of the crossing angle θ and time delay t between the pump (I_1 and I_2) and probe (I_3) pulses. The exponent β depends on the nature of the photon process. If the resonant excitation is generated by a one-photon process, $\beta=1$ for $I_1=I_2$. Therefore, determination of β from the dependence of diffraction efficiency on the pulse power I, gives information on the nature of photon processes creating material excitation. The time evolution of the grating provides information on the build up and decay of the excitation. The decay due to spatial migration of excitation will be dependent on the grating spacing and, therefore, on the crossing angle θ. Therefore, an angle dependence study of the grating decay will provide information on the nature of excitation migration (dispersive vs diffusive, etc).

For nonresonant processes, $\frac{\partial n}{\partial x}$ simply relates to the intensity dependent refractive index coefficient n_2. The nonlinear refractive index n_2, derived from the π-electron nonlinearity in conjugated polymers, makes the largest non-resonant electronic contributions and has the fastest response time. In anisotropic liquids, orientational nonlinearity can also make a significant non-resonant contribution, but its response is considerably slow.

In the case of resonant processes, local nonradiative relaxations produce local heating which results in a change of temperature and also launches density waves leading to counterpropagating ultrasonic waves of wave length equal to the grating spacing [22]. Because of these two effects, diffracted signals are also observed where $\frac{\partial n}{\partial x}$ is $\frac{\partial n}{\partial T}$ or $\frac{\partial n}{\partial \rho}$. In order to separate these contributions, one requires ultrashort pulses of several picoseconds or less. In our laboratory, we can select pulses of durations from 50 picoseconds to less than 100 femtoseconds. The 50 picoseconds pulse laser system has already been described above. The shortest pulse laser system which can produce pulses as short as 50 femtoseconds utilizes a state-of-the-art amplified CPM (colliding pulse mode-locked) laser. In this system, the CPM laser is pumped by an argon ion laser.

The pulses from the CPM laser are amplified in a multipass dye amplifier which is pumped with an 8 kHz pulsed copper vapor laser (Metalaser). For this design a crucial aspect is the synchronization of the CPM and the copper vapor laser pulses which was achieved by building our own fast electronic device. The pulse width of the CPM is within 40-80 fs range, typically 50 fs whereas an amplified pulse of ~40 μJ/pulse is somewhat broader, 60-100 fs, and depends on the amplification and compression in a prism pulse compensator. The output wavelength is practically fixed at 620 nm, which is the best condition to run the CPM and the amplifier. The peak power available is very high, of the order of 0.1-1 GW in an unfocused beam.

Another laser system with subpicosecond to picosecond (0.3-4 ps) pulse duration is the most widely used system in our laboratory. The oscillator consists of an actively mode-locked cw Nd-Yag laser coupled with a fiber-grating pulse compressor, the output from which is frequency doubled and stabilizied by an acousto-optical stabilization loop. The stabilized green output is used to synchronously pump a dye laser. Depending on the use of the pulse compressor and the dye laser tuning element, the output might be either within the subpicosecond range, 300-400 fs, or picosecond range, 1.5-4 ps. The oscillator pulses are amplified in a three stage dye amplifier to the pulse energy of 0.4-0.5 mJ. The amplifier is pumped with nanosecond frequency doubled pulses (30 Hz repetition rate) from a Nd-Yag (DCR-2A) laser.

The DFWM experiment is usually performed in our laboratory in the backward wave geometry (phase conjugation arrangement) in which two beams I_1 and I_2 cross at an angle; the third beam I_3 is counterpropagating to I_1 and the signal beam I_4 is generated as a phase conjugate to I_2 in a direction counterpropagating to it. This arrangement is described in detail by Dr. Hall in this book.

For studying the dispersion effect on optical nonlinearity one needs to increase the useful wavelength range. In order to extend tunability of the subpicosecond laser systems we utilize the generation of a continuum. The continuum is generated by focusing the output from the amplifier on a cell filled with water, or on an ethylene glycol jet. For higher power, a selected wavelength of the generated continuum can be passed through a second dye amplifier. Additional wavelength range can be reached by using difference frequency generation in a parametric mixing nonlinear crystal.

For the third harmonic generation studies, one can simply use a nanosecond laser system. To achieve a non-resonant wavelength range, one may have to shift the wavelength into near IR. For this purpose, we either use Raman shifting of the fundamental of the Nd-Yag laser or we generate a tunable IR using difference frequency mixing between the IR and the dye beam or between the two dye beams.

5. Solution Measurements

For soluble materials, solution measurements conveniently yield information on microscopic nonlinearity. In addition, solution measurements yield the sign of optical nonlinear coefficient as well as the information on whether it is real or complex. For second order nonlinear optical effect, the method used is electric field induced second-harmonic generation in which a dc (or low frequency ac) field is applied to align the dipoles and create a noncentrosymmetric arrangement of the nonlinear solute molecule in an otherwise isotropic solvent [12]. For third-order effect, both third-harmonic generation and degenerate four wave mixing has been used [14,23,24].

The solution behavior will be discussed here with an example of third-order optical nonlinearity measured in degenerate four wave mixing. The optical response observed is proportional to the absolute square of $\chi^{(3)}$ of solution which under the assumption of a noninteracting solute-solvent system can be written using equation 11 as

$$\chi^{(3)}_{solution} = F^4[N_{solvent}\langle\gamma\rangle_{solvent} + N_{solute}\langle\gamma\rangle_{solute}] \,. \tag{16}$$

In the above equation $N_{solvent}$ and N_{solute} correspond respectively to the number densities of the solvent and the solute; $\langle\gamma\rangle_{solvent}$ and $\langle\gamma\rangle_{solute}$ are the orientationally averaged hyperpolarizabilities of the solvent and the solute. We now choose a solvent for which γ is positive and real. In such a case, the optical response I (proportional to $|\chi^{(3)}|^2$) is given as

$$I \alpha F^8[|N_{solvent}\langle\gamma\rangle^R_{solvent} + N_{solute}\langle\gamma\rangle^R_{solute}|^2 + |N_{solute}\langle\gamma\rangle^{Im}_{solute}|^2] \,. \tag{17}$$

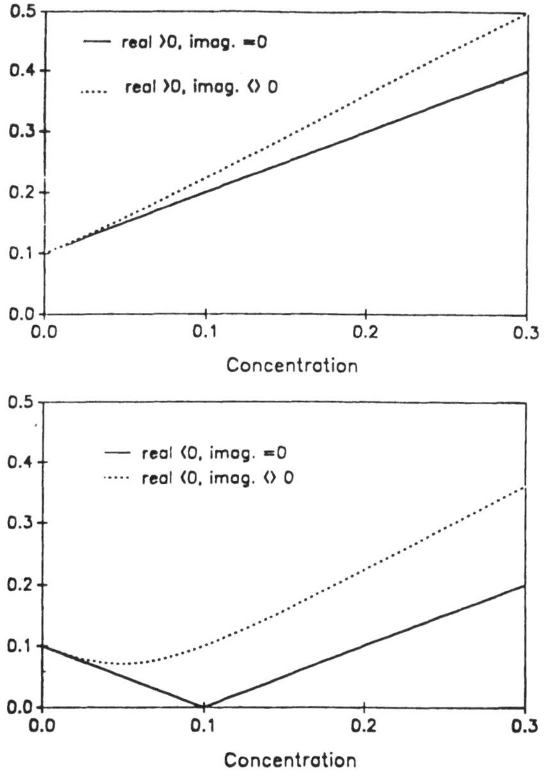

Figure 2. Various concentration dependence behaviors of $\chi^{(3)}$ of a dilute solution for which the solvent has a positive $\chi^{(3)}$.

The concentration dependence of $|\chi^{(3)}|$ in such a case is determined by the sign of the real part, $\langle\gamma\rangle^{R}_{solute}$, of the solute as well as by the contribution of the imaginary part, $\langle\gamma\rangle^{Im}_{solute}$. Figure 2 shows the various concentration dependence behavior of $|\chi^{(3)}|$. Therefore, a concentration dependence study can readily yield the sign of $\langle\gamma\rangle^{R}_{solute}$ as well as information on the relative contribution of $\langle\gamma\rangle^{Im}_{solute}$. We have extensively used this type of study on sequentially built and systematically derivatized structures which we have investigated to build an understanding of the structure-property relationships [24,25].

6. Dynamics of Resonant Third-Order Nonlinear Processes

In inorganic materials such as multiple quantum wells, the resonant third-order nonlinear optical processes provide the most effective nonlinearity [4]. Consequently, a considerable amount of effort has been placed on understanding the dynamics of excitations in inorganic semiconductors. In constrast, studies of dynamics of resonant third-order nonlinear optical processes in organic systems have been very limited [26-28].

The dynamics of resonant third-order nonlinear optical processes can be conveniently investigated by the power dependence and the temporal behavior of the transient grating formed in a four-wave mixing experiment. In the event that the excited state dynamics involves bimolecular decay, saturation of absorption, two-photon absorption or diffusion of excitation, the transient grating is expected to show deviations from a pure sinusoidal form. The higher Fourier components of the grating as discussed below should then produce higher-order diffractions even in a counterpropagating degenerate four-wave mixing (DFWM) arrangement.

We have found that the effective $\chi^{(3)}$ under resonance conditions can be dependent on the intensity of the laser pulse as well as on its pulse width. Also the time-response of resonant $\chi^{(3)}$ is dependent on the intensity in a case where excited state decays by a bimolecular mechanism [29]. The excited state dynamics is, therefore, expected to play an important role. We have found that the study of temporal behavior of higher order diffraction signals in the DFWM study provides very useful information on the excited state dynamics. A brief description of higher order diffraction is presented here. The usual expansion for the polarization for a symmetric system can be recast as

$$P = \chi^{(1)} \cdot E + \chi^{(3)} \vdots \, EEE + \chi^{(5)} \vdots \, EEEEE + \ldots \qquad (18)$$
$$= (\chi^{(1)} + \Delta\chi)\, E = \chi(E) \cdot E \; .$$

In the above equation $\Delta\chi$ is the change in susceptibility due to nonlinear terms. In the case where only $\chi^{(3)}$ terms are important, $\Delta\chi$ is proportional to $\chi^{(3)}E^2$ or $\chi^{(3)}I$ where I is the light intensity. The corresponding equation for refractive index can also be written as

$$n = n_0 + \Delta n = n_0 + n_2 I \qquad (19)$$

where n_0 relates to $\chi^{(1)}$, Δn relates to $\Delta\chi$, and n_2 is the nonlinear refractive index coefficient.

Close to a resonance χ and n are complex. The term $\Delta\chi$ (or Δn) is derived from the changes in both the real and imaginary parts of the susceptibility due to excitation. In the presence of processes (two-photon absorption, saturation of absorption, diffusion of excitons, bimolecular decay) which produce a nonsinusoidal deviation of the grating, the change in susceptibility can be expressed as a Fourier series [29]:

$$\Delta\chi(r,t) = c_0(t) + c_1(t)\cos(\Delta k \cdot r) + c_2(t)\cos(2\Delta k \cdot r) + c_3(t)\cos(3\Delta k \cdot r) + \ldots \; . \qquad (20)$$

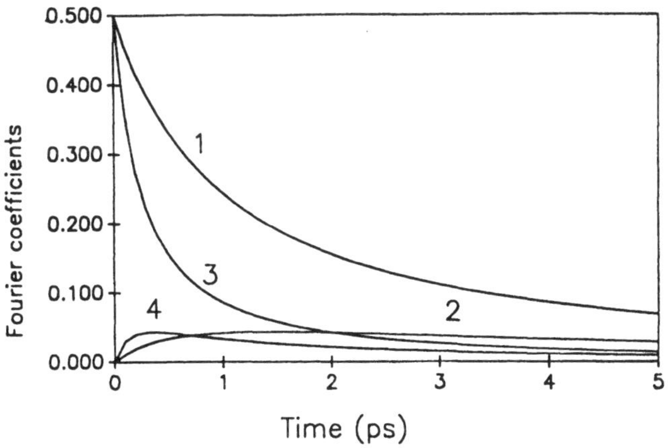

Figure 3. Temporal behavior of the Fourier coefficients in the case of a δ-function excitation and a pure bimolecular decay of the excited states. Curves 1 and 2 show the first and second Fourier coefficients, respectively, at some intensity, while curves 3 and 4 show these coefficients for a pump intensity four times higher.

In the above equation Δk is the grating vector which is the difference between the wave vectors of the pump beams forming the grating. In the above equation, $c_1(t)$ describes the usual sinusoidal grating (first-order diffraction). The case of two-photon absorption is the simplest, provided the excited states generated by a two-photon process decay according to an exponential law. One can easily prove that at any moment $c_1(t)/c_2(t)$ is a constant and thus the temporal behavior of the first and second order diffraction signals should be the same.

The general features of the expected results are demonstrated in Figure 3 for a case in which the exciting and reading pulses are δ functions and the excited state decay is bimolecular. The first Fourier coefficient $c_1(t)$ corresponding to the usual first-order diffraction shows a slightly distorted bimolecular decay behavior. Further coefficients, however, increase from a zero value to a maximum at longer times and show a much slower decay. Similar calculations for other choice of parameters also show that the maximum of $c_2(t)$ (and higher Fourier coefficients) will shift towards shorter times as the decay rate (i.e. the product of the bimolecular rate constant κ_2 and the density of excited states) increases. This characteristic behavior of the second order diffraction is the signature of a bimolecular decay process and can be used to distinguish it from other phenomena also giving rise to second-order diffraction such as saturation of absorption and two-photon absorption as well as from higher order diffractions due to geometric limitations like in the case of a thin grating [22]. We have successfully used this method for the investigation of dynamic nonlinearities of several organic systems which show bimolecular decay [29].

7. Role of Carriers

In organic systems a large variety of carriers (excitations) can be generated. Free carriers (electrons and holes) as well as excitons can be photogenerated in organic solids. Furthermore, these excitons can be of Frenkel type (small radius excitons) or Wannier type (large radius excitons), the latter existing in excited state donor-acceptor complexes. In conjugated polymeric structures which permit bond alternation, the excitation is strongly coupled with structural deformations which produce dressed excitations (by local relaxation) such as solitons and polarons [30]. In polaronic deformations, the polarons may condense at higher concentration densities to form bipolarons. These excitations can either be photoproduced or created by doping.

Electrochemical or chemical doping of conjugated polymers has shown enhancement of conductivity by more than eight orders of magnitude [30]. In the case of polyacetylene, the dopant has been suggested to introduce charged solitons on the polymer backbone, which are thought to be the dominant charge carriers giving rise to enhanced conductivity [30]. In the case of polythiophene, with a nondegenerate ground state, the doping has been suggested to create polarons under low dopant concentration and bipolarons under high dopant concentration [30].

We have conducted some preliminary investigation of the $\chi^{(3)}$ behavior of these polymers in the doped state [26,31]. Our results indicate that, for p-doping, $\chi^{(3)}$ drops drastically when only a small amount of dopant (and consequently positive charge on the polymer backbone) is introduced in the case of polythiophene either electrochemically or by iodine doping. The polaronic and bipolaronic contributions to $\chi^{(3)}$ have been theoretically computed by de Melo and Silbey [32]. However, our experimental results are not explained by the existing theoretical models.

8. Device Processes

In this section, two specific device processes will be discussed. These are: (i) nonlinear processes in optical waveguides, and (ii) optical bistability in a nonlinear Fabry-Perot etalon.

An optical waveguide structure provides the following attractive features: 1) high power densities which are feasible with relatively low total powers because of the light beam confinement by a small waveguide cross-section. 2) diffractionless propagation over considerable distances (cms) leading to long interaction lengths. An additional advantage offered by an optical waveguide structure is that standard integrated optical devices can operate in an all-optical mode waveguide built of a $\chi^{(3)}$ material. Devices operating in an all-optical mode have been analyzed theoretically and experimentally [33-36]. However, most of the experiments have been performed on inorganic systems such as semiconductors, photorefractive materials, multiple quantum well semiconductors and semiconductor doped glasses. Although some of them display very high resonant third-order optical nonlinearities, strong absorption makes them less suitable in designing all-optical integrated optical devices due to high losses and slowly integrating (thermal) nonlinear processes which always accompany a resonant process.

(i) Nonlinear processes in optical waveguides. There have been reports addressing optical nonlinearities in organic waveguide structures, the majority of which were focused on second order processes in poled polymer films [37,38]. We recently reported studies of electronic (Kerr) nonlinearities in organic polymers using many different experimental geometries [39].

For details of the theory of optical waveguide the readers are referred to the article by Dr. Boardman in this book. For guided wave fields, the nonlinear polarization may be expressed as (for TE polarization) [40]:

$$P_y^{NL}(z,\omega) = c\varepsilon_0^2 n^2 n_2 |E_y(z,\omega)|^2 E_y(z,\omega)|a(x)|^2 a(x) \; . \tag{21}$$

In this expression $|a(x)|^2$ describes the guided wave power. The term $E_y(z,\omega)$ is the electric field distribution; n_2, n, ε_0 and c are the nonlinear refractive index, linear refractive index, dielectric constant of air and the speed of light respectively.

We focus on devices in which power dependence of the guided wavevector, β, is important. The change in the guided wavevector ($\Delta\beta_0$) can be derived directly from the intensity dependent refractive index by averaging over spatial distribution of the guided fields. From the coupled mode theory one obtains [41]

$$\beta = \beta_0 + \Delta\beta_0 |a(x)|^2 \tag{22}$$

where

$$\Delta\beta_0 = \frac{\varepsilon_0 c}{2} \int_{-\infty}^{\infty} dydz \; P^{NL}(z,\omega)E(z,\omega)$$

is the guided wavevector derived from the nonlinear polarization.

Two kinds of distributed couplers can be used to transfer optical fields to the nonlinear waveguide: prism and grating. Light incident on the prism base (or grating) causes a guided wave to grow with the distance along the waveguide. Increase in the guided wave power induces change in the effective waveguide index, hence detuning the wavevector matching. Consequently, the coupling efficiency is reduced if the coupling angle is kept the same, since the synchronous coupling angle is changed.

In a nonlinear coupler experiment, the optimum coupling efficiency at low input powers is achieved by adjusting the coupling angle so that the power transmitted through the waveguide and eventually coupled out of the waveguide through another grating (or prism) is maximum. For the high input powers the change in the film index causes a wavevector mismatch, reducing thus coupling efficiency and changing the synchronous angle. One then observes a drop in the transmitted power. This effect results in a limiter action behavior. Reoptimization of the coupling angle leads then to a direct measure of the wavevector mismatch $\Delta\beta_0$. In addition to being important in relation to device configurations, this study also provides information on both the sign and magnitude of nonlinearity.

In conducting waveguide experiments, one has to be extremely careful because even a small residual absorption can give rise to significant thermal effects due to the long propagation length. The use of a long-pulse in the presence of any residual absorption can also show important manifestations due to thermal nonlinearity (refractive index change due to the change of temperature) which is integrating in nature due to its slow relaxation. In a recent work, we have shown that for a polymer (polyamic acid) even in the region of optical transparency, a waveguide nonlinear optical process (change of coupling angle with the change of intensity) observed for 10 nanosecond pulses is mainly thermal in nature [39]. However, the same experiment performed with picosecond and subpicosecond pulses shows manifestation of mainly electronic nonlinearity [39].

(ii) Optical Bistability Study. Optical bistability is described by the response of a device which yields two output states I_{out} for the same input intensity, I_{in}, over some range of input values [42]. The article by Dr. Wherrett in this book discusses optical bistability in detail. Therefore, its discussion here is very brief.

An optically nonlinear medium by itself cannot exhibit a bistable behavior. An external feedback is required to obtain bistability. The nature of this feedback depends on the device structure. One can use both dispersive and absorptive nonlinearities in a Fabry-Perot etalon geometry. A Fabry-Perot etalon or interferometer is an optical resonator formed by two parallel mirrors of high reflectivity which provide the feedback. The Fabry-Perot resonator is filled or partially filled with a nonlinear medium. In the case of absorptive bistability, this medium is a saturable absorber. When the input intensity is low (less than $I_{\uparrow}$), the beam is heavily attenuated by absorption inside the cavity. As a result, the output level is low. As the input intensity increases, at certain level $I_{\uparrow}$, the medium bleaches out due to saturation of absorption. The cavity then switches from the low to the high output state. In the reverse cycle, even when the input level is decreased below $I_{\uparrow}$, the local intensity (field) within the cavity remains high to keep the cavity in the high output state well below $I_{in}=I_{\uparrow}$. A switching back to the low output state occurs only when I_{in} is reduced now to below $I_{\downarrow}$. Thus an optically bistable behavior results.

For dispersive optical bistability, one uses a medium with intensity dependent refractive index in the Fabry-Perot cavity. Here depending on the sign of optical nonlinearity (n_2), the cavity is detuned in a certain direction and by a certain amount, the latter determined by the cavity parameters. At low input levels ($I_{in}<I_{\uparrow}$), the Fabry-Perot transmission is low. However as I_{in} increases, the refractive index changes in the appropriate direction to compensate for detuning. At $I_{in}=I_{\uparrow}$, this compensation now leads to cavity tuning whereby I_{out} now switches to the high output state. The intensity (field) within the cavity is high. In the reverse cycle, even though I_{in} drops below $I_{\uparrow}$, the intensity within the cavity being high maintains the compensation of detuning to keep I_{out} in the high output state. The resulting behavior again is optical bistability.

Optical bistability experiments need extreme care. The only meaningful correlation of the input and output powers can come from single pulse analysis. We have conducted a preliminary study of optical bistability in a polyimide (LARC-TPI) film using nanosecond single pulse analysis. For this type of experiment we have used a home assembled Fabry-Perot etalon which can be piezoelectrically scanned. The input and output pulses were recorded using a large bandwidth (350 MHz) scope and a digital camera.

Many reported studies of optical bistability using short pulses actually show a slowly switching hysteresis behavior derived from the integrating nature of thermal nonlinearity. The theoretical analysis conducted by Professor Stegeman's group at the University of Arizona has shown that in the case of thermal effect, the output pulse shape is very much power dependent [43].

9. Current Status and Future Directions of Research

For second order nonlinear optical effects, organic systems have exhibited nonlinear coefficients orders of magnitude larger than those of traditional inorganic salts (KDP, KTP etc). Organic crystals have been recognized for some time to have large d and r coefficients which describe the figures of merits for frequency doubling and electrooptic effect. The problem in their laboratory use has been materials quality and achieving phase matching for largest nonlinear coefficient [12]. Recently, polymeric structures with nonlinear groups (groups with large β) attached to their backbone as pendent side groups have become more promising. These structures are poled at higher temperatures to create a noncentrosymmetric alignment of the nonlinear side groups. The polymeric structures offer improved mechanical, thermal and environmental stabilities as well as higher optical damage threshold. However, such poled structures are thermodynamically unstable, and the long term relaxation behavior of the aligned side groups has to be understood. This is one area of research which is currently active and will continue to receive increased attention in future. Poled polymeric structures also offer the flexibility to make channeled waveguides for various device applications [37].

For third-order nonlinear optical effects, conjugated polymeric systems exhibit the largest non-resonant $\chi^{(3)}$ value [3]. However, the largest non-resonant value demonstrated is ~10^{-9} esu which is still at least an order of magnitude off from what one would need for a device application [3]. Resonance enhancement of $\chi^{(3)}$ organic system is not as large as one observes in multiple quantum well semiconductors. Therefore, at this time it appears unlikely that resonant third-order optical nonlinearity in organic materials will compete with those found for multiple quantum well systems.

The most suitable application of organics will be in guided-wave structures where non-resonant nonlinearities are needed [2]. Conjugated polymers in general do not form high optical quality (low optical loss) films or fibers. They tend to have high optical losses due to many factors such as (i) domain structures, (ii) impurities, and (iii) a distribution of conjugation lengths which give rise to residual absorption even in the so called non-resonant wavelength range. In summary, the improvements needed for device application utilizing third-order processes are both in the magnitude of $\chi^{(3)}$ and in the optical quality of materials.

In the opinion of this author the following areas offer opportunities for future research:

(i) Microscopic understanding of nonlinear optical processes in molecular systems. We need to develop this understanding in order to predict and synthesize structures with large nonlinearities. We need input from synthetic chemists who could design controlled and systematically derivatized structures for the study of structure-property relationship. We also need contribution from theorists in developing a more matured theory of microscopic nonlinearity.

(ii) Relation between bulk and microscopic nonlinearities. One often uses the Lorentz-Lorenz approximation for local field to relate microscopic nonlinearities with bulk susceptibilities. This approach ignores any many-body effect and the role of any bulk co-operative excitation. Solid state physicists can make important contributions on this issue.

(iii) Role of various excitations. As was discussed above, the roles of solitonic, polaronic and bipolaronic excitations are not well understood. The photophysics of these excitations as well as the nonlinearities in n- and p-doped conducting polymers offer challenging opportunities for basic research.

(iv) Improvement of material qualtiy. We need to have better optical quality materials for waveguide applications. Material characterization and processing control are also very important.

In conclusion, the area of nonlinear optics of molecular materials is truly multidisciplinary, which offers challenging opportunities for chemists, physicists, materials scientists, and engineers. The interest in this area is rapidly expanding worldwide.

10. Acknowledgements

This work was support by the Air Force Office of Scientific Research, the directorate of Chemical Sciences and Air Force Wright Aeronautical Laboratory - Polymer Branch under contract number F4962087C0042 and by the Office of Innovative Science and Technology - Defense Initiative Organization under contract number F4962087C0097. Partial support from NSF-Solid State Chemistry Program grant number DMR-8715688 is also acknowledged. The author thanks his group members Drs. J. Swiatkiewicz, M. Samoc, R. Burzynski, E. Perrin, A. Samoc, J. Klimovic, I. Kminek, Mr. He, Mr. Y. Pang, Mr. M. T. Zhao, Mr. M. Casstevens, and Ms. S. Ghosal.

References

1. B. Clymer and S. A. Collins, Jr., Opt. Eng. 24, 74 (1985).
2. G. I. Stegeman, Thin Solid Films 152, 231 (1987).
3. "Nonlinear Optical and Electroactive Polymers" Eds. P. N. Prasad and D. R. Ulrich, Plenum Press (New York, 1988).
4. "Optical Nonlinearities and Instabilities in Semiconductors" Ed. H. Huag, Academic Press (London, 1988).
5. Y. R. Shen "The Principles of Nonlinear Optics" Wiley & Sons (New York, 1984).
6. P. Chopra, L. Carlacci, H. F. King and P. N. Prasad, J. Phys. Chem. (in Press).
7. G. J. B. Hurst, M. Dupuis and F. Clementi, J. Chem. Phys. 89, 385 (1988).

8. M. Dupuis, J. Rys and H. F. King, J. Chem. Phys. 65, 111 (1976). M. Dupuis, J. D. Watts, H. O. Villar and B. J. Hurst, Hondo (7.0) available from QCPE, Indiana University.
9. H. Sekino and R. J. Bartlett, J. Chem. Phys. 85, 976 (1986).
10. J. Zyss and D. S. Chemla in "Nonlinear Optical Properties of Organic Molecules and Crystals" Eds. D. S. Chemla and J. Zyss, Academic Press (New York, 1987) p. 23.
11. A. F. Garito, K. Y. Wong and O. Zamani-Khamiri in reference 4, p. 13.
12. D. J. Williams, Angew. Chem. Int. Ed. 23, 690 (1984).
13. B. P. Singh, P. N. Prasad and F. E. Karasz, Polymer 29, 1940 (1988).
14. G. R. Meredith, J. Chem. Phys. 77, 5863 (1982); G. R. Meredith and B. Buchalter, J. Chem. Phys. 78, 1938 (1983).
15. E. Perrin, P. N. Prasad and M. Dupuis, to be published.
16. A. D. Buckingham and J. A. Pople, Proc. Phys. Soc. (London) 68, 905 (1955).
17. A. Samoc, M. Samoc, P. N. Prasad, C. Willand, and D. J. Williams, submitted to J. Phys. Chem.
18. "Electromagnetic Surface Excitations" Eds. R. F. Wallis and G. I. Stegeman, Springer-Verlag (Berlin, 1986).
19. X. Huang, R. Burzynski and P. N. Prasad, Langmuir 5, 325 (1989).
20. M. M. Carpenter, P. N. Prasad and A. C. Griffin, Thin Solid Films 161, 315 (1988).
21. G. H. Cross, I. R. Peterson, I. R. Girling, N. A. Cade, M. J. Goodwin, N. Carr, R. S. Sethi, R. Marsden, G. W. Gray, D. Lacey, A. M. McRoberts, R. M. Scrowston, and K. J. Toyne, Thin Solid Films 156, 39 (1988).
22. H. J. Eichler, P. Günter and D. W. Pohl, "Laser-Induced Dynamic Gratings" Springer (Berlin, 1986).
23. F. Kajzar in "Nonlinear Optical Effects in Organic Polymers" Eds. J. Messier, F. Kajzar, P. Prasad and D. Ulrich, NATO ASI Series, Vol. 162, Kluwer Academic Publishers (The Netherlands, 1989) p. 225.
24. M. T. Zhao, B. P. Singh and P. N. Prasad, J. Chem. Phys. 89, 5535 (1988).
25. M. T. Zhao, M. Samoc, B. P. Singh and P. N. Prasad, J. Phys. Chem. (in Press).
26. P. N. Prasad, J. Swiatkiewicz, and J. Pfleger, Mol. Cryst. Liq. Cryst. 160, 53 (1988).
27. Z. Z. Ho and N. Peyghambarian, Chem. Phys. Lett. 148, 107 (1988).
28. S. K. Ghoshal, P. Chopra, B. P. Singh, J. Swiatkiewicz, and P. N. Prasad, J. Chem. Phys. 90, 5078 (1989).
29. M. Samoc and P. N. Prasad, J. Chem. Phys. (in Press).

30. "Handbook of Conducting Polymers" Vols. 1 and 2, Ed. T. Skotheim, Marcel-Dekker (1987).

31. P. Logsdon, J. Pfleger and P. N. Prasad, Synthetic Metals 26, 369 (1988).

32. C. P. de Melo and R. Silbey, Chem. Phys. Lett. 140, 537 (1987).

33. S. M. Jensen, IEEE J. Quant. Electr., QE18, 1580 (1982).

34. A. Lattes, H. A. Hans, F. J. Leonberger and E. P. Ippen, IEEE J. Quantum Electr. QE19, 1718 (1983).

35. S. Wabnitz, E. M. Wright, C. T. Seaton and G. I. Stegeman, Appl. Phys. Lett. 49, 838 (1986).

36. S. Trillo and S. Wabnitz, Tech. Dig. CLEO 1987, Washington:OSA (1987) paper THK30, p. 260.

37. J. I. Thacara, G. F. Lipscomb, R. S. Lytel and A. J. Tickonor in "Nonlinear Optical Properties of Polymers" Eds. A. J. Heeger, J. Orenstein and D. R. Ulrich, Mat. Res. Soc. Symp. Proc. 109, 19 (1988).

38. R. DeMartino, D. Haas, G. Khanarian, T. Leslie, H. T. Man, J. Riggs, M. Sansone, J. Stamatoff, C. Teng and H. Yoon in "Nonlinear Optical Properties of Polymers" Eds. A. J. Heeger, J. Orenstein and D. R. Ulrich, Mat. Res. Soc. Symp. Proc. 109, 65 (1988).

39. R. Burzynski, B. P. Singh, P. N. Prasad, R. Zanoni and G. I. Stegeman, Appl. Phys. Lett. 53, 2011 (1988).

40. C. T. Seaton, G. I. Stegeman, W. M. Hetherington III and H. G. Winful, in Springer Ser. Opt. Sci., Vol. 48 (Integr. Optics), p. 179, (1985).

41. G. I. Stegeman, IEEE J. Quantum Electr. QE18, 1610 (1982).

42. H. M. Gibbs "Optical Bistability: Controlling Light with Light" Academic Press (New York, 1985).

43. G. Assanto, R. M. Fortenberg, C. T. Seaton and G. I. Stegeman, J. Opt. Soc. Am. B5, 432 (1988).

Nonlinear Optical Effects in Substituted Anilines

S. Rosenkilde [1], *A. Holm* [1], *T. Geisler* [2], *and P.S. Ramanujam* [3]

[1]Department of General and Organic Chemistry, The H.C. Ørsted Institute, Universitetsparken 5, University of Copenhagen, DK-2100 Copenhagen, Denmark
[2]Laboratory of Applied Mathematical Physics, Technical University of Denmark, DK-2800 Lyngby, Denmark
[3]Danish Institute of Fundamental Metrology, Bldg. 322, Lundtoftevej 100, DK-2800 Lyngby, Denmark

Abstract. A series of substituted anilines have been synthesized and examined for second harmonic generation (SHG) effects using Kurtz's powder technique and the results are compared with a urea reference. The materials are N-acyl-4-nitroaniline, N-alkyl-4-nitroaniline, N-acyl-2-methyl-4-nitroaniline and N-alkyl-2-methyl-4-nitroaniline. The influence of the methyl group, the chain length and the carbonyl group have been examined. N-octadecanoyl-4-nitroaniline shows a nonlinear activity comparable to that of urea.

1. INTRODUCTION

In recent years there has been an increasing interest in the development of organic materials showing large nonlinear optical responses. Some organic materials show nonlinear optical effects that far exceed those observed from their inorganic counterparts already having optoelectronic device application [1].

In most cases, the organic molecules in question feature electron-rich donor groups in resonance with acceptor groups through a conjugated molecular π-electron system [2]. The ability of a solid compound to generate second harmonic radiation depends not only on a large microscopic second order hyperpolarizability, β, but also demands that it possess a non-centrosymmetric structure, since in the electric dipole approximation the second-order susceptibility, $\chi^{(2)}$, vanishes for centrosymmetric structures. One way of getting around this problem is the use of Langmuir-Blodgett films [3,4], to orient the molecules.

Nitroanilines are among the most studied organic systems usually having large hyperpolarizabilities. However, as an example, pNA (4-nitroaniline) crystallizes in a centrosymmetric structure and has therefore no second harmonic activity. The centrosymmetric structure is destroyed by a methyl group in crystals of MNA (2-methyl-4-nitroaniline), making MNA a second harmonic generator.

Our primary interest lies in the preparation of thin solid films having extensive nonlinear optical properties, using the Langmuir-Blodgett technique. Since stable monolayers are formed by organic molecules with long aliphatic chains, it is of interest to investigate the nonlinear optical properties of such molecules.

We have synthesized four N-alkyl and four N-acyl derivatives of pNA and MNA respectively, with chain lengths of 8, 12, 18 and 22 carbon atoms respectively.

Springer Series in Wave Phenomena, Vol. 9 **Nonlinear Optics in Solids**
Editor: O. Keller

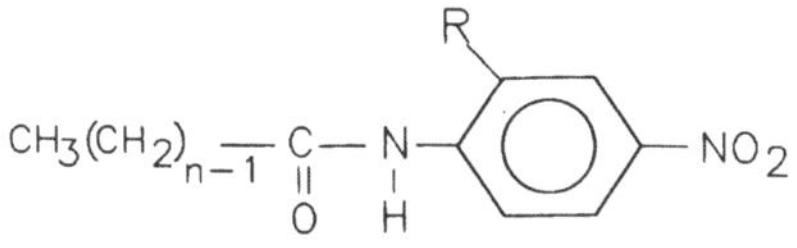

1

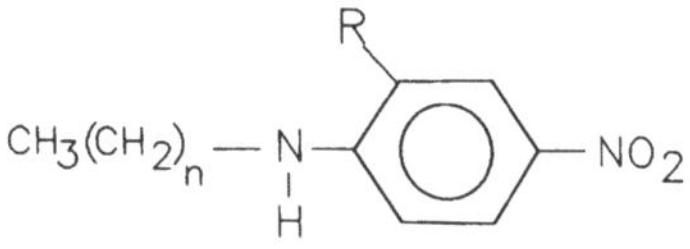

2

R = H, CH_3
n = 7,11,17 and 21

From measurements of the second harmonic generation (SHG) from powder samples of these molecules, the influence of the carbon chain length, the carbonyl group and that of the methyl group upon the nonlinear properties have been examined.

2. Syntheses and Experiment

2.1 Experimental Section

^{1}H-NMR spectra were recorded with a Bruker AM 250 spectrometer, IR spectra with a Perkin Elmer PE 580 spectrophotometer, UV-VIS spectra with a Perkin-Elmer 137 UV spectrophotometer. Mass spectra were measured with a VG Masslab VG 12-250 mass spectrometer.

The *N*-octadecanoyl derivatives of pNA and MNA were prepared from octadecanoyl chloride (Fluka, pract. 90-95% - purified by vacuum distillation; bp. 215 °C; 15 mmHg) and the corresponding anilines.

All the other *N*-acyl nitroanilines were prepared from the carboxylic acids (Aldrich Chem. Co., 98%) and the corresponding anilines *via* the acid chlorides.

4-Nitroaniline (Reidel-de Haën, purum 99%) was purified from a solution of acetone treated with charcoal and filtered through celite.

2-Methyl-4-nitroaniline (Aldrich Chem. Co., 97%) was re-crystallized from ethanol followed by sublimation.

2.2 General Procedures

***N*-Docosanoyl-2-methyl-4-nitroaniline.** Docosanoic acid (2.9 mmol) was treated with an excess of thionyl chloride (10 ml) at 80 °C for 30 min. Excess of thionyl chloride was removed by distillation *in vacuo*. A solution of 2-methyl-4-nitroaniline (6.9 mmol) in benzene (10 ml) was added to the docosanoyl chloride, and the reaction mixture refluxed for 30 min. After evaporation of the solvent *in vacuo* the crude material was recrystallized from ethanol and washed with cold water. Yield 1.27 g (93% based upon docosanoic acid) of the white crystalline *N*-docosanoyl MNA. The product was identified from the ^{1}H-NMR and IR spectra, and the mass spectrum exhibited a molecular ion at m/z = 474. Anal. $C_{29}H_{50}N_2O_3$: C,H,N. M.p. 121.0-121.5 °C. The remaining acyl derivatives were identified by similar analyses and yields in the range of 87-95% were obtained. The m.p. and λ_{max} are shown in table 1. To the best of our knowledge, the *N*-docosanoyl pNA and MNA have not previously been reported.

TABLE 1. SHG intensity from N-acylated nitroanilines.

Compound	Formula	SHG intensity rel. to urea	λ_{max} (nm) ($CHCl_3$)	m.p. (°C)
N-Octanoyl-pNA	1 (n=7)	* b	315	73.0
N-Octanoyl-MNA		* c	316	102.0
N-Dodecanoyl-pNA	1 (n=11)	* b	315	80-1
N-Dodecanoyl-MNA		0 c	316	108.0
N-Octadecanoyl-pNA	1 (n=17)	0.5 c 0.01 b	315	95-6
N-Octadecanoyl-MNA		0 c	316	116-7
N-Docosanoyl-pNA	1 (n=21)	0.01 c	316	104.0
N-Docosanoyl-MNA		0 c	316	121.0

N-Docosyl-2-methyl-4-nitroaniline. In order to reduce the carbonyl group in the N-acyl nitroanilines, we have applied the method of Kuehne and Shannon [5] to reduce amides and lactams by reaction with phosphorus oxychloride and sodium borohydride. Under such mild conditions, the carbonyl group was selectively reduced to a methylene group without conversion of the nitro group. Yields of 65-80% were obtained.

In a typical experiment, N-docosanoyl-2-methyl-4-nitroaniline (0.85 mmol) was added to phosphorus oxychloride at room temperature. The reaction mixture was stirred for 20 hr and excess of $POCl_3$ was evaporated at 20 °C (15 mmHg). To remove residual $POCl_3$, the solid white remanence was evacuated at 0.05 mmHg for 20 min and dissolved in tetrahydrofuran (THF, 30 ml). The solution was cooled in an ice/water bath and sodium borohydride (8.5 mmol) added gradually under vigorous stirring. Instantaneously the reaction mixture turned yellow. The ice bath was removed and the mixture stirred at room temperature for 2 hr. Hydrochloric acid (10%, 20 ml) was added dropwise at 0 °C, the solvent removed *in vacuo* and water (30 ml) added. The mixture was refluxed for 20 min, and made basic with sodium hydroxide (1.5 g). The basic solution was extracted with chloroform several times. The combined chloroform phases were dried over potassium carbonate and concentrated. Work-up by column chromatography and recrystallization from methanol gave a total yield of 285 mg (75% based upon N-docosanoyl-2-methyl-4-nitroaniline) of yellow crystals. The product was identified from the ^1H-NMR and IR spectra, and the mass spectrum exhibited a molecular ion at m/z = 460. Anal. $C_{29}H_{52}N_2O_2$: C,H,N. M.p. 92-93 °C. The remaining alkyl derivatives were identified by similar analyses. The m.p. and λ_{max} are shown in table 2.

TABLE 2. SHG intensity from N-alkylated nitroanilines.

Compound	Formula	SHG intensity rel. to urea	λ_{max} (nm) ($CHCl_3$)	m.p. (^oC)
N-Octyl-pNA		* a	378	51.0
N-Octyl-MNA	**2** (n=7)	0 b	383	82.0
N-Dodecyl-pNA		* b	378	65-6
N-Dodecyl-MNA	**2** (n=11)	0 b	382	70-1
N-Octadecyl-pNA		0 d	375	80-1
N-Octadecyl-MNA	**2** (n=17)	0 d	380	83-4
N-Docosyl-pNA		* d	375	87-8
N-Docosyl-MNA	**2** (n=21)	0 d	380	92-3

* weak SHG intensity (<<0.01 rel. Urea)
a recrystallized from pentane.
b recrystallized from cyclohexane:ether (1:1).
c recrystallized from ethanol.
d recrystallized from methanol.

2.3 SHG Powder Measurement

The samples were tested by measuring the SHG signal following the technique for screening of powdered materials for nonlinear activity, proposed by Kurtz and Perry [6]. The experimental setup for the SHG measurements are shown in fig. 1. The beam from a 500 Hz Q-switched Nd:YAG laser at 1,064 nm with a pulse duration of 120 ns and a pulse energy of approximately 0.4 mJ was weakly focused to an approximate diameter of 2.5 mm and incident normally on the sample to be tested. In order to collect the back-scattered SHG signal a parabolic reflector was inserted in front of the sample. The fundamental beam was removed by a filter (Schott KG1) and the SHG signal was then focused on a monochromator and detected with a photomultiplier situated behind the monochromator. The reduction in intensity as the detection setup was tuned away from 532 nm confirmed that the signal was due to SHG and not to any broadband fluorescence. The samples, reflector and most of the optics were placed inside a black box.

The powder was dense packed between two transparent glass plates with a sample thickness of 0.2 mm. The SHG powder efficiencies were calibrated with respect to the SHG signal from a reference urea powder (grain size 70-140 μm).

3. Results and Discussion

All the SHG measurements were done on ungraded powder and the results are shown in tables 1 and 2.

It must be noted that the powder technique is sensitive to average grain size. Furthermore the crystal structure might depend on the solvent used for recrystallization [7].

Chen et al. [8] have recently investigated a series of N-alkyl-4-nitroanilines, including unbranched alkyl chains of 3,4,5,6 and 8 carbon-atoms. They found that N-butyl-4-nitroaniline

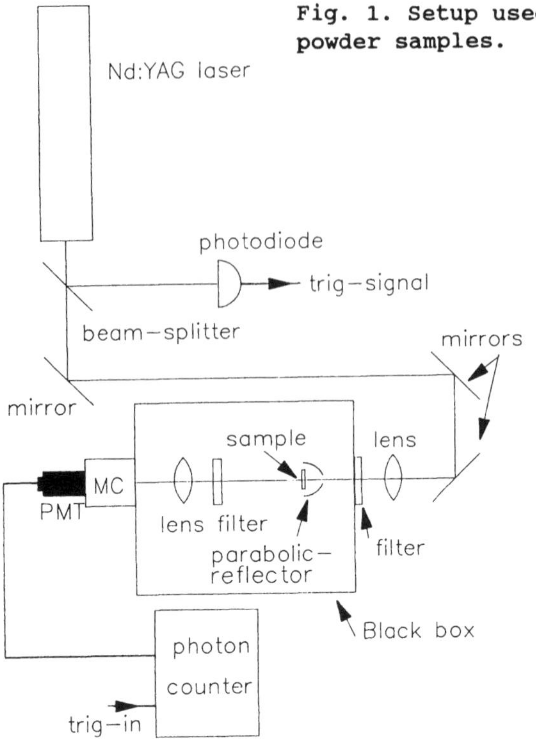

Fig. 1. Setup used for the study of SHG in powder samples.

(formula 2, n=3) produces a SHG intensity of 0.5 relative to urea when recrystallized from ethanol. On the other hand, changing to a less polar solvent (cyclohexane:ether), the SHG intensity was observed to be 14 times that of urea. An X-ray study verified differences in structure in the two samples. From N-octyl-4-nitroaniline (formula 2, n=7) only a weak signal was observed (0.05 times that of urea) when recrystallized from ethanol, and from the rest of the compounds examined there were zero or little effect.

We had undertaken a similar powder study of some long chained acyl and alkyl derivatives of pNA and MNA (formulas 1 and 2, n = 17 and 21). Chen's report induced us to extend our examinations to include the C_8- and the C_{12}- analogues.

Our results from N-octyl-4-nitroaniline (recrystallized from the nonpolar solvent pentane) are in accordance with those of Chen et al. as only a weak SHG intensity was observed. In general the N-alkyl-nitroanilines in our experiments showed no or very little effect.

The largest effect was observed from N-octadecanoyl-4-nitroaniline (ethanol) giving a SHG intensity 0.5 times that of urea. The intensity dropped to 0.01 if the less polar solvent cyclohexane:ether was used. N-docosanoyl-4-nitroaniline (ethanol) showed a small effect, 0.01 times that of urea. The other N-acyl-nitroanilines showed no or very little effect.

From a solvent of low polarity, molecules with a large ground-state dipole moment tend to line up in an antiparallel fashion. This leads to formation of centrosymmetric crystals when recrystallization occurs. In a solvent of stronger polarity, however, the molecules become more solvent separated, leading to weaker dipole-dipole interactions. Hence there is a greater probability of forming non-centrosymmetric crystals [7]. As expected from these arguments we observed a drop in SHG intensity from N-octadecanoyl-pNA when changing solvent from ethanol to cyclohexane:ether (table 1). However Chen et al. find the opposite solvent effect in case of N-butyl-pNA [8].

Reducing the N-acyl substituents to N-alkyl also results in a decrease in SHG intensity. Whereas the acyl substituents are electron-withdrawing, the alkyl substituents have an electron-donating effect. This leads to much smaller dipole moments in the N-acyl compounds as related to the N-alkyl analogues. Again molecules having large dipole moments more strongly interact in solution and more often form a centrosymmetric structure on recrystallization.

There is no clear dependence of the SHG intensity on chain length in the acyl- and in the alkyl series.

Further investigations of some of the long chained derivatives in Langmuir-Blodgett films are in progress in our laboratory.

4. Acknowledgments

The authors thank Dr.K.Schaumburg, Department of Physical Chemistry, The H.C.Ørsted Institute, University of Copenhagen for useful discussions and Dr.Charles Larsen, Department of General and Organic Chemistry, The H.C.Ørsted Institute, University of Copenhagen for recording the mass spectra. The Danish Technical Research Council and the Danish Metrology Council are acknowledged for financial support.

5. References

[1] D.S. Chemla and J. Zyss (ed.), "Nonlinear Optical Properties of Organic Molecules and Crystals", Academic Press, New York (1987).

[2] D.J. Williams, Angew. Chem. Int. Ed. Engl., **23**, 690 (1984).

[3] O.A. Aktsipetrov, N.N. Akhmediev, E.D. Mishina, V.R. Novak, JETP Lett., **37**, 207 (1983).

[4] D.B. Neal, M.C. Petty, G.G. Roberts, M.M. Ahmad, W.J. Feast, I.R. Girling, N.A. Cade, P.V. Kolinsky and I.R. Peterson, Electr. Lett., **22**, 460 (1986).

[5] M.E. Kuehne and P.J. Shannon, J. Org. Chem., **42**, 2082 (1977).

[6] S.K. Kurtz and T.T. Perry, J. Appl. Phys., **39**, 3798 (1968).

[7] H. Tabei, T. Kurihara and T. Kaino, Appl. Phys. Lett., **50**, 1855 (1987).

[8] D. Chen, N. Okamoto and R. Matsushima, Optics Comm., **69**, 425 (1989).

Part VII

Disordered Media

Weak Localization in Nonlinear Optics of Disordered Media

V.E. Kravtsov, V.M. Agranovich, and V.I. Yudson

Institute of Spectroscopy, USSR Academy of Sciences,
Troitsk, Moscow r-n, SU-142092, USSR

Abstract. The angular distribution of light generated in nonlinear processes of three- and four-wave mixing is studied theoretically for the case of a medium both disordered and nonlinear. The problems of special interest are weak localization effects in multiple scattering of a coherent light within the medium. It was shown that these effects result in sharp peaks in the angular distribution of generated light. The shape of the peaks is determined by the space correlations of the strongly scattered radiation within the medium and it is sensitive to the Anderson localization effects.

1. Introduction

Recently a problem of a coherent wave propagation in disordered media has attracted a considerable attention in connection with the study of the Anderson localization /1/. The point of special interest was coherent interference effects in multiple scattering of light which are referred to as "weak localization". Most investigations on the weak localization of photons dealt with a linear backscattering of light from a disordered media /2,3/. It was shown theoretically /2/ and verified experimentally (see review /3/) that the intensity in the backscattering direction is twice as large as that far beyond the narrow region of scattering angles close to 180°. The shape of this backscattering peak may give, especially in the time-domain experiments, detailed information about transport properties of photons in disordered media. Here we extend the ideas of weak localization to the consideration of the nonlinear optics in disordered media and summarize the main results obtained in /4,5/.

Springer Series in Wave Phenomena, Vol. 9 **Nonlinear Optics in Solids**
Editor: O. Keller

2. Second Harmonic Generation

In this section we consider the second harmonic generation (SHG) as an example for the discussion of those qualitative effects which result from the weak localization of photons in a nonlinear medium. In addition to its general nature this problem becomes one of special interest because doped semiconductors at low temperatures have been suggested /6/ as possible systems where the Anderson localization of photons may occur. They are just the systems which exhibit large nonlinearities especially near excitonic resonances.

First of all we consider a situation when only the generated light of the frequency 2ω undergoes a strong elastic scattering by imperfections. The corresponding elastic mean free path $l_{2\omega}$ is supposed to be much less than both the thickness of a sample L and the absorption length l_a which is determined by the imaginary part of the dielectric constant ϵ at a frequency 2ω. On the other hand we suppose that a criterion of weak localization $\lambda/2\pi l_{2\omega} \ll 1$ is fulfilled where λ is a wavelength. At such conditions the calculations /4/, made in the frame of the impurity diagram technique, give for the differential cross section of SHG (per unit area of a sample) the quantity

$$\frac{d\sigma}{d\Omega} = J_0\left(1+\beta\frac{f(\theta)}{f(0)}\right) . \qquad (1)$$

Here

$$J_0 \sim 6\pi k_2^2 l_{2\omega} L_0 \epsilon^{-4} |\chi E_\omega|^2, \qquad (2)$$

with E_ω being the amplitude of an incident light of the frequency ω ; χ being the nonlinear susceptibility. The effective thickness of a sample L_0 is given by

$$L_0 = \min\{L, (l_{2\omega} l_a)^{1/2}\}. \qquad (3)$$

The SH wave vector k_2 is supposed to be equal in magnitude to twice the wave vector k_1 of an incident light. The constant β in Eq.(1) is of the order of $l_{2\omega}/L_0$. The function $f(\theta)$ dependent on the angle θ between $-2\mathbf{k}_1$ and $\mathbf{k}_2$ for the normal incidence and $\theta \ll 1$ is given by

$$f(\theta) = \frac{1.7+0.7\,q l_{2\omega}}{(1+0.7\,q l_{2\omega})(1+2q l_{2\omega})^2}, \qquad q = (\theta^2 k_2^2 + L_0^{-2})^{1/2} . \qquad (4)$$

The expressions (1)-(4) show that effects of weak localization result in a sharp peak in the direction opposite to one determined by the phase matching condition $\mathbf{k}_2 = 2\mathbf{k}_1$. At $L_0 \gg l_{2\omega}$ the peak has a triangular form with the width being of the order of $\lambda/l_{2\omega}$. In this case the ratio peak/background β is small.

However at a finite elastic mean free path $l_\omega < L_0$ of the incident light one should replace L_0 by l_ω in the expression for β. Hence, at comparable values $l_{2\omega}$ and l_ω (e.g. for sizes of imperfections larger than λ) this nonlinear backscattering peak may be observable.

3. Phase Conjugation in Disordered Media

The backscattering peak in SHG discussed above is produced by SH photons generated in a small surface layer of the thickness $l_{2\omega}$. It is just the cause of the relative smallness of this peak. On the contrary, in the phase conjugation processes for the case of four-wave mixing one deals with the bulk effect of weak localization /5/. Moreover, the phase conjugation in a disordered media may occur in an experimental geometry with only one pump wave (of the frequency ω_p) and a probe wave (of the frequency ω_i). This occurs because a multiply scattered pump wave has components with all possible (e.g. counterpropagating) directions of the wave vector. In calculations /5/ of the angular distribution of the generated signal light (of the frequency $\omega_s = 2\omega_p - \omega_i$) it is supposed that all waves with close frequencies $(\omega_p \approx \omega_i \approx \omega_s)$ have the same elastic mean free path l $(\lambda \ll l \ll L_0)$. This calculations show that the angular distribution of the signal light exhibits a sharp peak in the direction opposite to the direction of incidence of a probe wave $(\theta_s \approx \theta_i)$. At $\omega_p - \omega_i \to 0$ the peak value of the differential cross-section (normalized per unit flux of a probe wave and unit sample area) is given by

$$\left(\frac{d\sigma}{d\Omega}\right)_{peak} \sim \frac{2\pi l L_0}{\lambda^2}(\delta n_{NL})^2 \tag{5}$$

where δn_{NL} is a nonlinear correction to the pump wave refractive index.

The proportionality between the peak intensity and the effective thickness of a sample L_0 allows one to consider the PC backscattering peak as a bulk effect. The shape of the peak has nothing to do with that for linear backscattering or SHG. For $\theta \equiv |\theta_i - \theta_s| \ll 1$ we find

$$\frac{d\sigma(\theta)}{d\Omega} \propto [\theta^2 + (\theta^4 + (\Delta\theta)^4)^{1/2}]^{-1/2} \quad (6)$$

where the width of the peak is given by

$$\Delta\theta = \lambda[2\pi L_0 \cos\theta_i]^{-1} . \quad (7)$$

The value of $\Delta\theta$ turns out to be much less than $\lambda(2\pi l)^{-1}$ because of the relatively small contribution made by photon trajectories corresponding to several scattering events.

The most striking feature of the PC backscattering peak is that the peak value may be some orders of magnitude larger (by the parameter $l/\lambda \gg 1$) than the diffuse background at the same frequency. This results from the phase matching condition which governs the four-wave mixing in a region of the size l (see /5/).

At not too small values of

$$\tau|\omega_p - \omega_i| \gtrsim l^2 L_0^{-2} \quad (8)$$

(τ is a time of the mean free path) one should replace L_0 in Eqs. (5), (7) by the value

$$L_{\Delta\omega} = l(|\omega_i - \omega_s|\tau)^{-1/2} . \quad (9)$$

As a result both the peak intensity and its width become dependent on a small frequency difference. The slope of the dependence of $\Delta\theta$ on $|\omega_i - \omega_p|^{1/2}$ may give the value of l while the crossover from the linear to constant behaviour of $\Delta\theta$ on $|\omega_i - \omega_p|^{1/2}$ allows one to find the absorption length l_a . It is also worth noting that when approaching the Anderson localization point the length L_0 in Eq.(8) should be replaced by the localization radius L_{loc} , providing that $L_{loc} < L_0$. If the experimental value of L_0 determined by the crossover condition (8) turns out to be much less than the value of $\min\{L, (ll_a)^{1/2}\}$ obtained independently, such an experiment may be considered as evidence in favour of the Anderson localization of photons.

References

1. P.A.Lee, T.V.Ramakrishnan, Rev.Mod.Phys., 57, 287, (1985).
2. V.I.Tatarski, "The effect of a turbulent atmosphere on wave propogation", (National Technical Information Service, Washington D.C., 1971).

3. E.Akkerman, P.E.Wolf, R.Maynard, G.Maret, J.Phys.France, 49, 77, (1988).
4. V.M.Agranovich, V.E.Kravtsov, Phys.Lett. A, 131, 378, 386 (1988).
5. V.E.Kravtsov, V.I.Yudson, V.M.Agranovich. Solid State Comm., 1989 (in press); Phys.Rev.B, 1990 (in press).
6. V.M.Agranovich, V.E.Kravtsov, I.V.Lerner, Phys.Lett.A, 125, 435 (1987).

Part VIII

Photon Echoes, Raman Line Shapes

Optical Data Storage and Image Processing in Rare Earth Ion Doped Crystals Using Stimulated Photon Echo

S. Kröll, E.Y. Xu, R. Kachru, and D.L. Huestis*

Molecular Physics Laboratory, SRI International,
Menlo Park, CA 94025, USA
*Permanent address: Dept. of Atomic Physics, Combustion Center,
Lund Institute of Technology, Box 118, S-221 00 Lund, Sweden

Abstract Three experiments relevant for optical data storage and image processing using the stimulated photon echo technique in rare earth ion doped crystals are described. These are: an apparent laser excitation energy dependence of the homogeneous dephasing time for these crystals, optical image processing on a nanosecond time scale using microjoule input energies and optical storage and recovery of multiple data bits.

1. Introduction

Storing digital information in different homogeneous components within an inhomogeneously broadened absorption line has been suggested as a method for high density optical data storage. The reading and writing of the digital information may be performed in the frequency domain, e.g. [1], or in the time domain [2]. The time domain approach utilizes the stimulated photon echo technique and has previously been given less attention but recently both storage and retrieval of several pulses [3] and storage during extended time periods (days) [4] have been performed in rare earth ion doped crystals at liquid helium temperatures. Three experiments connected to optical data storage and image processing using the time domain approach are briefly described below.

2. The homogeneous dephasing time

In the context of optical storage the homogeneous dephasing time puts a limit on both the number of bits that can be stored in a single spatial location (given by T_2/T_2^*, where T_2^* is the inhomogeneous dephasing time) and the duration of the input datapulse sequence (e.g. [3]). Recently an intensity dependence of the photon echo decay has been observed [5,6,7]. In the first two papers the decay rate predominantly changed depending on the intensity of the second laser pulse [5,6] and was therefore explained in terms of instantaneous spectral diffusion [8]. In our measurements on the $^3P_0$-$^3H_4$ transition in a 1.27 cm long sample of 0.01% Pr^{3+}:YAG the effect of the first pulse intensity was comparable to the effect of the second pulse intensity indicating that additional mechanisms may contribute to the observed intensity dependence [7]. We have now observed the homogeneous dephasing time, T_2^+, as a function of total laser input energy on the $^1D_2$-$^3H_4$ transition in a 0.1 cm long 0.1% Pr^{3+}:$YAlO_3$ crystal (Fig. 1). (We denote the measured dephasing time T_2^+ to empasize that it may be different from

Springer Series in Wave Phenomena, Vol. 9 **Nonlinear Optics in Solids**
Editor: O. Keller

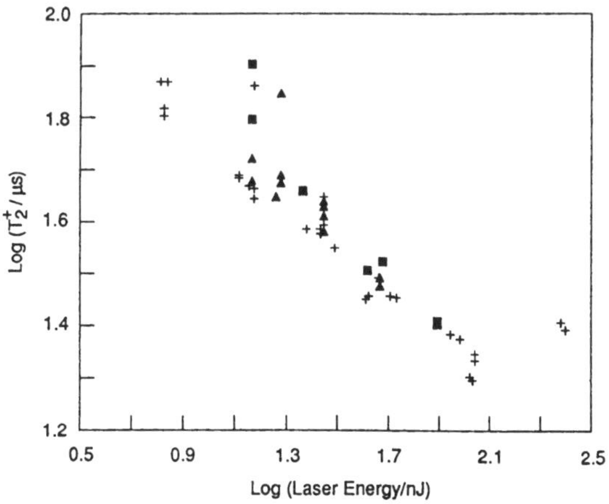

Fig. 1. The logarithm of the observed homogeneous decay time, T_2^+, versus the logarithm of the total laser input energy. +, second pulse has twice the energy of the first pulse, ■, energy of first pulse is 5 nJ, ▲ energy of second pulse is 10 nJ.

the intrinsic homogeneous dephasing time, T_2.) Unfortunately the optical density of this sample is ~0.8 and optical density effects [9] may here be a major cause of the intensity dependent decay time. However, we note that the homogeneous dephasing time in the limit of weak intensity ,70 µs, is more than a factor of two larger than the literature value for T_2 of 35 µs [10,11]. This may have profound implications for the modelling of nuclear spin exchange and other interactions that determine the homogeneous dephasing time in this crystal at temperatures of 2 K and below. The magnetic field dependence for this transition have been investigated previously by Macfarlane et al. [10]. They noted that in the region 0-100 Gauss T_2^+ increased rapidly with applied magnetic field saturating at 40 Gauss. Our data show a partially different magnetic field dependence (Fig. 2), the cause of which is as yet unknown to us.

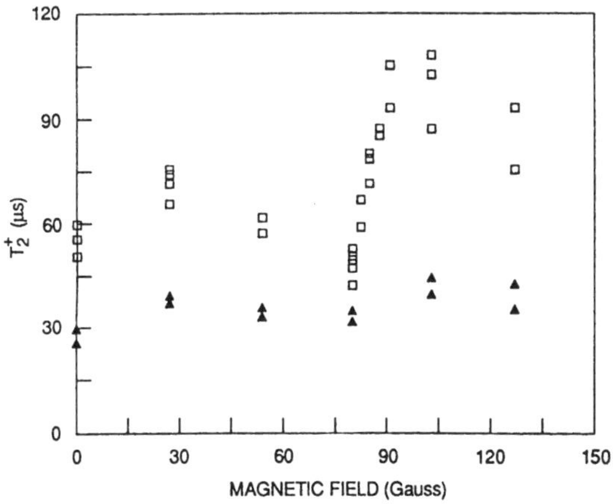

Fig. 2. Observed homogeneous dephasing time as a function of magnetic field along the crystal c axis. Laser energy is ~100 nJ (▲) or ~5 nJ (□).

3. Image processing

We have performed image processing on a nanosecond timescale using μJ input energies on the $^3P_0$ - $^3H_4$ transition in 0.1% Pr^{3+}:LaF_3 [12]. The input beams are propagated through image masks and sequentially focussed into the crystal. Due to the focusing the generated nonlinear polarization in the crystal is proportional to the product of the spatial Fourier transforms of the input masks. As the generated echo propagates through a lens it is itself Fourier transformed and is then the spatial convolution or correlation of the input images.

4. Multibit data storage

Utilizing the $^3P_0$-$^3H_4$ transition in a 0.01% Pr^{3+}:YAG crystal we have stored and recalled sequences of about 40 data bits (Fig. 3). The data bit sequence was generated by acousto optically modulating the output of a cw dye laser. Write and read pulses were provided by a YAG laser pumped dye laser. A gated Pockels cell between two crossed polarizers blocked scattered light from the input pulses but transmitted the recalled data sequence. The 4-5 μs opening time of the Pockels cell in conjunction with a maximum modulation rate of ~10 MHz for the dye laser output limited the maximum number of stored and retrieved data pulses to about 40. However, the maximum number of bits as limited by the ratio of the inhomogeneous and homogeneous linewidths is as large as 10^6 for this transition [11]. The sequence in Fig. 3 is a 32 shot average but single shot data with good signal-to-noise could be obtained. Finally, we emphasize that several bits are stored within the laser focal volume, therefore by storing information at spatially different locations very high storage densities may, in principle, be obtained.

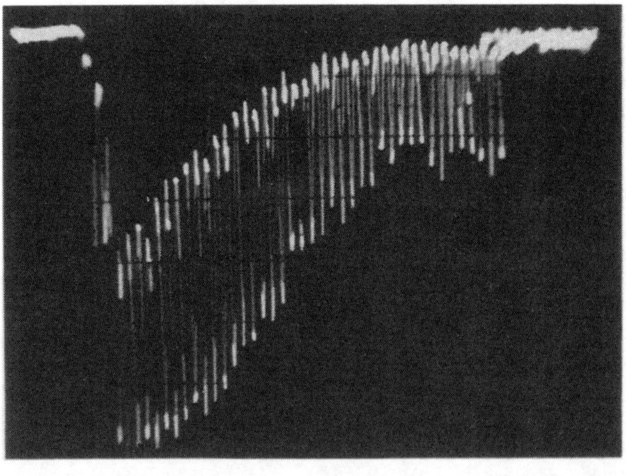

Fig. 3. Trace showing the retrieval of a stored data pulse sequence of 40 bits. Scale is 2 mV and 500 ns per division.

Acknowledgements

This work was supported by Nippon Telegraph and Telephone Corporation (NTT). S.K. acknowledges the support from the Swedish Board of Technical Developments (STU).

References

[1] F.M. Schellenberg, W. Lenth and G.C. Bjorklund, Appl. Opt. **25,** 3207 (1986)
[2] T.W. Mossberg, Opt. Lett. **7**, 77 (1982).
[3] M.K. Kim and R.Kachru, Appl. Opt. **28**, 2186 (1989).
[4] W.R. Babitt, A. Lezama and T.W. Mossberg, Phys. Rev. B39, 1987 (1989)
[8] D.R. Taylor and J.P. Hessler, Phys. Lett. **50A**, 205 (1974).
[5] G.K. Liu, N.F. Joubert, R.L. Cone and B. Jaquier, J. of Luminescence **38**, 34 (1987).
[7] S. Kröll, E.Y. Xu, M.K. Kim, M. Mitsunaga and R. Kachru, submitted to Phys. Rev. Lett.
[6] J. Huang, J.M. Zhang, A. Lezama and T.W. Mossberg, Phys. Rev. Lett. **63**, 78 (1989).
[9] R.W. Olson, H.W.H. Lee, F.G. Patterson and M.D. Fayer, J. Chem. Phys. **76**, 31 (1982).
[10] R.M. Macfarlane, R.M. Shelby and R.L. Shoemaker, Phys. Rev. Lett. **43**, 1726 (1979).
[11] R.M. Macfarlane and R.M. Shelby, in "Spectroscopy of solids containing rare earth ions", Eds. A.A. Kaplyanskii and R.M. Macfarlane, (North-Holland, 1987) Chap. 3.
[12] E.Y. Xu, S. Kröll, D.L. Huestis, R. Kachru and M.K. Kim, submitted to Opt. Lett.

A Practical Analysis for Coherent Raman Scattering Spectra by the Fourier Series Method

E.M. Vartiainen[1] *and K.-E. Peiponen*[2,*]

[1]Vaisala Laboratory, Department of Physics, University of Joensuu, P.O. Box 111, SF-80101 Joensuu, Finland
[2]Department of Engineering, Kyoto Sangyo University, Kamigamo, Kyoto 603, Japan
*On leave of absence from Vaisala Laboratory

Abstract. A simple method for computing the lineshapes of the real part and the squared modulus of the Raman susceptibility and the relative magnitude of the nonresonant background susceptibility with the aid of a coherent Raman scattering spectrum is presented. The method is valid when the spectrum results from a single Raman resonance.

1. Introduction

Coherent Raman spectroscopy is a collection of nonlinear optical spectroscopy techniques, which are applications of stimulated Raman scattering. Well-known examples of these techniques are coherent anti-Stokes Raman scattering spectroscopy (CARS) and Raman indused Kerr-effect spectroscopy (RIKES).

The spectra are usually measured by mixing two laser-beams at frequency ω_1 and ω_2 in the sample to generate an output at frequency $\omega = \omega_1 \pm \omega_1 \mp \omega_2$. One of the frequencies is fixed while the other one is scanned. The output intensity is proportional to the squared modulus of the total third-order nonlinear susceptibility $\chi^{(3)}$ [1]. In condensed phases $\chi^{(3)}$ can be written as the sum of a resonant part (or Raman susceptibility) χ_R and a nonresonant part χ_{NR}. Only the Raman term is resonantly enhanced when $|\omega_1 - \omega_2|$ approaches a Raman active transition frequency ω_r. Thus, χ_R is a complex quantity whereas χ_{NR} is usually assumed to have a real value. In this case the spectrum consists of three components, which can be written as follows:

$$|\chi^{(3)}|^2 = \chi_{NR}^2 + 2\chi_{NR} Re(\chi_R) + |\chi_R|^2, \qquad (1)$$

where $Re(\chi_R)$ is the real part of χ_R. Assuming a Lorentzian lineshape for a single Raman resonance [2 - 4], the Raman susceptibility can be written as

$$\chi_R = \frac{R}{\omega_R - (\omega_1 - \omega_2) - i\Gamma_R}, \qquad (2)$$

Springer Series in Wave Phenomena, Vol. 9 **Nonlinear Optics in Solids**
Editor: O. Keller

where ω_R and Γ_R are the frequency and the linewidth (HWHM) of a Raman mode, respectively. R is assumed to have negligible dispersion in the vicinity of ω_R. Fleming and Johnson [4] calculated the components of (1) using the theoretical model of (2) for CARS spectra. In this paper we present a method for numerical calculations of the components of (1) for more general lineshape than that of (2).

2. Method

Consider a case where a coherent Raman scattering spectrum of $|\chi^{(3)}|^2$ is measured within an interval $-\delta_N \leq \delta \leq \delta_N$, where $\delta = \omega_1 - \omega_2 - \omega_R$. We note that $\chi_R(\delta)$ is now virtually a complex function of one variable, and can be approximated with the polynome

$$\chi_R(\theta) = \sum_{n=0}^{N} c_n \exp(in\theta), \tag{3}$$

where $c_n = a_n + ib_n$ and $\theta = \pi\delta/\delta_N$. From (3) we obtain

$$\begin{aligned} |\chi_R(\theta)|^2 &= \left\{\sum_{n=0}^{N} c_n \exp(in\theta)\right\}\left\{\sum_{n=0}^{N} c_n^* \exp(-in\theta)\right\} \\ &= A_0' + 2\sum_{n=1}^{N}(A_n' \cos n\theta + B_n' \sin n\theta), \end{aligned} \tag{4}$$

where

$$A_n' = \sum_{k=0}^{N-n} (a_k a_{k+n} + b_k b_{k+n}), \quad n = 0, 1, ..., N$$

$$B_n' = \sum_{k=0}^{N-n} (a_{k+n} b_k - a_k b_{k+n}), \quad n = 1, 2, ..., N-1.$$

If we assume that $Re(\chi_R)$ is an odd function of δ like in the case of (2), the coefficients $a_n \equiv 0$ because the Fourier expansion of $Re(\chi_R)$ involves only sine terms. Thus, we can write

$$\chi_R(\theta) = \sum_{n=0}^{N} ib_n \exp(in\theta), \tag{5}$$

$$Re[\chi_R(\theta)] = -\sum_{n=1}^{N} b_n \sin n\theta, \tag{6}$$

$$|\chi_R(\theta)|^2 = A_0 + 2\sum_{n=1}^{N} A_n \cos n\theta, \tag{7}$$

where

$$A_n = \sum_{k=0}^{N-n} b_k b_{k+n}, \quad n = 0, 1, ..., N.$$

According to the theory of the Fourier series [5] the series of (6) and (7) can be recognized as the finite Fourier series of an odd and an even function, respectively, in a discrete set of points

$$\theta_m = \pi\delta_m/\delta_N = \pi m/N, \quad m = -N+1, ..., -1, 0, 1, ..., N. \tag{8}$$

By substituting (6) and (7) into (1) in the set of points of (8) we obtain

$$|\chi^{(3)}(\theta_m)|^2 = \alpha_0 + 2\sum_{n=1}^{N-1}(\alpha_n \cos n\theta_m + \beta_n \sin n\theta_m) + \alpha_N \cos \pi m, \tag{9}$$

where

$$\begin{aligned} \alpha_0 &= \chi_{NR}^2 + A_0, \quad \alpha_N = 2A_N, \\ \alpha_n &= A_n, \qquad n = 1, 2, ..., N-1, \\ \beta_n &= -\chi_{NR} b_n, \quad n = 1, 2, ..., N-1. \end{aligned} \tag{10}$$

The coefficients α_n and β_n can be computed with the aid of measured data by using discrete Fourier transformation formulas, as follows

$$\alpha_n = \frac{1}{2N} \sum_{m=-N+1}^{N} |\chi^{(3)}(\theta_m)|^2 \cos n\theta_m, \quad n = 0, 1, ..., N, \tag{11}$$

$$\beta_n = \frac{1}{2N} \sum_{m=-N+1}^{N} |\chi^{(3)}(\theta_m)|^2 \sin n\theta_m, \quad n = 1, 2, ..., N-1. \tag{12}$$

Finally it follows from (6),(7) and (9) that we get the approximations

$$|\chi_R(\theta_m)|^2 + \chi_{NR}^2 = \alpha_0 + 2\sum_{n=1}^{N-1} \alpha_n \cos n\theta_m + \alpha_N \cos \pi m, \tag{13}$$

$$\chi_{NR} Re\left[\chi_R(\theta_m)\right] = \sum_{n=1}^{N-1} \beta_n \sin n\theta_m. \tag{14}$$

Since $|\chi_R(\delta)|^2$ converges rapidly to zero as δ increases, the value of χ_{NR} can also be obtained by extrapolating it from the lineshape of (13).

3. Summary

We have presented a simple method for calculating the lineshapes of $|\chi_R|^2$ and $Re(\chi_R)$ when the lineshape of $|\chi^{(3)}|^2$ is known. The method is based on the use of the finite Fourier series expansion, and it can be applied to the study of coherent Raman scattering spectra. The method presented here is valid provided the nonresonant susceptibility has a real value and the spectrum results from a single Raman resonance. This model is somewhat more general than that of [4] because we are not restricted to any special lineshape model. We only assumed that $Re(\chi_R)$ is an odd function of the variable δ.

References

1. Here the susceptibility $\chi^{(3)}$ is a so-called effective susceptibility, which is usually a tensor-element $\chi^{(3)}_{ijkl}$ or a linear combination of two elements of the third-order nonlinear susceptibility tensor $\bar{\chi}^{(3)}$.
2. Y. R. Shen: *The Principles of Nonlinear Optics*, (Wiley, New York 1984) p. 268
3. J. J. Song, M. D. Levenson: J. Appl. Phys. **48**, 8 (1977)
4. J. W. Fleming, C. S. Johnson Jr.: J. Raman Spectrosc. **8**, 5 (1979)
5. R. W. Hamming: *Numerical Methods for Scientists and Engineers*, (McGraw-Hill, Tokyo 1962)

Index of Contributors

Printed by Publishers' Graphics LLC